THE ENGLISH-CHINESE ENCYCLOPEDIA OF PRACTICAL TRADITIONAL CHINESE MEDICINE

Chief Editor   Xu Xiangcai
Assistants   You Ke   Kang Kai
                 Bao Xuequan   Lu Yubin

英汉实用中医药大全

主　编　　徐象才
主编助理　尤　可　　康　凯
　　　　　鲍学全　　路玉滨

Higher Education Press
高等教育出版社

19 急 症 学

	中文		英文	
主　编	邵念方		朱忠宝	
副主编	卢尚岭	包培荣	陶锦文	
编　者：	曹晓岚	戴法轩	李震生	王洁华
	张　伟		陈宝荣	田开宇
			申　光	李广荣

EMERGENTOLOGY

	English	Chinese
Chief Editor	Zhu Zhongbao	Shao Nianfang
Deputy Chief Editors	Tao Jinwen	Lu Shangling
		Bao Peirong
Editors	Li Zhensheng	Cao Xiaolan
	Wang Jiehua	Dai Faxuan
	Chen Baorong	Zhang Wei
	Tian Kaiyu	
	Shen Guang	
	Li Guangrong	

(京) 112号

The English-Chinese
Encyclopedia of Practical TCM
Chief Editor   Xu Xiangcai

19
EMERGENTOLOGY
English Chief Editor   Zhu Zhongbao
Chinese Chief Editor   Shao Nianfang

英汉实用中医药大全

主编 徐象才

19
急 症 学

中文主编   邵念方
英文主编   朱忠宝

*

高等教育出版社出版
新华书店总店科技发行所发行
国防工业出版印刷厂印

*

开本 850×1168 1/32 印张 16.625 字数 430 000
1994年9月第1版 1994年9月第1次印刷
印数 0001—5 040
ISBN7-04-004587-7/R·27
定价 ▇▇ 元

The Leading Commission of Compilation and Translation
编译领导委员会

Honorary Director 名誉主任委员	Hu Ximing 胡熙明		
Honorary Deputy Directors 名誉副主任委员	Zhang Qiwen 张奇文	Wang Lei 王镭	
Director 主任委员	Zou Jilong 邹积隆		
Deputy Director 副主任委员	Wei Jiwu 隗继武		
Members 委员 (以姓氏笔划为序)	Wan Deguang 万德光	Wang Yongyan 王永炎	Wang Maoze 王懋泽
	Wei Guikang 韦贵康	Cong Chunyu 丛春雨	Liu Zhongben 刘中本
	Sun Guojie 孙国杰	Yan Shiyun 严世芸	Qiu Dewen 邱德文
	Shang Chichang 尚炽昌	Xiang Ping 项平	Zhao Yisen 赵以森
	Gao Jinliang 高金亮	Cheng Yichun 程益春	Ge Linyi 葛琳仪
	Cai Jianqian 蔡剑前	Zhai Weimin 翟维敏	
Advisers 顾问	Dong Jianhua 董建华	Huang Xiaokai 黄孝楷	Geng Jianting 耿鉴庭
	Zhou Fengwu 周凤梧	Zhou Ciqing 周次清	Chen Keji 陈可冀

The Commission of Compilation and Translation
编译委员会

Director 主任委员	Xu Xiangcai 徐象才

Deputy Directors
副主任委员

Zhang Zhigang 张志刚	Zhang Wengao 张文高	Jiang Zhaojun 姜兆俊
Qi Xiuheng 亓秀恒	Xuan Jiasheng 宣家声	Sun Xiangxie 孙祥燮

Members
委员
(以姓氏笔划为序)

Yu Wenping 于文平	Wang Zhengzhong 王正忠	Wang Chenying 王陈应
Wang Guocai 王国才	Fang Tingyu 方廷钰	Fang Xuwu 方续武
Tian Jingzhen 田景振	Bi Yongsheng 毕永升	Liu Yutan 刘玉檀
Liu Chengcai 刘承才	Liu Jiaqi 刘家起	Liu Xiaojuan 刘晓娟
Zhu Zhongbao 朱忠宝	Zhu Zhenduo 朱振铎	Xun Jianying 寻建英
Li Lei 李磊	Li Zhulan 李竹兰	Xin Shoupu 辛守璞
Shao Nianfang 邵念方	Chen Shaomin 陈绍民	Zou Jilong 邹积隆
Lu Shengnian 陆胜年	Zhou Xing 周行	Zhou Ciqing 周次清
Zhang Sufang 张素芳	Yang Chongfeng 杨崇峰	Zhao Chunxiu 赵纯修
Yu Changzheng 俞昌正	Hu Zunda 胡遵达	Xu Heying 须鹤瑛
Yuan Jiurong 袁久荣	Huang Naijian 黄乃健	Huang Kuiming 黄奎铭
Huang Jialing 黄嘉陵	Cao Yixun 曹贻训	Lei Xilian 雷希濂
Cai Huasong 蔡华松	Cai Jianqian 蔡剑前	

Preface

I am delighted to learn that THE ENGLISH−CHINESE ENCYCLOPEDIA OF PRACTICAL TRADITIONAL CHINESE MEDICINE will soon come into the world.

TCM has experienced many vicissitudes of times but has remained evergreen. It has made great contributions not only to the power and prosperity of our Chinese nation but to the enrichment and improvement of world medicine. Unfortunately, differences in nations, states and languages have slowed down its spreading and flowing outside China. At present, however, an upsurge in learning, researching and applying Traditional Chinese Medicine (TCM) is unfolding. In order to maximize the effect of this upsurge and to lead TCM, one of the brilliant cultural heritages of the Chinese nation, to the world for it to expand and bring benefit to the people of all nations, Mr. Xu Xiangcai called intellectuals of noble aspirations and high intelligence together from Shandong and many other provinces in China and took charge of the work of both compilation and translation of THE ENGLISH−CHINESE ENCYCLOPEDIA OF PRACTICAL TRADITIONAL CHINESE MEDICINE. With great pleasure, the medical staff both at home and abroad will hail the appearance of this encyclopedia.

I believe that the day when the world's medicine is fully

developed will be the day when TCM has spread throughout the world.

I am pleased to give it my preface.

Prof. Dr. Hu Ximing
    Deputy Minister of the Ministry of Public Health of the People's Republic of China,
    Director General of the State Administrative Bureau of Traditional Chinese Medicine and Pharmacology,
    President of the World Federation of Acupuncture—Moxibustion Societies,
    Member of China Association of Science & Technology,
    Deputy President of All—China Association of Traditional Chinese Medicine,
    President of China Acupuncture & Moxibustion Society.

December, 1989

Preface

The Chinese nation has been through a long, arduous course of struggling against diseases and has developed its own traditional medicine—Traditional Chinese Medicine and Pharmacology (TCMP). TCMP has a unique, comprehensive, scientific system including both theories and clinical practice. Some thousand years since ito-beginnings, not only has it been well preserved but also continuously developed. It has special advantages, such as remarkable curative effects and few side effects. Hence it is an effective means by which people prevent and treat diseases and keep themselves strong and healthy.

All achievements attained by any nation in the development of medicine are the public wealth of all mankind. They should not be confined within a single country. What is more, the need to set them free to flow throughout the world as quickly and precisely as possible is greater than that of any other kind of science. During my more than thirty years of being engaged in Traditional Chinese Medicine(TCM), I have been looking forward to the day when TCMP will have spread all over the world and made its contributions to the elimination of diseases of all mankind. However it is to be deeply regretted that the pace of TCMP in extending outside China has been unsatisfactory due to the major difficulties in expressing its concepts in foreign languages.

Mr. Xu Xiangcai, a teacher of Shandong College of TCM, has sponsored and taken charge of the work of compilation and translation of The English—Chinese Encyclopedia of Practical Traditional Chinese Medicine—an extensive series. This work is a great project, a large—scale scientific research, a courageous effort and a novel creation. I deeply esteem Mr. Xu Xiangcai and his compilers and translators, who have been working day and night for such a long time, for their hard labor and for their firm and indomitable will displayed in overcoming one difficulty after another, and for their great success achieved in this way. As a leader in the circles of TCM, I am duty—bound to do my best to support them.

I believe this encyclopedia will be certain to find its position both in the history of Chinese medicine and in the history of world science and technology.

Mr. Zhang Qiwen
Member of the Standing Committee of
All—China Association of TCM,
Deputy Head of the Health Department
of Shandong Province.
March, 1990

Publisher's Preface

Traditional Chinese Medicine(TCM) is one of China's great cultural heritages. Since the founding of the People's Republic of China in 1949, guided by the farsighted TCM policy of the Chinese Communist Party and the Chinese government, the treasure house of the theories of TCM has been continuously explored and the plentiful literature researched and compiled. As a result, great success has been achieved. Today there has appeared a world-wide upsurge in the studying and researching of TCM. To promote even more vigorous development of this trend in order that TCM may better serve all mankind, efforts are required to further it throughout the world. To bring this about, the language barriers must be overcome as soon as possible in order that TCM can be accurately expressed in foreign languages.

Thus the compilation and translation of a series of English-Chinese books of basic knowledge of TCM has become of great urgency to serve the needs of medical and educational circles both inside and outside China.

In recent years, at the request of the health departments, satisfactory achievements have been made in researching the expression of TCM in English. Based on the investigation into the history and current state of the research work mentioned above, the English-Chinese Encyclopedia of Practical TCM has been published to meet the needs of extending the knowledge of TCM around the world.

The encyclopedia consists of twenty-one volumes, each dealing with a particular branch of TCM. In the process of compilation, the distinguishing features of TCM have been given close attention and great efforts have been made to ensure that the content is scientific, practical, comprehensive and concise. The chief writers of the Chinese manuscripts include professors or associate professors with at least twenty years of practical clinical and / or teaching experience in TCM. The Chinese manuscript of each volume has been checked and approved by a specialist of the relevant branch of TCM. The team of the translators and revisers of the English versions consists of TCM specialists with a good command of English professional medical translators, and teachers of English from TCM colleges or universities. At a symposium to standardize the English versions, scholars from twenty-two colleges or universities, research institutes of TCM or other health institutes probed the question of how to express TCM in English more comprehensively, systematically and accurately, and discussed and deliberated in detail the English versions of some volumes in order to upgrade the English versions of the whole series. The English version of each volume has been re-examined and then given a final checking.

Obviously this encyclopedia will provide extensive reading material of TCM English for senior students in colleges of TCM in China and will also greatly benefit foreigners studying TCM.

The assiduous efforts of compiling and translating this encyclopedia have been supported by the responsible leaders of the State Education Commission of the People's Republic of China, the State Administrative Bureau of TCM and Pharmacy, and the Education Commission and Health Department of Shandong

Province. Under the direction of the Higher Education Department of the State Education Commission, the leading board of compilation and translation of this encyclopedia was set up. The leaders of many colleges of TCM and pharmaceutical factories of TCM have also given assistance.

We hope that this encyclopedia will bring about a good effect on enhancing the teaching of TCM English at the colleges of TCM in China, on cultivating skills in medical circles in exchanging ideas of TCM with patients in English, and on giving an impetus to the study of TCM outside China.

Higher Education Press
March, 1990

Foreword

The English—Chinese Encyclopedia of Practical Traditional Chinese Medicine is an extensive series of twenty—one volumes. Based on the fundamental theories of traditional Chinese medicine(TCM) and with emphasis on the clinical practice of TCM, it is a semi—advanced English—Chinese academic works which is quite comprehensive, systematic, concise, practical and easy to read. It caters mainly to the following readers: senior students of colleges of TCM, young and middle—aged teachers of colleges of TCM, young and middle—aged physicians of hospitals of TCM, personnel of scientific research institutions of TCM, teachers giving correspondence courses in TCM to foreigners, TCM personnel going abroad in the capacity of lecturers or physicians, those trained in Western medicine but wishing to study TCM, and foreigners coming to China to learn TCM or to take refresher courses in TCM.

Because Traditional Chinese Medicine and Pharmacology is unique to our Chinese nation, putting TCM into English has been the crux of the compilation and translation of this encyclopedia. Owing to the fact that no one can be proficient both in the theories of Traditional Chinese Medicine and Pharmacology and the clinical practice of every branch of TCM, as well as in English, to ensure that the English versions express accurately the inherent meanings of TCM, collective translation measures have been taken. That is, teachers of English familiar with TCM, professional medical translators, teachers or physicians of TCM and even teachers of palaeography with a strong command of English were all invited together to co-translate the Chinese manuscripts and, then, to co-deliberate and discuss the English versions. Finally English-speaking foreigners studying TCM or teaching English in China were asked to polish the English versions. In this way, the skills of the above translators and foreigners were merged to ensure the quality of the English versions. However, even using this method, the uncertainty that the English versions will be wholly accepted still remains. As for the Chinese manuscripts, they do reflect the essence, and give a general picture, of traditional Chinese medicine and pharmacology. It is not asserted, though, that they are perfect, I whole-heartedly look forward to any criticisms or opinions from readers in order to make improvements to future editions.

More than 200 people have taken part in the activities of compiling, translating and revising this encyclopedia. They come from twenty-eight institutions in all parts of China. Among these institutions, there are fifteen colleges of TCM:Shandong, Beijing, Shanghai, Tianjin, Nanjing, Zhejiang, Anhui, Henan, Hubei, Guangxi, Guiyang, Gansu, Chengdu, Shanxi and Changchun, and scientific research centers of TCM such as China Academy of TCM and Shandong Scientific Research Institute of TCM.

The Education Commission of Shandong province has included the compilation and translation of this encyclopedia in its scientific research projects and allocated funds accordingly. The Health Department of Shandong Province has also given financial aid together with a number of pharmaceutical factories of TCM. The subsidization from Jinan Pharmaceutical Factory of

TCM provided the impetus for the work of compilation and translation to get under way.

The success of compiling and translating this encyclopedia is not only the fruit of the collective labor of all the compilers, translators and revisers but also the result of the support of the responsible leaders of the relevant leading institutions. As the encyclopedia is going to be published, I express my heartfelt thanks to all the compilers, translators and revisers for their sincere cooperation, and to the specialists, professors, leaders at all levels and pharmaceutical factories of TCM for their warm support.

It is my most profound wish that the publication of this encyclopedia will take its role in cultivating talented persons of TCM having a very good command of TCM English and in extending, rapidly, comprehensive knowledge of TCM to all corners of the globe.

Chief Editor   Xu Xiangcai
Shandong College of TCM
March, 1990

Content

Notes 1

1 First-aid for Imminent Symptoms and Signs 1
1.1 High Fever 1
1.2 Shock 20
1.3 Coma 25

2 Cardiovascular Diseases 35
2.1 Angina Pectoris 35
2.2 Acute Myocardial Infarction 41
2.3 Acute Pulmonary Heart Disease 48
2.4 Acute Infectious Endocarditis 52
2.5 Viral Myocarditis 56
2.6 Arrythmia 61
2.7 Heart Failure 68

3 Respiratory Diseases 72
3.1 Acute Bronchitis 72
3.2 Lobar Pneumonia 76
3.3 Acute Aspiration Pulmonary Abscess 82

4 Neurogenic and Psychogenic Diseases 85
4.1 Trigeminal Neuralgia 85
4.2 Migraine 88
4.3 Acute Polyneuritis 93
4.4 Acute Infective Polyneuroradiculitis 96
4.5 Hypertensive Cerebral Hemorrhage 100

4.6	Arteriosclerotic Cerebral Infarction	106

5 Digestive Diseases — 118

5.1	Acute Gastroenteritis	118
5.2	Gastroduodenal Ulcerative Bleeding	121
5.3	Acute Gastric Dilatation	126
5.4	Hepatic Coma	129
5.5	Biliary Ascariasis	133
5.6	Acute Cholecystitis and Cholelithiasis	136
5.7	Acute Pancreatitis	140
5.8	Acute Appendicitis	143
5.9	Gastroduodenal Ulcer and Perforation	147
5.10	Acute Peritonitis	151
5.11	Pseudomembranous Enteritis	154

6 Diseases of the Urinary System — 157

6.1	Acute Renal Failure	157
6.2	Uremia	162
6.3	Acute Urinary Tract Infection	166
6.4	Acute Urine Retention	169

7 Diseases of the Endocrine and Metabolic System — 175

7.1	Acute Suppurative Thyroiditis	175
7.2	Acute Hypoadrenocorticism	177
7.3	Spontaneous Hypoglycemia	180

8 Diseases of Hematopoietic System — 183

8.1	Aplastic Anemia	183
8.2	Acute Hemorrhagic Anemia	187
8.3	Allergic Purpura	190

9 Diseases of the Connective Tissues and Allergic Reactions — 195

9.1	Systemic Lupus Erythematosus	195

- 9.2 Allergic Subacute Septicemia ... 198
- **10 Infectious Diseases** ... 202
  - 10.1 Influenza ... 202
  - 10.2 Epidemic Cerebrospinal Meningitis ... 205
  - 10.3 Epidemic Encephalitis B ... 208
  - 10.4 Epidemic Hemorrhagic Fever ... 213
  - 10.5 Acute Fatal Hepatitis ... 221
- **11 Poisoning and Physicochemical Traumatic Diseases** ... 226
  - 11.1 Alcoholism ... 226
  - 11.2 Aconite Poisoning ... 229
  - 11.3 Dog Button Poisoning ... 230
  - 11.4 Heliosis ... 231
- **12 Common Emergent Cases of Gynecology** ... 234
  - 12.1 Dysmenorrhea ... 234
  - 12.2 Acute Pelvic Inflammation ... 238
  - 12.3 Functional Uterine Bleeding ... 241
- **13 Common Emergent Cases of Paediatrics** ... 250
  - 13.1 Neonatal Pneumonia ... 250
  - 13.2 Neonatal Septicemia ... 255
  - 13.3 Toxic Dysentery ... 257
- **14 Acute Diseases of Ear, Nose and Throat** ... 260
  - 14.1 Acute Nasosinusitis ... 260
  - 14.2 Acute Tonsillitis ... 264
  - 14.3 Acute Purulent Otitis Media ... 267
  - 14.4 Meniere's Disease ... 271
- **15 Common Emergent Cases of Dermatology** ... 277
  - 15.1 Herpes Zoster ... 277
  - 15.2 Acute Erysipelas ... 281

16 The Main Approaches of Emergency Treatment in TCM ... 284

- 16.1 Emetic and Purgative Method ... 284
- 16.2 Acupuncture–Massage Therapy ... 286
- 16.3 Venous Transfusion ... 290
- 16.4 Aerosol Inhalation ... 292
- 16.5 Administration Per Anus ... 293

Index of Recipes ... 294

The English–Chinese Encyclopedia of Practical TCM (Booklist) ... 509

Notes

"EMERGENTOLOGY" is the nineteenth volume of The English-Chinese Encyclopedia of Practical Traditional Chinese Medicine.

Traditional Chinese medicine has a long history and rich experiences in treating emergency cases. It is substantial in content and unique in theory.

This volume covers sixteen chapters concerning first-aid for imminent symptoms and signs and for emergency cases of cardiovascular diseases, respiratory diseases, neurogenic and psychogenic diseases, digestive diseases, diseases, of the urinary system, diseases of the endocrine and metabolic system, diseases of the hematopoietic, system, diseases of the connective tissues and allergic reactions, infectious diseases, poisoning and physicochemical traumatic injuries, common emergent cases of gynecology, common emergent cases of paediatrics, common acute cases of ear, nose and throat, common energent cases of dermatology, as well as the main approaches taken by traditional Chinese medicine for their treatment.The book adopts the western way in naming the diseases, whereas its contents is presented in the form of treatment based upon differentiaion of syndrome according to the theory of traditional Chinese medicine.Each disease is introduced in terms of name of the disease, etiology and pathogenesis, essentials of diagnosis and treatment based upon differentiation of syndrome.The dosage for each disease, with the exception of chapter thirteen which is on paediatrics, is prescribed for grown-ups

Some of the materials of traditional chinese medicine referred to in this volume come from rare animals which are near extinction and of which trapping and killing has been prohibited by lau. So, they have lost their sources. But they havetheir substitutes with the same effects in both hospitals and pharmaceutical factories. Why we deal with these materials in this volume is just to the academic convenience.

The Chinese version of the book is checked and approved by Professor Wang Yongyan, deputy director of All-China Association of Traditional Chinese Medicine.

The Editors

1 First-aid for Imminent Symptoms and Signs

Imminent symptoms and signs in Western medicine correspond to critical syndromes in traditional Chinese medicine. The common ones are as follows.

1.1 High Fever

Clinically, high fever, a critical condition in enormous emergent and serious diseases, may reflect various kinds of pathogenesis. In the literature of traditional Chinese medicine, high fever falls into *Fare*(fever), *Zhuangre*(sthenic fever), *Chaore*(tidal fever) and *Hanre*(chills and fever).

Etiology and Pathogenesis

Exogenous high fever: Mostly caused by the struggle between vital *Qi* and the six exogenous pathogenic factors or other pestilential factors.

Endogenous high fever: The result of preponderance of *Yang-heat* due to either dysfunction of *Zang-Fu* or consumption of *Qi*, blood and body fluid caused by overstrain and overstress, improper diet, mental depression, internal blood stasis, damp-heat retention, etc.

Essentials of Diagnosis

1. Sudden rise of body temperature, usually over 39 °C, or instant high fever resulted from a low one, characterized by ab-

rupt onset, high body temperature, short course and rapid progress.

2. The main clinical features include excessive thirst, feverish sensation of the body and rapid pulse.

3. Chills and avertion to wind in the early stage of exogenous high fever, usually accompanied by other exopathic symptoms;no chills in case of endogenous high fever, which, however, is always complicated by other symptoms due to endogenous functional impairment of the internal organs.

Treatment Based on Differentiation of Syndromes

1. The Exterior Syndrome

(1) The exterior-cold syndrome

Symptoms and signs: Chills, fever, no sweating, headache with rigidity of nape, general aching, lumbago, arthralgia, thin and white tongue coating, and superficial and tense pulse.

Principle of treatment: Relieving the exterior with drugs pungent in flavor and warm in property.

Recipe: *Jing Fang Jie Biao Tang*(1)
Ingredients:

Herba Schizonepetae	9 g
Radix Ledebouriellae	12 g
Rhizoma seu Radix Notopterygii	9 g
Radix Angelicae Dahuricae	9 g
Bulbus Allii Fistulosi	1 cun(3.33 cm)
Rhizoma Zingiberis Recens	3 slices
Radix Glycyrrhizae	6 g

Decoct the above ingredients in water for oral dose.

(2)The exterior-heat syndrome

Symptoms and signs: Fever with slight avertion to wind and

cold, little or no sweating, headache, nasal obstruction, cough, slight thirst, redness, swelling or pain in the throat, redness of the tip and border of the tongue, thin and white or thin and yellow tongue coating, and superficial and rapid pulse.

Principle of treatment: Relieving the exterior with drugs pungent in flavor and cool in property.

Recipe: *Yin Qiao San* (2) with modification

Ingredients:

Flos Lonicerae	30 g
Fructus Forsythiae	12 g
Radix Platycodi	9 g
Herba Menthae	9 g
Semen Armeniacae Amarum (stir-fried)	9 g
Spica Schizonepetae	9 g
Herba Lophatheri	9 g
Radix Glycyrrhizae	3 g

Decoct the above ingredients in water for oral dose, 1 or 2 doses a day.

Modification: In case of high fever, add Radix Isatidis, Radix Scutellariae and Gypsum Fibrosum.

In case of obvious headache, add Folium Mori 12 g and Flos Chrysanthemi 12 g

In case of frequent cough, add Folium Eriobotryae 12 g, Cortex Mori Radicis 12 g and Bulbus Fritillariae Cirrhosae 12 g.

In case of swelling and pain of the throat, add Fructus Arctii 12 g (stir-fried) and Radix Sophorae Subprostratae 12 g.

(3) The exterior-dampness syndrome

Symptoms and signs: Chills, recessive fever or high fever in the afternoon, headache as if being tightly bound, heaviness sensation and lassitude of the body and limbs, sticky sweat, fullness and stuffiness in the chest and epigastrium, loss of appetite, thirst with no desire for drinking due to subsequent nausea, white or yellow and sticky tongue coating, and soft and rapid pulse.

Principle of treatment: Expelling pathogenic factors from the exterior by means of aromatics, and clearing away heat and promoting diuresis.

Recipe: *Huo Xiang Zheng Qi San* (3) with modification
Ingredients:

Herba Agastaches	12 g
Folium Perillae	9 g
Rhizoma Atractylodis Macrocephalae	12 g
Pericarpium Arecae	12 g
Poria	9 g
Massa Pinelliae Fermentatae	9 g
Pericarpium Citri Reticulatae	9 g
Cortex Magnoliae Officinalis	9 g
Radix Platycodi	9 g
Fructus Forsythiae	12 g
Herba Menthae	9 g
Semen Plantaginis (to be wrapped and then decocted)	12 g
Radix Glycyrrhizae	3 g

Decoct the above ingredients in water for oral dose.

(4) The syndrome of dryness of the lung

Symptoms and signs: Fever, thirst, dry nose and lips, dry cough or cough with blood-stained sputum in severe cases, red tongue with thin and yellow coating and little saliva, and superficial, slippery and rapid pulse.

Principle of treatment: Moistening the lung with drugs pungent in flavor and cool in nature, and clearing away heat and promoting the production of body fluid.

Recipe: *Sang Xing Tang*(4) and *Sha Shen Mai Men Dong Tang*(5) with modification.

Ingredients:

Folium Mori	12 g
Semen Armeniacae Amarmum (stir-fried)	9 g
Radix Adenophorae Strictae	24 g
Radix Ophiopogonis	15 g
Bulbus Fritillariae	9 g
Fructus Gardeniae	9 g
Rhizoma Polygonati Odorati	12 g
Radix Trichosanthis	12 g
Exocarpium Pyrus	9 g

Decoct the above ingredients in water for oral dose.

(5) The syndrome of summer-heat

Symptoms and signs: Fever and chills with or without sweating, headache and heaviness sensation in the head, irritability, thirst without desire for drinking, fullness in epigastrium, nausea and vomiting, scanty and deep yellow urine, red tongue with white and sticky coating, and soft and rapid or soft and superficial pulse.

Principle of treatment: Clearing away summer-heat, eliminating dampness and relieving the exterior heat.

Recipe: *Xin Jia Xiang Ru Yin*(6) with modification.

Ingredients:

Herba Elsholtziae seu Moslae	9 g
Flos Dolichoris (fresh)	24 g
Cortex Magnoliae Officinalis	12 g
Flos Lonicerae	30 g
Fructus Forsythiae	12 g
Herba Agastachis	12 g
Folium Eupatorii	9 g
Semen Coicis	24 g
Six to One Powder	15 g

Decoct the above ingredients in water for oral dose.

2. The Exterior−interior Syndrome

(1) The half exterior and half interior syndrome

Symptoms and signs: Alternate attacks of chills and fever, bitter mouth and dry throat, vertigo, fullness and discomfort in the chest and hypochondrium, mental restlessness and nausea, loss of appetite, white and slippery tongue coating and wiry pulse.

Principle of treatment: Mediating the exterior and interior.

Recipe: *Xiao Chai Hu Tang*(7) with modification.

Ingredients:

Radix Bupleuri	30 g
Radix Scutellariae	15 g
Rhizoma Pinelliae	9 g
Radix Ginseng	9 g
Rhizoma Zingiberis Recens	3 slices
Fructus Ziziphi Jujubae	3 pieces
Radix Glycyrrhizae	3 g

Decoct the above ingredients in water for oral dose.

(2) The syndrome of exterior cold and interior heat

Symptoms and signs: High fever, cough with dyspnea, shortness of breath, thirst, yellow or white sticky expectoration which could be blood stained, or rusty sputum, chest pain due to coughing, white—and—yellow tongue coating, and superficial and rapid pulse.

Principle of treatment: Clearing away heat to promote the dispersing function of the lung.

Recipe: *Ma Xing Shi Gan Tang*(8) with modification.

Ingredients:

Herba Ephedrae	6 g
Semen Armeniacae Amarmum (stir—fried)	9 g
Gypsum Fibrosum	30 g
Radix Glycyrrhizae	6 g
Cortex Mori Radicis	12 g
Folium Eriobotryae	12 g
Radix Scutellariae	9 g
Herba Houttuyniae	18 g
Radix Platycodi	9 g
Cortex Lycii Radicis	18 g

Decoct the above ingredients in water for oral dose.

Modification: In case of haemoptysis, add Radix Rehmanniae 15 g and Cortex Moutan Radicis 12 g.

(3)The syndrome of heat in both exterior and interior

Symptoms and signs: Fever, sore throat, thirst, constipation, red tongue with yellow coating, and rapid pulse.

Principle of treatment: Expelling heat from the exterior and clearing out heat from the interior.

Recipe: *Liang Ge San*(9) with modification.

Ingredients:

Radix et Rhizoma Rhei	6–9 g
Radix Scutellariae	9 g
Fructus Gardeniae	9 g
Fructus Forsythiae	15 g
Herba Menthae	9 g
Natrii Sulfas	6 g
Radix Glycyrrhizae	3 g

Decoct the above ingredients in water for oral dose.

3. The Interior Syndrome

(1) The syndrome of excessive heat in the *Qi* system

Symptoms and signs: High fever, excessive thirst, sweating, or accompanied by foul breating, swelling and pain of gingiva with erosion or bleeding, red tongue with dry and yellow coating, and large and full pulse.

Principle of treatment: Clearing away heat and promoting the production of body fluid.

Recipe: *Bai Hu Tang*(10) with modification.

Ingredients:

Gypsum Fibrosum	60–120 g
Rhizoma Anemarrhenae	30 g
Radix Trichosanthis	30 g
Radix Pseudostellariae	30 g
Radix Bupleuri	30–60 g
Radix Isatidis	45–60 g
Rhizoma Paridis	15 g
Radix Rehmanniae	30 g
Radix Glycyrrhizae	6 g

Decoct the above ingredients in water for oral dose, 1 or 2 doses a day.

Modification: In case of excessive noxious heat, add Flos lonicerae 30 g and Fructus Forsythiae 15 g;

In case of coma, delirium and constipation, add Radix et Rhizoma Rhei 10 g and Natrii Sulfas 6 g;

In case of high fever accompanied by skin eruptions, add powder of Cornu Rhinoceri 3 g, Radix Scrophulariae 30 g and Cortex Moutan Radicis 12 g;

In case of foul breathing and severe swelling of gingiva, add Rhizoma Coptidis 10 g and Fructus Gardeniae 12 g.

(2)The syndrome of accumulation of heat in the stomach and intestines

Symptoms and signs: High fever that becomes worse in the afternoon, constipation or watery diarrhea, abdominal pain that is aggravated by pressure, fullness and discomfort in the chest and epigastrium, dysphoria and delirium, or even mental confusion in severe cases, red thorny tongue with yellow and dry or charred brown coating, and wiry and rapid or deep and strong pulse.

Principle of treatment: Promoting bowel movement to purge away the intestinal heat.

Recipe: *Da Cheng Qi Tang*(11)with modification.

Ingredients:

Radix et Rhizoma Rhei (to be decocted later)	9 g
Natrii Sulfas	9 g
Fructus Aurantii Immaturus	12 g
Cortex Magnoliae Officinalis	12 g

Decoct the above ingredients in water for oral dose.

Modification: In case of deficiency of *Yin* due to accumulation of heat with constipation, add Radix Rehmanniae 30 g, Radix Ophiopogonis 30 g and Radis Scrophulariae 30 g;

In case of dysphoria with feverish sensation in the chest and costal region and oral ulceration, add Fructus Gardeniae 12 g, Radix Scutellariae 15 g, Fructus Forsythiae 12 g, Herba Menthae 10 g and Herba Lophatheri 10 g.

(3)The syndrome of retention of noxious heat in the lung

Symptoms and signs: Fever, sweating, cough with dyspnea, expectoration, thirst, shortness of breath, red tongue with yellow coating, and slippery and rapid pulse.

Principle of treatment: Removing heat from the lung and dissolving phlegm, and promoting discharge of pus and eliminating toxin.

Recipe: *Qian Jin Wei Jing Tang* (12) with modification.

Ingredients:

Rhizoma Phragnitis (fresh)	30 g
Semen Coicis	30 g
Semen Benincasae	12 g
Semen Persicae	9 g
Radix Platycodi	24 g
Radix Glycirrhizae	6 g
Flos Lonicerae	30 g
Caulis Sargentodoxae	30 g
Herba Houttuyniae	24 g
Radix Scutellariae	9 g
Cortex Mori Radicis	12 g
Fructus Trichosanthis	24 g
Herba Taraxaci	24 g

Herba Violae	18 g

Decoct the above ingredients in water for oral dose.

Modification: In case of difficulty in lying flat due to cough with shortness of breath, fullness in the chest and profuse expectoration, add Semen Lepidii seu Descurainiae 12 g;

In case of haemoptysis, add Cortex Moutan Radicis 12 g, Fructus Gardeniae 12 g, Rhizoma Bletillae 15 g, Nodus Nelumbinis Rhizomatis 30 g, and powder of Radix Notoginseng 3 g (to be swallowed with water);

In case of chest pain and dyspnea, add Rhizoma Cyperi 10 g and Radis Curcumae 10 g;

In case of excessive thirst, add Radix Trichosanthis 15 g and Rhizoma Anemarrhenae 15 g.

(4) The syndrome of invasion of the spleen by damp-heat

Symptoms and signs: Recessive fever which becomes worse in the afternoon, stuffy chest, fullness in epigastrium, drowsiness, sticky tongue coating, and soft and rapid pulse.

Principle of treatment: Dispersing and dissolving damp-heat.

Recipe: *Wang Ren Tang*(13) with modification

Ingredients:

Semen Coicis	30 g
Semen Armeniacae Amarum (stir-fried)	9 g
Semen Amomi Cardamomi	6 g
Talcum	24 g
Cortex Magnoliae Officinalis	9 g
Rhizoma Pinelliae	6 g
Medulla Tetrapanacis	9 g

Herba Lophatheri	9 g

Decoct the above ingredients in water for oral dose.

(5) The syndrome of damp—heat in the liver and gall bladder

Symptoms and signs: Fever, thirst, bitter mouth, hypochondriac pain, fullness and distention in epigastrium and abdomen, nausea or vomiting of yellow fluid, scanty and deep yellow urine, red tongue with yellow and sticky coating, and wiry, slippery and rapid pulse.

Principle of treatment: Relieving the depressed liver and gall bladder, regulating *Qi* and clearing away damp—heat.

Recipe: *Hao Qin Qing Dan Tang*(14) with modification.

Ingredients:

Herba Artemisiae Chinghao	24 g
Radix Scutellariae	12 g
Pericarpium Citri Reticulatae	9 g
Caulis Bambusae in Taeniam	9 g
Rhizoma Pinelliae	9 g
Poria Rubra	12 g
Fructus Aurantii	9 g
Rhizoma Coptidis	6 g
Herba Eupatorii	6 g
Rhizoma Acori Graminei	9 g
Green Jade Powder	10 g

Decoct the above ingredients in water for oral dose.

(6) The syndrome of damp—heat in the urinary bladder

Symptoms and signs: Frequent, urgent, difficult, painful and dribbling urination, fever, lumbago, red tongue with yellow and sticky or yellow and thin coating, and slippery and rapid pulse.

Principle of treatment: Clearing away damp—heat to relieve

the lower *Jiao*.

Recipe: *Ba Zheng San* (15) with modification.

Ingredients:

Flos Lonicerae	30 g
Fructus Forsythiae	15 g
Herba Plantaginis	30 g
Herba Polygoni Avicularis	15 g
Herba Dianthi	18 g
Caulis Akebiae	9 g
Fructus Gardeniae	9 g
Talcum	24 g
tip of Radix Glycyrrhizae	9 g
Medulla Junci	3 g
Radix et Rhizoma Rhei (to be decocted later)	6 g

Decoct the above ingredients in water for oral dose, 1 or 2 doses a day.

Modification: In case of deep yellow urine, add Rhizoma Imperatae 30 g, Herba Cephalanoploris 30 g and Herba Leonuri 30 g;

In case of turbid urine, add Folium Pyrrosiae 24 g and Rhizoma Dioscoreae Hypoglaucae 24 g.

(7) The syndrome of damp—heat giving rise to dysentery

Symptoms and signs: Fever, abdominal pain, stool with white and red mucous, tenesmus or spouting diarrhea, burning sensation of the anus, nausea and vomiting, redtongue with sticky coating, and slippery and rapid pulse.

Principle of treatment: Clearing away toxic heat, drying dampness and stopping dysentery.

Recipe: *Bai Tou Weng Tang*(16) with modification.

Ingredients:

Radix Pulsatillae	24 g
Rhizoma Coptidis	9 g
Cortex Phellodendri	9 g
Cortex Fraxini	12 g
Radix Aucklandiae	9 g
Semen Arecae	9 g
Radix Paeoniae	30 g
Radix Glycyrrhizae	10 g
Fructus Crataegi	15 g

Decoct the above ingredients in water for oral dose.

(8) The syndrome of intensive heat in both *Qi* and *Ying* systems

Symptoms and signs: High fever, thirst, sweating, insomnia, coma, deep red tongue with yellow and dry coating, and rapid pulse.

Principle of treatment: Clearing away heat from the *Qi* system and cooling the *Ying* system.

Recipe: *Yu Nü Jian*(17) with modification.

Ingredients:

Gypsum Fibrosum	90—120 g
Radix Rehmanniae	30 g
Radix Ophiopogonis	30 g
Rhizoma Anemarrhenae	30 g
Radix Scutellariae	15 g
Fructus Gardeniae	9 g
Rhizoma Coptidis	9 g
Radix Isatidis	30 g

Rhizoma Phragmitis	30 g
Semen Cardamomi Rotundi	9 g

Decoct the above ingredients in water for oral dose.

Modification: In case of bleeding, add powder of Cornu Rhinoceri 3 g(to be swallowed with water), Radix Paeoniae Rubra 12 g and Cordex Moutan Radicis 12 g.

In case of coma, add powder of Cornu Antelonis 6 g(to be swallowed with water).Rhizoma Acori Graminei 10 g and Radix Curcumae 10 g.

In case of heart failure, add Radix Panacis Quinquefolii 15 g.

(9) The syndrome of invasion of the *Ying* system by pathogenic heat.

Symptoms and signs: Fever which is higher at night, mental restlessness, insomnia, thirst with no desire for drinking, and coma, delirium and faint eruption in severe cases, red tongue with little coating and fluid, and thready and rapid pulse.

Principle of treatment: Clearing heat from the *Ying* System.

Recipe: *Qing Ying Tang*(18) with modification.

Ingredients:

Cornu Rhinoceri	10 g
or Cornu Bubali	
(to be decocted first)	30 g
Radix Rehmanniae	15 g
Radix Scrophulariae	15 g
Radix Ophiopogonis	30 g
Radix Salviae Miltiorrhizae	15 g
Rhizoma Coptidis	9 g
Flos Lonicerae	30 g
Fructus Forsythiae	15 g

leaves bud of Herba Lophatheri   9 g

Decoct the above ingredients in water for oral dose.

(10) The syndrome of the invasion of the heart by pathogenic heat.

Symptoms and signs: Flaming fever, coma, delirium or mental mist with no speech, restlessness, deep red tongue with dry and yellow coating, and thready and rapid pulse.

Principle of treatment: Removing pathogenic heat from the heart to cause resuscitation.

Ripe: *Qing Gong Tang*(19) with modification.

Ingredients:

Cornu Rhinoceri	10 g
or Cornu Bubali	
(to be decocted first)	30 g
Radix Scrophulariae	15 g
Radix Ophiopogonis	30 g
Plumula Nelumbinis	9 g
Plumula Forsythiae	9 g
leaves bud of Herba Lophatheri	9 g
Rhizoma Acori Graminei	12 g

Decoct the above ingredients in water for oral dose.

Modification: In case of tense syndrome manifested by high fever and coma accompanied with trismus, opened eyes and clenching of fists, administer *An Gong Niu Huang Wan* (20) or *Niu Huang Qing Xin Wan*(21) immediately, one pill each time, 3 or 4 times daily by nasal feeding, or put *Qing Kai Ling* 20−40ml or *Xing Nao Jing* 10−20 ml in 5% glucose solution for intravenou drip, 1 or 2 times daily.

In case of retention of profuse phlegm, give *Zhu Li Shui*

(Succus Gambusae) 1 ampule each time and 3 times a day. Artifician *Niu Huang Fen*(powder of Calculus Bovis)3 g can be added following its infusion. Besides, *Zao Jiao Fen*(Fructus Gleditsiae Abnormalis powder) 1 g or *She Xiang Fen*(Moschus powder)0.01 g or *Hou Zao San*(Calculus Macacae powder)0.1 g can be given by nasal insufflation.

(11) The syndrome of extreme heat giving rise to endogenous wind (convulsive syndrome caused by high fever).

Symptoms and signs: High fever, excessive thirst, headache and vertigo, clonic convulsion, rigidity of the neck, upward staring of the eyes, trismus, and even opisthotonos, coma and cold limbs in severe cases, deep red tongue with yellow coating, and wiry and swift pulse.

Principle of treatment: Clearing away pathogenic heat from the liver, nourishing the blood and calming the endogenous wind.

Recipe: *Ling Yang Gou Teng Tang*(22) with modification.

Ingredients:

Ingredient	Amount
Cornu Antelopis (to be decocted first)	6 g
Ramulus Uncariae cum Uncis	15 g
Folium Mori	12 g
Flos Chrysanthemi	15 g
Radix Rehmanniae	15 g
Radix Paeoniae Alba	12 g
Bulbus Fritillariae Cirrhosae	9 g
Poria	12 g
Caulis Bambusae in Taeniam	9 g
Radix Glycyrrhizae	6 g

Decoct the above ingredients in water for oral dose, 1 or 2

doses daily.

Modification: *Qing Kai Ling* 40-60 ml added to 500 ml of 5% glucose solution can be given by intravenous drip, 1 or 2 times daily, or *Zhi Jing San* 1.5 g taken after infusion 1 or 2 times daily, or *An Gong Niu Huang Wan*, 1 pill each time, 3 or 4 times daily by oral administration.

(12) The Syndrome of stirring-up of endogenous wind due to deficiency of *Yin*.

Symptoms and signs: Low fever, athetosis, dry mouth and tongue, sinking eyes and blurred vision, deep red tongue with little coating, and thready, rapid and forceless pulse.

Principle of treatment: Nourishing *Yin* and blood, soothing the liver *Yang* and dispelling wind.

Recipe: *Da Ding Feng Zhu*(23) with modification.

Ingredients:

Radix Rehmanniae	24 g
Radix Ophiopogonis	30 g
Radix Paeoniae Alba	12 g
Concha Ostreae	30 g
Carapax Trionycis	18 g
Plastrum Testudinis	18 g
Glycyrrhizae Praeparata	9 g
Fructus Cannabis	6 g
Fructus Schisandrae	6 g
Colla Corii Asini (to be melted)	9 g
fresh egg yolks	2 pieces

Decoct the above ingredients in water for oral dose.

(13) The syndrome of skin eruption due to heat in the blood.

Symptoms and signs: Sthenic fever, delirium, irritability, mania, obvious skin eruption, or hematemesis, hemoptysis, epistaxis, hematochezia and hematuria, deep red tongue with dry and yellow coating, and rapid pulse.

Principle of treatment: Clearing heat, cooling the blood, eliminating toxin and dispersing macula.

Recipe: *Hua Ban Tang*(24) and *Xi Jiao Di Huang Tang*(25) with modification.

Ingredients:

Cornu Rhinoceri (to be decocted first)	6–10 g
Gypsum Fibrosum	30–90 g
Rhizoma Anemarrhenae	15–30 g
Radix Rehmanniae	30 g
Radix Scrophulariae	30 g
Cortex Moutan Radicis	12 g
Radix Paeoniae Alba	12 g
Semen Oryzae Sativae	9 g
Radix Glycyrrhizae	6 g

Decoct the above ingredients in water for oral dose.

Injection: *Qing Kai Ling* 30–40 ml in 100 ml of 5% glucose solution is given by intravenous drip, 1 or 2 times daily.

(14) The syndrome of deficiency of *Yin* and collapse of *Yang* (collapse syndrome due to high fever)

Symptoms and signs: Coma, dry throat, sinking eyes, restlessness, cold limbs, cold or greasy sweat, red tongud with no coating, and fading pulse.

Principle of treatment: Tonifying *Qi*, nourishing *Yin*, retoring *Yang*, and saving the collapse.

Recipe: *Sheng Mai San*(26) and *Si Ni Tang*(27) with modification.

Ingredients:
Radix Ginseng	9–30 g
Radix Ophiopogonis	30 g
Fructus Schisandrae	6 g
Radix Aconiti Praeparata	9 g
Rhizoma Zingiberis	6 g
Radix Glycyrrhizae	6 g

Decoct the above ingredients in water for oral dose.

1.2 Shock

According to its manifestations, shock due to various causes falls into the category of *Jue Tuo*(syndrome of syncope and prostration) in traditional Chinese medicine. It should therefore be taken as syncope and prostration in differentiation and treatment.

Etiology and Pathogenesis

The disease is chiefly caused by invasion of the *Ying* and blood systems by the six exogenous pathogenous factors, or by exhaustion of *Qi* and impairment of *Yin* affecting the five *Zang* organs due to severe pain, fear and fright, loss of body fluid, bleeding, consumption of vital essence, allergy, poisoning and lingering illness, which lead to disturbance of the flow of *Qi* and blood, imbalance between *Yin* and *Yang*, and adverseness of *Qi* activities.

Essentials of Diagnosis

1. This disease is characterized by sudden onset and rapid progress and transmission.

2. As a secondary syndrome of different kinds of illness, it has its primary causes and the corresponding manifestations.

3. This disease is characterized by such clinical manifestations as cold limbs, pallor or flushed face or cynosis, frequent cold sweat, dysphoria or apathetic expression, feeble or short and coarse breathing, deep and thready pulse, or even fading and impalpable pulse.

4.Drop of blood pressure: Systolic pressure is often lower than 80 mmHg.The difference of pulse pressure is less than 20 mmHg.Systolic pressure in patients with hypertention could be one third or 30 mmHg lower than the usual level.

Treatment Based on Differentiation of Syndromes

1. Exuberance of Noxious Heat

Symtoms and signs: Fever without chills, cold limbs, fidgets, thirst, derilium, burning sensation in the chest and abdomen, scanty and deep yellow urine, constipation or foul stools, red tongue with yellow and dry coating, and deep, thready and rapid pulse.

Principle of treatment: Purging away the heat and removing the toxin.

Recipe: *Ren Shen Bai Hu Tang*(28) with modification.

Ingredients:

Radix Ginseng	15—30 g
Gypsum Fibrosum	30—90 g
Rhizoma Anemarrhenae	15 g
Semen Oryzae Sativa	15 g
Radix Glycyrrhizae	6 g
Radix Scutellariae	15 g
Rhizoma Coptidis	9 g

| Herba Taraxaci | 30—60 g |

Decoct the above ingredients in water for oral dose.

Modification: In case of constipation due to obstruction of *Qi* of the *Fu* organs, add Radix and Rhizoma Rhei 6 g, Natrii Sultas 6 g, Fructus Aurantii Immaturus 12 g and Cortex Magnoliae Officinalis 12 g.

In case of rattle in the throat due to stagnation of *Qi* and accumulation of phlegm, add Rhizoma Pinelliae, Pericarpium Citri Reticulatae, Rhizoma Arisaematis, Fructus Aurantii Immaturus, etc., 12 g each.

Injection: 300—400 ml of *Qing Qi Jie Du* Injection for intravenous drip, 1 or 2 times daily.

Add 20—40 ml of *Qing Kai Ling* Injection to 250 ml of 5% glucose solution for intravenous drip, 1 or 2 times daily.

2. Deficiency of Both *Qi* and *Yin*

Symptoms and signs: Spiritlessness and lassitude, cold limbs, thirst, sweating, weakness of breath, red or light red tongue with thin and white coating, and thready and rapid pulse.

Principle of treatment: Tonifying *Qi* and nourishing *Yin*.

Recipe: *Sheng Mai San*(26) with modification.

Ingredients:

Radix Ginseng	15 g
Radix Astragali seu Hedysari	24 g
Radix Ophiopogonis	30 g
Fructus Schisandrae	9 g

Decoct the above ingredients in water for oral dose.

Injection: Add 50—100 ml of *Shen Mai* Injection to 250—500 ml of 5% glucose solution for intravenous drip, which should not be discontinued until the patient's condition is improved. Or add

20 ml of *Shen Mai* Injection to 20 ml of 25% glucose solution for intravenous injection, once every 10—15 minutes for successively 3—5 times. After the blood pressure goes up and becomes stable, 50—100 ml of *Shen Mai* Injection should be put in 500 ml of *Zeng Ye* or *Yang Yin* Injection for intravenous drip.

3. Sudden Collapse of *Yang Qi*

Symptoms and signs: Cold limbs, low body temperature or hypothemia, profuse sweating, short feeble breathing, pale tongue, and fading or impalpable pulse.

Principle of treatment: Restoring *Yang* and saving the collapse.

Recipe: *Shen Fu Tang*(29) or *Si Ni Tang*(27) with modification.

Ingredients:

Radix Ginseng Rubra	15—30 g
Radix Aconiti	
(to be decocted first)	15—30 g
Rhizome Zingiberis	9 g
Glycyrrhizae Praeparata	9 g

Decoct the above ingredients in water for oral dose.

4. Exhaustion of Kidney *Yin*

Symptoms and signs: Trance, flushed face, fever, mental restlessness, fear and fright, palpitation, thirst, scanty and yellow urine, cold limbs, dry and glossy tongue without coating, deficient and rapid or knotted pulse.

Principle of treatment: Nourishing *Yin*, supplementing body fluid and saving the collapse.

Recipe: *San Jia Fu Mai Tang* (30) with modification.

Ingredients:

Concha Ostreae	30 g
Carapax Trionycis	
(to be decocted first)	30 g
Plastrum Testudinis	
(to be decocted first)	30 g
Radix Rehmanniae	30 g
Radix Ophiopogonis	30 g
Fructus Corni	15 g
Fructus Schisandrae	9 g
Radix Glycyrrhizae Praeparata	9 g

Decoct the above ingredients in water for oral dose.

5. Insufficiency of Heart *Qi*

Symptoms and signs: Palpitation, restlessness, shortness and weakness of breath, asthenia, pale tongue with thin and white coating, and thready and rapid or knotted pulse.

Principle of treatment: Tonifying and invigorating the *Qi* of the heart.

Recipe: *Zhi Gan Cao Tang*(31) with modification.

Ingredients:

Radix Glycyrrhizae Praeparata	9–15 g
Radix Ginseng	9–30 g
Ramulus Cinnamomi	9 g
Radix Ophiopogonis	30 g
Os Draconis	
(to be decocted first)	15 g
Concha Ostreae	
(to be decocted first)	30 g

Decoct the above ingredients in water for oral dose.

6. Stagnation of *Qi* and Stasis of Blood

Symptoms and signs: Cyanosis, ecchymosis of skin, abdominal distention, dark purple tongue, and deep, thready and hesitant pulse.

Principle of treatment: Activating blood circulation, dispersing blood stasis, regulating *Qi* and saving the collapse.

Recipe: *Xue Fu Zhu Yu Tang*(32) with modification.

Ingredients:

Radix Bupleuri	9 g
Fructus Aurantii Immaturus	9 g
Pericarpium Citri Reticulatae Viride	9 g
Radix Paeoniae Alba	9 g
Rhizoma Ligustici Chuanxiong	9 g
Flos Carthami	6 g
Radix Salviae Miltiorrhizae	15 g
Radix et Rhizoma Rhei Praeparata	9 g
Radix Glycyrrhizae	6 g

Decoct the above ingredients in water for oral dose.

Injection: Add 20–30 ml of *Dan Shen* Injection(Injection Salviae Miltiorrhizae) to 250 ml of 5% glucose solution for intravenous drip.

1.3 Coma

Coma is one of the common emergencies in internal medicine. According to its clinical manifestations, it corresponds to *Shen Hun, Hun Meng, Hun Jue* and *Hun Kui* in traditional Chinese medicine.

Etiology and Pathogenesis

Coma is caused by the obstruction or lack of nourishment of

the heart and brain. It may be resulted from either *Qi* obstruction in the brain due to the invasion of epidemic noxious heat and the up-stirring of the wind-phlegm and the blood or lack of nourishment of the brain or mental derangement brought about by the comsumption of *Qi* and blood in the interior or internal injuries, all of which may lead to the isolation of the depleted *Yin* and *Yang*.

Essentials of Diagnosis

1. The onset of the disease is abrupt.

2. The main clinical feature is unconsciousness with no response to arousing or failure to recognize acquaintances.

3. Coma is a kind of critical condition resulted from deterioration of various kinds of diseases and is mostly associated with the symptoms of the primary diseases.

Treatment Based upon Differentiation of Syndromes

1. Coma of Excess Type

(1) Impairment of heart-nutriment by pathogenic heat

Symptoms and signs: High fever, coma, dysphoria, delirium, or unconsciousness with no response to call, accompanied with macular eruption, epistaxis, frequent clonic convulsion, or opisthotonos, deep red tongue with yellow coating and little fluid, and slippery and rapid or thready and rapid pulse.

Principle of treatment: Clearing heat away from the heart for resuscitation and expelling heat to recuperate *Yin*.

Recipe: *Qing Ying Tang*(18) with modification.

Ingredients:

Corni Rhinoceri	
(to be decocted first)	10 g
Radix Rehmanniae	24 g

Scrophulariae	18 g
Leaves bud of Herba Lophatheri	3 g
Plumula Nelumbinis	3 g
Fructus Forsythiae	15 g
Radix ophiopogonis	30 g
Flos Lonicerae	30–60 g
Rhizoma Coptidis	6 g

Decoct the above ingredients in water for oral dose.

Modification: In case of severe coma, add Rhizoma Acori Graminei 12 g and Radix Curcumae 12 g, and give *An Gong Niu Huang Wan* or *Zhi Bao Dan* orally or nasally, one pill each time and 4–6 times a day.

In case of clonic convulsion, add Cornu Antelopis 6 g(to be prepared into powder and taken with water)and Ramulus Uncariae cum Uncis 30 g and Lumbricus 12 g, and give *Zi Xue Dan* orally or nasally, one pill each time and 3–4 times a day.

(2) Dyspnea and mist of the heart by phlegm.

Symptoms and signs: Dementia, alternate unconsciousness and consciousness, delirium or semiconsciousness, accompanied with stuffy chest, nausea, cough with dyspnea, shortness of breath, accumulation of phlegm, recessive fever which is higher in the afternoon, white and sticky or yellow, thick and dirty tongue coating, and soft, slippery and rapid pulse.

Principle of treatment: Resolving phlegm for resuscitation, clearing away heat and eliminating dampness to calm asthma.

Recipe: *Chang Pu Yu Jin Tang*(33) and *San Zi Yang Qin Tang*(34)with modification.

Ingredients:

Rhizoma Acori Graminei	12 g

Radix Curcumae	12 g
stir-fried Fructus Gardeniae	12 g
Fructus Forsythiae	15 g
Herba Lophatheri	9 g
Fructus Arctii	12 g
Rhizoma Pinelliae prepared with ginger juice	9 g
Poria	12 g
Pericarpium Citri Reticulatae	9 g
Semen Sinapis Albae	6 g
Fructus Perillae	9 g
Semen Raphani	12 g
Cortex Moutan Radicis	12 g
Flos Chrysanthemi	12 g
Talcum	15 g

Decoct the above ingredients in water for oral dose.

Meantime, 15-30 ml of *Xian Zhu Li* (Succus Bambusae) should be administered following its infusion 2 or 3 times daily.

Modification: In case of severe coma, add *Su He Xiang Wan*, one pill each time and 3 or 4 times a day.

(3) The excess syndrome of *Fu* organs due to accumulation of dryness.

Symptoms and signs: Coma, delirium, fidgets, restlessness, high fever or afternoon fever, fullness, distention and pain in the abdomen, constipation, red tongue with yellow dry or thorny coating, and deep, excess and rapid pulse.

Principle of treatment: Eliminating constipation to purge away heat.

Recipe: *Da Cheng Qi Tang*(11).

Ingredients:

Radix et Rhizoma Rhei (to be decocted later)	9–30 g
Natrii Sulfas (to be melted)	6–9 g
Fructus Aurantii Immaturus	12 g
Cortex Magnoliae Officinalis	12 g

Decoct the above ingredients in water for oral dose.

Modification: In case of thirst with excessive drinking, add Gypsum Fibrosum 30–60 g and Rhizoma Anemarrhenae 15 g.

In case of delirium mania, add *Zi Xue Dan* 0.15–0.3 g, 2 or 3 times a day.

(4) Acute jaundice due to damp—heat

Symptoms and signs: Coma, delirium, or spells of unconsciousness, restlessness, deepening jaundice, macular eruption and epitaxis or apparent abdominal distention, deep red tongue with sticky coating, and wiry and rapid pulse.

Principle of treatment: Removing dampness by diuresis and clearing pathogenic heat, restoring consciousness by cooling the blood.

Recipe: *Yin Chen Hao Tang*(35) with modification.

Ingredients:

Herba Artemisiae Capillaris	30 g
stir—fried Fructus Gardeniae	12 g
Cornu Bubali (to be filed into powder and taken following its infusion)	15 g
Radix et Rhizoma Rhei	9 g
Radix Rehmanniae	15 g

Cortex Moutan Radicis	12 g
Radix Scrophulariae	15 g
Rhizoma Acori Graminei	12 g
Herba Dendrobii	12 g

Decoct the above ingredients in water for oral dose.

Shen Xi Dan may also be given 3 g each time, 3 or 4 times daily.

Injection: Add 10–20 ml of *Xing Nao Jing* to 250 ml of 5% glucose solution for intravenous drip, 1 or 2 times a day.

(5) Stagnation of blood and heat obstructing the orifice.

Symptoms and signs: Coma, delirium or mania, high fever that gets worse at night, thirst with preference of drinking, lower abdominal fullness, distention and pain, constipation, cyanosis, deep red or dark purple tongue, and wiry and rapid pulse.

Principle of treatment: Restoring consciousness by clearing away heat and dispersing blood stasis.

Recipe: *Xi Di Qing Luo Yin*(36) with modification.

Ingredients:

Cornu Rhinoceri	
(to be cut into slcies and decocted first)	15–30 g
Radix Rehmanniae	30 g
Radix Paeoniae Rubra	15 g
Cortex Moutan Radicis	12 g
Fructus Forsythiae	30 g
Semen Persicae	12 g
Rhizoma Acori Graminei	12 g
Radix et Rhizoma Rhei	6 g
Powder of Succinum	
(to be taken after being infused)	2 g

Decoct the above ingredients in water for oral dose.

Modification: In case coma with high fever, add *Zi Xue Dan* 0.15—0.3 g each time, 2 or 3 times daily, or *An Gong Niu Huang Wan*, 1 pill each time, 2 or 3 times daily.

In case of flaming of heart fire, add Rhizoma Coptidis 10 g and Fructus Gardeniae 10 g.

In cases of haematemesis or epistaxis, add Cacumen Biotae, Herba Ecliptae and Rhizoma Imperatae 30 g each.

(6) Apoplexy due to hyperactivity of *Yang*.

Symptoms and signs: Sudden syncope with no consciousness, trismus, clenched fists, discontinuance of defecation and urination, rigidity and convulsion of limbs, hemiplegia, accompanied with flushed face, coarse breathing, frequent snore, red tongue with yellow and dry coating, and wiry, slippery and rapid pulse.

Principle of treatment: Soothing the liver *Yang*, dispelling wind and restoring consciousness.

Recipe: *Ling Yang Gou Teng Tang* (22) with modification.

Ingredients:

Cornu Antelopis (to be filed into powder and taken following its infusion)	3 g
Ramulus Uncariae cum Uncis	15 g
Radix Rehmanniae	15 g
Cortex Moutan Radicis	12 g
Plastrum Testudinis	18 g
Spica Prunellae	30 g
Concha Haliotidis	30 g
Radix Paeoniae Alba	12 g

Radix Bupleuri	9 g
Herba Menthae	9 g
Flos Chrysanthemi	15 g
Concha Ostreae (to be decocted first)	30 g

Decoct the above ingredients in water for oral dose.

Modification: In case of coma due to accumulation of phlegm, add Arisaema cun Bile 12 g, Fructus Trichosanthis 30 g, Concretio Silicea Bambusae 12 g, Succus Bambusae 30 g and *Sheng Jiang Zhi*(ginger juice)12 g.

In case of clonic convulsion, add Scorpio 10 g, Scolopendra 3 pieces and Bombyx Batryticatus 12 g.

In case of constipation, foul breathing and abdominal distention, add Radix et Rhizoma Rhei 10 g, Natrii Sulfas 6 g and Fructus Aurantii Immaturus 12 g.

Injection: Add 20–40 ml of *Qing Kai Ling* Injection to 250 ml of 5% glucose solution for intravenous drip, 1 or 2 times daily.

(7) Sunstroke in summer

Symptoms and signs: Dizziness, headache, stuffiness in the chest, fever, flushed face, followed by loss of consciousness and sudden fall, cold limbs, or coma with delirium, red and dry tongue, and full and rapid pulse.

Principle of treatment: Eliminating summer–heat and supplementing *Qi*, and clearing heat from the heart for resuscitation.

Recipe: First administer either one pill of *Wan Shi Niu Huang Qing Xin Wan* or one pill of *Zi Xue Dan*, and then modified *Bai Hu Tang*(10).

Bai Hu Tang with modification.

Ingredients:

Gypsum Fibrosum	30–60 g
Rhizoma Anemarrhenae	12 g
Radix Glycyrrhizae	3 g
Herba Dendrobii	15 g
Radix Pseudostellariae	30 g
Petiolus Nelumbinis	15 g
Rhizoma Coptidis	6 g
Radix Ophiopogonis	30 g
Herba Lophatheri	12 g
Exocarpium Citrulli	30 g

Decoct the above ingredients in water for oral dose.

Modification: In case of clonic convulsion, add Cornu Antelopis 5 g (to be taken following its infusion), Ramulus Uncariae cum Uncis 30 g and Scorpio 10 g.

2. Coma of Deficiency Type

(1) Coma caused by exhaustion of *Yin*.

Symptoms and signs: Coma, sweating, flushed face, fever, red and dry lips and tongue, and deficient and rapid pulse.

Principle of treatment: Recuperating the depleted *Yin* and astringing *Yang*.

Recipe: *Sheng Mai San*(26) with modification.

Ingredients:

Radix Ginseng Rubra	
(to be decocted alone)	6–30 g
Radix Ophiopogonis	30 g
Fructus Schisandrae	6 g
Fructus Corni	15 g
Rhizoma Polygonati	30 g
Os Draconis	30 g

Concha Ostreae 30 g

Decoct the above ingredients in water for oral dose.

Modification: In case of stiff tongue and aphasia caused by blockage of the orifice due to phlegm-heat, add Bulbus Fritillariae Cirrhosae, Succus Bambusae, Arisaema cum Bile and Concretio Bambusae.

(2) Coma caused by collapse of *Yang*.

Symptoms and signs: Pallor, mental confusion with no speech, closed eyes, open mouth, feeble breathing, cold limbs with flaccidity of hands, profuse sweating, incontinence of urination and bowels, pale and moist or bluish purple lips and tongue, fading pulse.

Principle of treatment: Recuperating the depleted *Yang* and rescuing the patient from collapse.

Recipe: *Shen Fu Tang*(29) with modification.

Ingredients:

Radix Ginseng Rubra	12 g
Radix Aconiti Praeparata	30 g
Os Draconis	30 g
Concha Ostreae	30 g
Rhizoma Acori Graminei	12 g

Decoct the above ingredients in water for oral dose.

Modification: In case of preponderance of *Yin* and cold in the interior, add Rhizoma Zingiberis 10 g and Radix Glycyrrhizae 10 g.

2 Cardiovascular Diseases

Cardiovascular diseases in modern medicine correspond to the diseases of the heart and vessel system in traditional Chinese medicine. Its common clinical emergencies are as follows.

2.1 Angina Pectoris

Angina pectoris in modern medicine is defined as *Xin Tong*(precordial pain), *Xiong Tong*(chest pain), *Xin Bi*(obstruction of the heart *Qi*)and *Xiong Zhong Tong*(pain in the chest).

Etiology and Pathogenesis

The onset of this disease is resulted from dysfunction of *Zang—Fu*, stagnation of *Qi*, obstruction of the heart and vessels and malnourishment of the heart due to improper food—intake, mental depression, over—strain and over—stress giving rise to internal injury, invasion of the interior by pathogenic cold, etc.

Essentials of Diagnosis

1. People over 40 years old are at the higher risk for the disease.

2. Chest pain radiating to the left back and shoulder or medial side of the left arm.

3. Its sudden onset is often induced by overfatigue, excessive eating, cold or excitement, usually last for 3—6 minutes and can be alleviated by rest or medication.

4. Chronic coronary insufficiency or acute myocardial

ischemia can be found in E.C.G.

Treatment Based on Differentiation of Syndromes

1. Stagnation of the heart blood

Symptoms and signs: Fixed stabbing pain in the chest which gets worse at night, accompanied by stuffiness in the chest, shortness of breath, palpitation, dark purple tongue, and deep and hesitant pulse.

Principle of treatment: Activating blood circulation, dispersing stagnation, removing obstruction from the channel and arresting pain.

Recipe: *Xue Fu Zhu Yu Tang*(32) with modification.

Ingredients:

Radix Angelicae Sinensis	12 g
Radix Paeoniae Rubra	12 g
Rhizoma Ligustici Chuanxiong	10 g
Semen Persicae	10 g
Flos Carthami	10 g
Radix Bupleuri	10 g
Radix Curcumae	10 g
Fructus Aurantii (stir-fried)	10 g
Rhizoma Corydasis (stir-fried)	10 g

Decoct the above ingredients in water for oral dose.

2. Stagnation of phlegm

Symptoms and signs: Stuffiness in the chest, chest pain radiating to the shoulder and back, obesity, shortness of breath, asthma, cough with much expectoration, puffy tongue greasy coating, and wiry and slippery pulse.

Principle of treatment: Removing obstruction of the channel by eliminating the turbid and activating *Yang* by resolving phlegm.

Recipe: *Gua Lou Xie Bai Ban Xia Tang*(37) with modification.

Ingredients:

Fructus Trichosanthis	24 g
Bulbus Allii Macrostemi	15 g
Rhizoma Pinelliae	12 g
Fructus Aurantii Immaturus (stir-fried)	10 g
Rhizoma Zingiberis	6 g
Semen Amomi Cardamomi	10 g
Pericarpium Citri Reticulatae	10 g

Decoct the above ingredients in water for oral dose.

Modification: In case of complication of blood stasis, add *Fu Fang Dan Shen Pian* 4 tablets each time and 3 times a day.

3. Stagnation of *Yin* and cold in the interior

Symptoms and signs: Chest pain radiating to the back which is aggravated by cold, accompanied with stuffiness in the chest, shortness of breath, palpitation, pallor, cold limbs, white and slippery tongue coating, and deep and thready pulse.

Principle of treatment: Activating *Yang* with the drugs pungent in flavour and warm in property, removing obstruction from the channel and dispelling cold.

Recipe: *Gua Lou Xie Bai Bai Jiu Tang*(38) with modification.

Ingredients:

Fructus Trichosanthis	24 g
Bulbus Allii Macrostemi	15 g
Radix Aconiti Praeparata	12 g

Ramulus Cinnamomi	10 g
Fructus Aurantii Immaturus (stir-fried)	10 g
Radix Glycyrrhizae Praeparata	6 g
Lignum Santali	6 g
White Spirit (to be used as a guiding drug)	10 g

Decoct the above ingredients in water for oral dose.

Modification: In case of cough with expectoration, add Pericarpium Citri Reticulatae 12 g, Poria 12 g and stir-fried Semen Armeniacae Amarum 12 g.

In case of severe pain difficult to alleviate, add *Su He Xiang Wan* 1 pill each time and twice a day.

4. Deficiency of both *Qi* and *Yin*

Symptoms and signs: Stuffiness and dull pain in the chest, palpitation, shortness of breath, lassitude, dislike of speaking, dizziness and vertigo, slightly red tongue, and thready and forceless or knotted pulse.

Principle of treatment: Tonifying *Qi*, nourishing *Yin*, activating blood circulation and removing obstruction from the channel.

Recipe: *Sheng Mai San*(26) with modification.

Ingredients:

Radix Ginseng	12 g
Radix Ophiopogonis	24 g
Fructus Schisandrae	10 g
Rhizoma Atractylodis Macrocephalae	12 g
Poria	12 g
Rhizoma Polygonati	15 g

| Radix Salviae Miltiorrhizae | 15 g |
| Radix Glycyrrhizae Praeparata | 6 g |

Decoct the above ingredients in water for oral dose.

Modification: In case of obvious chest stuffiness and pain, add Radix Curcumae 10 g and Radix Notoginseng 6 g.

In case of persistent severe chest pain, add 2ml of *Fu Fang Dan Shen* Injection to 20 ml of 5% glucose solution for intravenous injection.

5. Deficiency of *Yin* of both heart and kidney

Symptoms and signs: Stuffiness and pain in the chest, dizziness, tinnitus, mental restlessness, insomnia, soreness and weakness of lumbus and knees, red tongue with possible purple spots, and thready and rapid pulse.

Principle of treatment: Nourishing the heart and kidney, removing obstruction from the channel and calming the mind.

Recipe: *Zuo Gui Yin*(39) with modification.

Ingredients:

Radix Rehmanniae Praeparata	15 g
Fructus Corni	12 g
Fructus Lycii	15 g
Rhizoma Dioscoreae	12 g
Poria	12 g
Cortex Moutan Radicis	12 g
Radix Salviae Miltiorrhizae	20 g
Radix Ophiopogonis	15 g
Semen Ziziphi Spinosae (stir-fried)	20 g

Decoct the above ingredients in water for oral dose.

Modification: In case of significant mental restlessness and insomnia, add Fructus Gardeniae 10 g and Bulbus Lilii Praeparata 24 g.

In case of dizziness and vertigo complicated with obvious flushed face, add Ramulus Uncariae cum Uncis 30 g, Concha Haliotidis 30 g, Radix Polygoni Multiflori Praeparata 30 g and Radix Achyranthis Bidentatae 15 g.

6. Stasis of blood due to deficiency of *Qi*

Symptoms and signs: Stuffiness in the chest, shortness of breath, stabbing pain in the chest, aggravated by exertion and accompanied by general lassitude, palpitation, light red tongue with stagnant spots, and thready, forceless and hesitant pulse.

Principle of treatment: Tonifying *Qi*, nourishing the heart, activating blood circulation and calming the mind.

Recipe: *Bu Yang Huan Wu Tang*(40) with modification.
Ingredients:

Radix Astragali seu Hydysari	60 g
Radix Codonopsis Pilosulae	15 g
Tail of Radix Angelicae Sinensis	6 g
Semen Persicae	6 g
Flos Carthami	6 g
Rhizoma Ligustici Chuanxiong	6 g
Stir-fried Semen Ziziphi Spinosae	20 g
Radix Glycyrrhizae Praeparata	6 g

Decoct the above ingredients in water for oral dose.

Modification: In case of significant chest pain, add *Guan Xin No.2* six tablets each time and three times a day, or *Qi Li San* one gram each time and two or three times a day.

2.2 Acute Myocardial Infarction

Acute myocardial infarction in modern medicine corresponds to *Zhen Xin Tong*(angina pectoris), *Cu Xin Tong*(sudden attack of cardiac pain) and *Jue Xin Tong*(cardiac pain with cold limbs) in traditional Chinese medicine.

Etiology and Pathogenesis

The heart *Yang* is suppressed and the heart channel is obstructed by either phlegm, or blood stasis, or *Qi* stagnation, or stagnation of cold, which deprives the heart channel of nourishment and injures the heart, giving rise to sudden attack.

Essentials of Diagnosis

1. With intermittent attacks of chest pain, the patient may suddenly have squeezing pain, palpitation, profuse sweating, cold limbs, and fading pulse.

2. The attack is often induced by overfatigue, excessive enotional changes, sudden change of the weather and excessive food-intake and alcohol drinking.

3. People over 40 tend to have more opportunity to have this disease.

4. The ECG sequence can tell the pathological changes of myocardiac infarction.

5. Laboratorial examination will give the result that CPK, GOT, LOH and HBDH are higher than normal.

Treatment Based on Differentiation of Syndromes

Acute stage(within one week)

1. Exhaustion of *Qi* of the heart.

Symptoms and signs: Sudden chest pain followed by cold limbs and coma complicated with pallor, cold sweating, pale tongue with white and slippery coating, and feeble and thready pulse.

Principle of treatment: Tonifying *Qi*, activating *Yang* and saving the collapse for resuscitation.

Recipe: *Sheng Mai San*(26) with modification.
Ingredients:

Radix Ginseng	30 g
Radix Astragali seu Hedysari	30 g
Radix Ophiopogonis	30 g
Ramulus Cinnamomi	12 g
Fructus Schisandrae	9 g
Moschus (wrapped in a piece of silk to be decocted)	0.15 g

Decoct the above ingredients for oral administration, 1 or 2 doses daily.

Recipe: *Bu Qi Gu Tuo San*(experienced prescription)
Ingredients:

Radix Ginseng	6 portions
Cortex Cinnamomi	1.5 portions
Moschus	0.05 portion

Mix the drugs according to the proportion mentioned above, and grind them into very fine powder for oral or nasal administration 9 g each time, 2 or 3 times daily.

Acupuncture treatment: Puncture point Nei Guan(PC 6) and Jiu Wei(RN 15)with strong stimulation and no retention of needles.

Recipe: *Yi Xin Wan*, 1 pill for oral administration or nasal feeding.

2. Collapse of Heart *Yang*

Symptoms and signs: Very severe chest pain, profuse cold sweating, aversion to cold, cold limbs, pallor, listlessness, bluish purple lips and tongue, and fading pulse.

Principle of treatment: Restoring *Yang* and astringing *Yin* to rescue the patient from collapse.

Recipe: *Ren Shen Gui Zhi Tang*(41) with modification.

Ingredients:

Radix Ginseng	15 g
Ramulus Cinnamomi	12 g
Radix Aconiti Praeparata	12 g
Fructus Corni	15 g
Rhizoma Zingiberis	10 g
Radix Glycyrrhizae Praeparata	10 g

Decoct the above ingredients in water for oral dose.

Modification: In case of persistent sweating, add Os Draconis 30 g and Concha Ostreae 30 g to astringe sweat and save collapse.

Ear acupuncture: Puncture or press ear seeds on the auricular points Heart, Sympathetic Nerve and Subcortex.

3. Attack of the heart by the retained fluid.

Symptoms and signs: Sudden attack of cardiac pain, severe stuffiness in the chest, cough with frothy expectoration, aversion to cold, cold limbs, listlessness, shortness of breath, spontaneous sweating, palpitation, facial edema, puffy tongue with tooth marks on the border, and deep, slow and feeble pulse.

Principle of treatment: Warming the *Yang*, dispersing the retained fluid, removing obstruction and opening the mind.

Recipe: *Wen Yang Hua Yin Tang*(experienced prescription)

Ingredients:

Ramulus Cinnamomi	15 g
Poria	30 g
Rhizoma Zingiberis	12 g
Semen Lepidii seu Descurainiae	30 g
Rhizoma Acori Graminei	24 g
Radix Curcumae	10 g
Herba Ephedrae Praeparata	6 g

Decoct the above ingredients in water for oral dose.

Modification: In case of significant edema and palpitation, add Radix Aconiti Praeparata 12 g, Rhizoma Alismatis 24 g and Semen Plantaginis 24 g.

In case of obvious cough with expectoration, add Fructus Trichosanthis 24 g, Radix Peucedani 12 g, Rhizoma Pinelliae 12 g and Herba Asari 6 g.

Su Bin Di Wan could be swallowed or sucked 2—4 pills each time and 2 or 3 times daily.

Kuan Xiong Wan can also be taken 1 pill each time and 3 times daily.

Ear acupuncture: Puncture auricular points Heart, Subcortex, Adrenal, Lung, etc.

Remission stage.

1. Stagnation of *Qi* and stasis of blood

Symptoms and signs: Fixed stabbing pain in the chest combined with stuffiness in the chest, fullness and distention in hypochondriac region, irritability, insomnia, dream-disturbed sleep, red border with stagnant spots on the tongue, and wiry pulse.

Principle of treatment: Activating circulation of *Qi* and

blood, removing obstruction from the channel and calming the mind.

Recipe: *Xue Fu Zhu Yu Tang*(32) with modification.

Ingredients:

Radix Bupleuri	12 g
Radix Curcumae	12 g
Radix Salviae Miltiorrhizae	30 g
stir-fried Fructus Aurantii Immaturus	10 g
Rhizoma Ligustici Chuanxiong	10 g
Cortex Moutan Radicis	10 g
Flos Rosae Rugosae	12 g
Flos Carthami	10 g
Rhizoma Cyperi	8 g

Decoct the above ingredients in water for oral dose.

Modification: In case of severe paroxysmal pain radiating to the right hypochondriac area, add stir-fried Rhizoma Corydalis 12 g and Fructus Meliae Toosendan 10 g.

In case of mental restlessness and insomnia, add Radix Ophiopogonis 24 g, Succinum powder 1 g and Cinnabaris powder 1 g(to be taken following its infusion).

2. Chest-*Bi* syndrome caused by phlegm(see Angina Pectoris).

3. Stagnation of cold in the heart channel(see Angina Pectoris).

4. Stagnation of phlegm and retention of food.

Symptoms and signs: Stuffiness and pain in the chest, fullness and distention in epigastrium and abdomen, nausea, poor appetite, constipation, palpitation, insomnia, red tongue with light yellow, thick and sticky coating, and wiry and slippery pulse.

Principle of treatment: Regulating the stomach, removing obstruction from the *Fu* organ, resolving phlegm and promoting circulation in the channel.

Recipe: *Ban Xia Xie Xin Tang*(42) with modification.

Ingredients:

Rhizoma Pinelliae	12 g
stir-fried Fructus Citri Aurantii Immaturus	10 g
Rhizoma Coptidis	6 g
Rhizoma Zingiberis	3 g
Radix et Rhizoma Rhei Praeparata	6 g
Lignum Dalbergiae Odoriferae	10 g
Fructus Trichosanthis	24 g
Radix Salviae Miltiorrhizae	24 g
Fructus Crataegi	24 g

Decoct the above ingredients in water for oral dose.

Modification: In case of fullness and distention in epigastrium and abdomen combined with poor appetite, add Fructus Amomi 10 g, and Endothelium Corneum Gigeriae Galli powder 3 g(to be taken following its infusion).

5. Deficiency of heart *Qi*

Symptoms and signs: Paroxysmal dull pain in the chest combined with emptiness sensation, shortness of breath lassitude and palpitation which get worse after exertion, light red tongue with thin and white coating, and thready and forceless pulse.

Principle of treatment: Tonifying the *Qi* of the heart, removing obstruction from the channel and calming the mind.

Recipe: *Bao Dan Yin*(experienced prescription)

Ingredients:

Radix Ginseng	12 g
Radix Astragali seu Hydysari	30 g
Radix Salviae Miltiorrhizae	24 g
Radix Angelicae Sinensis	12 g
Semen Persicae	6 g
Ramulus Cinnamomi	6 g
Lignum Santali	6 g
Radix Glycyrrhizae Praeparata	3 g

Decoct the above ingredients in water for oral dose.

Modification: In case of deficiency of *Yin* add Radix Ophiopogonis 21 g and Fructus Schisandrae 6 g.

In case of deficiency of *Yang*, add Radis Aconiti Praeparata 6 g.

6. Deficiency of heart *Yang*.

Symptoms and signs: Stuffiness and pain in the chest, palpitation, aversion to cold, cold limbs, spontaneous sweating, shortness of breath, worse after exertion, light red tongue with white and slippery coating, and deep, slow, thready and forceless pulse or knotted pulse.

Principle of treatment: Warming *Yang*, tonifying *Qi*, resolving retained fluid and removing obstruction from the channel.

Recipe: *Ren Shen Si Ni Tang*(43) with modification.

Ingredients:

Radix Ginseng	12 g
Radix Aconiti Praeparata	10 g
Poria	24 g
Rhizoma Atractylodis Macrocephala	12 g

Radix Glycyrrhizae　　　　　　　　　　　　　　　　3 g

Decoct the above ingredients in water for oral dose.

Modification: In case of upward attack of excessive fluid giving rise to nausea and vomiting or spitting of phlegm-fluid, add Rhizoma Pinelliae Prepared with ginger Juice 10 g, Semen Lepidii seu Descurainiae 15 g and Pericarpium Citri Reticulatae 10 g.

In case of obvious palpitation, add Os Draconis 30 g and Concha Ostreae 30 g.(see Collapse of Heart *Yang*)

7. Deficiency of both *Qi* and *Yin*(see Angina Pecotris).

8. Deficiency of *Yin* of both heart and kidney(see Angina Pectoris).

2.3 Acute Pulmonary Heart Disease

Acute pulmonary heart disease in modern medicine is involved in *Ke Chuan*(cough and asthma), *Ka Xue*(hemoptysis) and *Shui Zhong*(edema)of traditional Chinese medicine.

Etiology and Pathogenesis

It is resulted from prolonged cough and asthma which lead to consumption of antipathogenic *Qi*, accumulation of phlegm and stasis of blood in the interior, perversion of *Qi* and upward attack of retained fluid, affecting the heart and lung.

Essentials of Diagnosis

1. It is characterized by prolonged cough and asthma accompanied by sudden worsening, shortness of breath and difficulty in lying flat.

2. Recent profuse expectoration or epigastric and abdominal fullness and distention and edema of the lower limbs.

3. Severe cough with suffocation sensation, hemoptysis.

Treatment Bassed on Differentiation of Syndromes

1. Retention of phlegm—fluid affecting the heart and lung.

Symptoms and signs: Cough and asthma with profuse frothy expectoration, difficulty in lying flat, palpitation, edema of lower limbs, dark red and puffy tongue, and slippery and rapid pulse.

Principle of treatment: Warming *Yang*, resolving phlegm, dispersing the lung and calming the mind.

Recipe: *Ling Gui Zhu Gan Tang*(44) with modification.

Ingredients:

Poria	30 g
Ramulus Cinnamomi	12 g
Rhizoma Atractylodis Macrocephalae	24 g
Semen Lepidii seu Descurainiae	18 g
Rhizoma Acori Graminei	24 g
Cortex Mori Radicis	15 g
Radix Peucedani	12 g
Semen Persicae	6 g
Radix Glycyrrhizae	3 g
Semen Plantaginis (wrapped in a piece of cloth to be decocted)	24 g

Decoct the above ingredients in water for oral dose.

Modification: In case of sticky expectoration, add Fructus Trichosanthis 24 g and Radix Platycodi 10 g.

In case of constipation, add Radix et Rhizoma Rhei 6 g and stir-fried Semen Raphani 15 g.

Acupuncture: Puncture point Feishu(BL 13), Chize(LU 5), Yinlingquan(SP 9)towards Yanglingquan(GB 34), Shuifen (RN 9), Fuliu (KI 7) and Zusanli(ST 36).

2. Deficiency of *Yang* of the heart and kidney leading to ac-

cumulation of phlegm and stasis of blood in the interior.

Symptoms and signs: Palpitation, shortness of breath, edema of the face and lower limbs, soreness and weakness of the lumbus and knees, cough and asthma with stuffiness in the chest, fullness and distention in epigastrium and abdomen, nausea, poor appetite, bluish purple lips and tongue, yellow and sticky tongue coating, and deep, thready and forceless pulse.

Principle of treatment: Tonifying *Qi*, warming *Yang*, resolving phlegm and dispersing blood stasis.

Recipe: *Zhen Wu Tang*(45) with modification.

Ingredients:

Radix Aconiti Praeparata	12 g
Poria	30 g
Rhizoma Atractylodis Macrocephalae	30 g
Radix Paeniae Alba	12 g
Radix Paeniae Rubra	12 g
Rhizoma Zingiberis	10 g
Semen Lepidii seu Descurainia	15 g
Cortex Acanthopancis Radicis	15 g
Semen Persicae	6 g
Herba Lycopi	10 g
stir-fried Semen Raphani	15 g
Fructus Crataegi	16 g

Decoct the above ingredients in water for oral dose.

Modification: In case of fullness and distention in epigastrium and abdomen and palpable mass in the right hypochondriac region, add Rhizoma Sparganii 6 g, Rhizoma Zedoariae 6 g and Endothelium Corneum Gigeriae Galli 12 g.

In case poor appetite, add Fructus Amomi 10 g and Massa

Fermentata Medicinalis 12 g.

Shen Fu Injection may be taken for intramuscular injection 4ml each time and 2 times a day.

3. Obstruction of the channel by phlegm and stasis of blood leading to injury of vessels and bleeding.

Symptoms and signs: Severe cough, chest pain radiating to the neck, dark bloody expectoration, stuffiness in the chest with suffocation sensation, dyspnea, bluish purple lips and tongue, and thready and hesitant pulse.

Principle of treatment: Clearing heat from the lung, resolving phlegm, removing blood stasis and stopping bleeding.

Recipe: *Xi Jiao Di Huang Tang*(25) with modification.

Ingredients:

powder of Cornu Rhinoceri (to be taken following its infusion)	3 g
Radix Rehmanniae	30 g
Cortex Moutan Radicis	12 g
Radix Paeoniae Rubra	12 g
Rhizoma et radix Rhei powder (to be taken following its infusion)	15 g
powder of Rhizoma Bletillae (to be taken following its infusion)	15 g
Bulbus Fritillariae Thunbergii	12 g
Cortex Cori Radicis	15 g

Decoct the above ingredients in water for oral dose.

powder of Rhizoma et Radix Rhei could be swallowed 6–12

g each time and 3 times a day.

Grind powder of Radix Notoginseng 6 g, Crinis Carbonisatus 6 g, Bulbus Fritillariae Thunbergii 6 g and Ophicalcitum 24 g together into fine powder which is divided into 4 portion to be taken with boiled water, 1 portion each time.

Yun Nan Bai Yao could be taken with warm boiled water 0.5 g each time and 3 times a day.

2.4 Acute Infectious Endocarditis

Acute infectious endocarditis belongs to epidemic febrile disease in traditional Chinese medicine.

Etiology and Pathogenesis

This disease is caused by the transmission of exogenous toxic factor into the interior which injures the *Ying* and blood systems, giving rise to various symptoms and signs.

Essentials of Diagnosis

1. It has the history of exterior syndrome.
2. It has constant irregular fever or lower grade fever.
3. It is often accompanied with injury of the five *Zang* organs and stagnant spots or patches on the skin.

Treatment Based on Differentiation of Syndromes

1. Invasion of the exterior by wind—heat

Symptoms and signs: Fever, slight aversion to wind and cold, headache, nasal obstruction, yellow nasal discharge, sorethroat, slightly thirst, red tongue with thin and yellow coating, and superficial and rapid pulse.

Principle of treatment: Relieving the exterior syndrome with drugs pungent in flavour and cool in property, clearing away heat and eliminating toxin.

Recipe: *Yin Qiao San*(2).
Ingredients:

Flos Lonicerae	30 g
Fructus Forsythiae	15 g
Flos Chrysanthemi	12 g
Herba Schizonepetae	10 g
Herba Menthae	10 g
Fructus Arctii	12 g
Rhizoma Phragmitis	30 g
Radix Scutellariae	12 g
stir-fried Armeniacae Amarum	12 g
Radix Glycyrrhizae	3 g

Decoct the above ingredients in water for oral dose.

Grind 24 tablets of *Ling Yang Jie Du Pian* into powder to be taken following its infusion, 2 or 3 times a day.

2. Preponderance of heat in the lung and stomach

Symptoms and signs: High fever, asthma, cough with yellow and sticky expectoration, thirst with preference of cold drinks, deep yellow urine, swelling and pain of the gums, constipation, red tongue with yellow coating, and full and rapid pulse.

Principle of treatment: Clearing away heat, reducing fire, eliminating constipation and soothing asthma.

Recipe: *Bai Hu Tang*(10) and *Xie Bai San*(46).
Ingredients:

Gypsum Fibrosum	30 g
Rhizoma Anemarrhenae	12 g
Cortex Mori Radicis	15 g
Cortex Lycii Radicis	12 g
stir-fried Semen Armeniacae	

Amarum	12 g
Radix Scrophulariae	30 g
Rhizoma et Radix Rhei	8 g
Herba Lophatheri	10 g

Decoct the above ingredients in water for oral dose.

3. Invasion of the *Ying* and blood systems by pathogenic heat.

Symptoms and signs: Fever which is worse at night, thirst with no desire of drinking, restlessness, coma, delirium, skin eruption, or hematemesis, epistaxis, hematuria, deep red tongue with yellow and dry coating, and thready and rapid pulse.

Principle of treatment: Clearing away heat and toxin, purging heat from the blood.

Recipe: *Xi Jiao Di Huang Tang*(25) with modification.

Ingredients:

powder of Cornu Rhinoceri (to be taken following its infusion)	3 g
Radix Rehmanniae	30 g
Cortex Moutan Radicis	12 g
Radix Paeoniae Rubra	12 g
Flos Lonicerae	30 g
powder of Cornu Antelopis (to be taken following its infusion)	6 g
Rhizoma Imperatae	45 g
Herba Lophatheri	6 g

Decoct the above ingredients in water for oral dose.

Modification: In case of significant epistaxis and

hematochezia, add Herba Agrimoniae 30 g and powder of Radix Notoginseng 3 g(to be taken following its infusion).

In case of significant hemoptysis, add powder of Rhizoma et Radix Rhei and Rhizoma Bletillae powder 15 g each to be taken following its infusion.

In case of coma and constipation, add *Zi Xue San*(to be taken following its infusion) 3 g each time and 2 or 3 times daily, or either *An Gong NiuHuang Wan* or *Zhi Bao Dan* 1 pill each time and 1 or 2 times daily.

Add 90 ml of *Qing Kai Ling* Injection to 500 ml of 10% glucose solution for intravenous drip, 1 or 2 times daily.

4. Extreme heat stirring up the endogenous wind.

Symptoms and signs: High fever, coma, convulsion, neck rigidity, or even opisthotonos, clenching of teeth, or convulsion of one side of the body which occurs several times a day and finally leads to hemiplegia, deep red tongue, and wiry and rapid pulse.

Principle of treatment: Clearing away heat from the liver, dispelling wind and checking convulsion.

Recipe: *Ling Yang Gou Teng Tang*(22) with modification.

Ingredients:

powder of Cornu Antelopis (to be taken following its infusion)	5 g
powder of Carapax Eretmochelydis (to be taken following its infusion)	3 g
Concha Haliotidis	30 g
Ramulus Uncariae cum Uncis	30 g
Radix Achyranthis Bidentatae	15 g
Radix Rehmanniae	30 g
Fructus Gardeniae	12 g

Indigo Naturalis	6 g
Scorpio	12 g
Lumbricus	12 g
Radix Paeoniae Alba	15 g
Flos Chrysanthemi	12 g
Folium Mori	12 g
Poria cum Ligno Hospite	15 g

Decoct the above ingredients in water for oral dose.

Modification: In case of significant convulsion, add *Zhi Jing San* 1.5 g(to be taken following its infusion), 1 or 2 times daily.

In case of persistent high fever, add *An Gong Niu Huang Wan* 1 pill each time and 3 times a day, or *Niu Huang Qing Xin Wan*, or *Zixue San*, or pricking point Shixuan (EX-UE11)to cause bleeding.

2.5 Viral Myocarditis

Viral myocarditis corresponds to *Xin Bi*(Heart-*Bi*), *Xin Tong*(Cardiac Pain)and *Xin Ji*(Palpitation) in traditional Chinese medicine.

Etiology and Pathogenesis

This disease is resulted from obstruction of the heart channel, malnourishment of the heart and dysfunction of the heart in housing mind due to invasion of exogenous pathogenic factors.

Essentials of Diagnosis

1. Fresh history of common cold within last two weeks, sudden occurrence of palpitation, stuffiness in the chest and shortness of breath, etc.

2. The patient often suffers from fever, chest pain, etc.

3. There is often knotted pulse.

Treatment Based on Differentiation of Syndromes

1. Blood stasis in the heart channel

Symptoms and signs: Persistent fixed stabbing pain in the Chest which gets worse at night, accompanied with palpitation, dark purple tongue with possible purple spots, and deep and hesitant pulse.

Principle of treatment: Activating blood circulation, dispersing blood stasis, removing obstruction from the channel and calming the mind.

Recipe: *Xue Fu Zhu Yu Tang*(32) with modification.

Ingredients:

Radix Angelicae Sinensis	12 g
Rhizoma Ligustici Chuanxiong	10 g
Radix Paeoniae Rubra	10 g
Cortex Moutan Radicis	12 g
Semen Persicae	6 g
Flos Carthami	6 g
Flos Rosae Rugosae	12 g
stir-fried Fructus Aurantii	10 g
Radix Bupleuri	10 g
Radix Glycyrrhizae	3 g

Decoct the above ingredients in water for oral dose.

Modification: In case of significant palpitation, omit stir-fried Fructus Aurantii and add Rhizoma Nardostachyos 12 g and Radix Ophiopogonis 24 g.

In case of severe chest pain, add stir-fried Rhizoma Corydalis 12 g and Lignum Dalbergiae Odoriferae 10 g.

In case of complication of phlegm, add Fructus

Trichosanthis 18 g and Rhizoma Pinelliae 10 g.

Acupuncture: Puncture Point Neiguan (PC6), Xinshu (BL15) and Danzhong(RN17) with strong stimulation and no needle retention.

2. Stagnation of cold in the heart channel.

Symptoms and signs: Stabbing pain in the chest radiating to the back which is worse with cold, accompanied by stuffy chest, shortness of breath, palpitation, pallor, cold limbs, white and slippery tongue coating, and deep and slow pulse.

Principle of treatment: Warming the channel, dispelling cold, removing obstruction from the channel and arresting pain.

Recipe: *Ma Huang Fu Zi Xi Xin Tang*(47) with modification.

Ingredients:

Herba Ephedrae Praeparata	12 g
Radix Aconiti Praeparata	10 g
Herba Asari	6 g
Radix Angelicae Sinensis	15 g
Rhizoma Ligustici Chuanxiong	12 g
Lignum Santali	10 g
Radix Glycyrrhizae Praeparata	12 g
Ramulus Cinnamoni	6 g

Decoct the above ingredients in water for oral dose.

Modification: In case of persistent severe pain, add *Su He Xiang Wan* 1 pill each time and 3 times a day, or *Guan Xin Su He Wan*.

In case of preponderance of *Yin* and cold, add Cortex Cinnamomi 6 g.

3. Hyperactivity of fire due to deficiency of *Yin*

Symptoms and signs: Palpitation, night sweating, lower

grade fever, mental restlessness, insomnia, poor memory, dull pain in the chest, dizziness, tinnitus, red tongue with little fluid, and thready and rapid pulse.

Principle of treatment: Tonifying *Yin*, clearing away fire, nourishing the heart and calming the mind.

Recipe: *Tian Wang Bu Xin Dan*(48) with modification.

Ingredients:

Radix Rehmanniae	20 g
Radix Scrophulariae	21 g
Radix Asparagi	15 g
Radix Ophiopogonis	15 g
Radix Salviae Miltiorrhizae	20 g
Radix Panacis Quinquefolii	10 g
Radix Sophorae Flavescentis	10 g
Poria	15 g
stir-fried Semen Ziziphi Spinosae	30 g
Caulis Polygoni Multiflori	30 g
Radix Polygalae Praeparata	10 g

Decoct the above ingredients in water for oral dose.

Modification: In case of significant stuffiness and pain in the chest, add Rhizoma Ligustici Chuanxiong 12 g and Radix Curcumae 12 g.

In case of obvious palpitation and vertigo, add Radix Polygoni Multiflori Praeparata 30 g, Concha Haliotidis 30 g, Fructus Ligustri Lucidi 18 g and Ramulus Uncariae cum Uncis 30 g.

Acupuncture: Puncture Point Neiguan(PC6) and Shen men (HT7) with medium stimulation and no retention of needles.

Add 20 ml of *Shen Mai* Injection to 20 ml of 50% glucose solution for intravenous injection, 3 times a day.

Take 2 ml of *Ku Shen Jian* Injection for intramuscular injection 2 or 3 times a day, or take *Ku Shen Jin Gao Pian* and *Dan Shen Pian* 3 tablets each every time orally and 2 times a day.

4. Weakness of heart *Yang*

Symptoms and signs: Palpitation, shortness of breath which become worse with exertion, pallor, lassitude, spiritlessness, aversion to cold, spontaneous sweating, pale tongue with white coating, and deep, forceless and thready pulse.

Principle of treatment: Tonifying *Qi*, invigorating *Yang*, removing obstruction from the channel and calming the mind.

Recipe: *Shen Fu Tang*(29) with modification.

Ingredients:

Radix Ginseng	12 g
Radix Aconiti Praeparata	10 g
Ramulus Cinnamomi	10 g
Radix Glycyrrhizae Praeparata	12 g
Poria	20 g
Rhizoma Zingiberis	10 g
Fructus Ziziphi Jujubae	6 pieces

Decoct the above ingredients in water for oral dose.

Modification: In case of cold limbs and fading pulse, add Radix Ginseng Rubra 24 g(instead of Radix Ginseng), Fructus Corni 24 g, Os Draconis 30 g and Concha Ostreae 30 g.

In case of edema and palpitation, add Radix Stephaniae Tetrandrae 12 g and Semen Plantaginis 24 g.

5. Deficiency of both *Qi* and *Yin*

Symptoms and signs: Intermittent stuffiness with dull pain

and palpitation which get worse in the afternoon, accompanied by pallor, lassitude, dislike of speaking, dizziness and vertigo especially during exertion, red tongue, and thready, weak and knotted pulse.

Principle of treatment: Tonifying *Qi*, nourishing *Yin*, eliminating palpitation and calming the mind.

Recipe: *Zhi Gan Cao Tang*(31) with modification.

Ingredients:

Radix Glycyrrhizae Praeparata	12 g
Ramulus Cinnamomi	12 g
Radix Ginseng	10 g
Semen Biotae	15 g
Radix Rehmanniae	30 g
Radix Ophiopogonis	24 g
Colla Corii Asini (to be melted)	12 g
stir-fried Semen Ziziphi Spinosae	30 g
Radix Salviae Miltiorrhizae	21 g
Rhizoma Zingiberis	8 g
Fructus Ziziphi Jujubae	30 pieces

Decoct the above ingredients in water for oral dose.

2.6 Arrhythmia

Arrhythmia corresponds to palpitation in traditional Chinese medicine.

Etiology and pathogenesis

This disease could be caused by either malnourishment of the heart or invasion of the heart by pathogenic factors which leads to timidity due to deficiency of the heart, insufficiency of *Qi* and blood of the heart, weakness of heart *Yang*, hyperactivity of fire due to deficiency of *Yin*, retention of fluid in the interior or obstruction of the channel by stasis of blood, etc. giving rise to mental restless and palpitation.

Essentials of Diagnosis

1. The patient has palpitation with fright, or even can not act on his(her) own. The pulse could be rapid, slow, slow with irregular missing beats, slow with regular missing beats, hesitant or abrupt.

2. Attack of the disease is usually induced by excessive emotional changes or overfatigue.

3. It is often complicated with insomnia, poor memory, vertigo and tinnitus.

4. Arrhysthmia could be found with E.C.G.

Treatment Based on Differentiation of Syndromes

1. Stagnation of *Qi* and stasis of blood

Symptoms and signs: Palpitation, Stabbing pain in the chest, which is fixed and worse with anger, dark purple tongue with possible purple spots, and deep and hesitant pulse.

Principle of treatment: Activating circulation of *Qi* and blood, removing obstruction from the channel and arresting pain.

Recipe: *Xue Fu Zhu Yu Tang*(32) with modification.

Ingredients:

Radix Angelicae Sinensis	12 g
Rhizoma Ligustici Chuanxiong	10 g
Semen Persicae	12 g
Flos Carthami	12 g
Radix Bupleuri	12 g
stir-fried Fructus Aurantii	10 g

Rhizoma Nardostachyos	12 g
Radix Salviae Miltiorrhizae	21 g
Flos Rosae Rugosae	10 g

Decoct the above ingredients in water for oral dose.

Modification: In case of obvious insomnia, add Caulis Polygoni Multiflori 30 g.

In case of palpitation, add powder of Succinum 3 g(to be taken following its infusion).

In case of severe chest pain, add *Guan Xin Su He Wan* 1 pill each time and twice a day.

Add 8 ml of *Fu Fang Dan Shen* Injection to 250 ml of 50% glucose solution for intravenous drip.

2. Obstruction of the channel by phlegm.

Symptoms and signs: Palpitation, stuffiness in the chest, cough with profuse expectoration, chest pain radiating to the back, dizziness and vertigo, obesity, light red tongue with white and slippery coating, and slippery or knotted pulse.

Principle of treatment: Warming *Yang*, resolving phlegm, and removing obstruction from the channel.

Recipe: *Gua Lou Xie Bai Ban Xia Tang* (37) with modification.

Ingredients:

Fructus Trichosanthis	24 g
Bulbus Allii Macrostemi	12 g
Rhizoma Pinelliae	12 g
stir-fried Fructus Aurantii Immaturus	12 g
Cortex Magnoliae Officinalis	10 g
Semen Amomi Cardamomi	10 g

Exoparpium Citri Grandis	12 g
Poria	12 g

Decoct the above ingredients in water for oral dose.

Take *Gua Lou Pian* 4 tablets each time and 3 times a day.

Take *Su Bing Di Wan* 2 pills each time and twice a day.

3. Attack of the heart by retained fluid.

Symptoms and signs: Palpitation, vertigo, fullness and stuffiness in the chest and epigastrium, nausea, avertion to cold, cold limbs, edema of lower limbs, oliguria, thirst with no desire of drinking, white and slippery tongue coating, and wiry and slippery pulse.

Principle of treatment: Invigorating *Yang* of the heart and promoting transformation of *Qi* for diuresis.

Recipe: *Ling Gui Zhu Gan Tang*(44) with modification.

Ingredients:

Poria	30 g
Ramulus Cinnamomi	12 g
Rhizoma Atractylodis Macrocephalae	15 g
Radix Glycyrrhizae	3 g
Radix Stephaniae Tetrandrae	12 g
Semen Lepidii seu Descurainiae	15 g
Radix Peucedani	12 g
ginger juice–prepared Rhizoma Pinelliae	12 g

Decoct the above ingredients in water for oral dose.

Modification: In case of significant nausea and vomiting, add Precarpium Citri Reticulatae 10 g and Rhizoma Zingiberis 10 g.

In case of edema of lower limbs and dysuria, add Radix Aconiti Praeparata 12 g and Semen Plantaginis 24 g(wrapped to be decocted).

4. Timidity due to deficiency of the heart.

Symptoms and signs: Palpitation, susceptibility to fear and fright, insomnia, dream-disturbed sleep, light red tongue with thin and white coating, and tremulous and rapid pulse.

Principle of treatment: Arrestomg fright, calming the mind, nourishing the heart and settling the will.

Recipe: *An Shen Ding Zhi Wan*(49) with modification.

Ingredients:

Radix Ginseng	10 g
Radix Glycyrrhizae Praeparata	10 g
stir-fried Semen Ziziphi Spinosae	30 g
Fructus Schisandrae	6 g
Poria cum Ligno Hospite	20 g
Radix Polygalae Praeparata	10 g
Dens Draconis Praeparata	30 g
powder of Succinum	
(to be taken following its infusion)	1 g
powder of Cinnabaris	
(to be taken following its infusion)	1 g
Magetitum	15 g

Decoct the above ingredients in water for oral dose.

Modification: In case of palpitation complicated with fight and mental restlessness, subtract Fructus Schisandrae and Dens Draconis, add Rhizoma Pinelliae 10 g, Exocarpium Citri Grandis 10 g and Rhizoma Coptidis 6 g.

5. Deficiency of heart blood.

Symptoms and signs: Palpitation, dizziness, pallor, lassitude, dislike of speaking, light red tongue, and thready and forceless or knotted pulse.

Principle of treatment: Tonifying blood, calming the mind, invigorating *Qi* and nourishing the heart.

Recipe: *Gui Pi Tang*(50) with modification.

Ingredients:

Radix Ginseng	10 g
Radix Astragali seu Hedysari Praeparata	12 g
Poria	12 g
Rhizoma Atractylodis Macrocephala	12 g
Radix Glycyrrhizae Praeparata	10 g
Radix Angelicae Sinensis	10 g
Arillus Longan	12 g
Fructus Amomi	6 g
Radix Aucklandiae	6 g
stir-fried Semen Ziziphi Spinosae	24 g
Radix Polygalae Praeparata	10 g
Radix Rehmanniae	12 g

Decoct the above ingredients in water for oral dose.

Modification: In case of obvious knotted pulse, use modified *Zhi Gan Cao Tang*.

If the patient has mental restlessness due to deficiency of heart *Yin*, add Radix Ophiopogonis 24 g and Fructus Schisandrae 10 g.

6. Deficiency of *Yin* of both heart and kidney.

Symptoms and signs: Palpitation, insomnia, dysphoria with

feverish sensation in the chest, palms and soles, dizziness and vertigo, tinnitus, soreness of the lumbus, red tongue with little coating, and thready and rapid or knotted pulse.

Principle of treatment: Tonifying *Yin*, reducing fire, nourishing the heart and calming the mind.

Recipe: *Tian Wang Bu Xin Dan* (48) with modification.

Ingredients:

Radix Rehmanniae	30 g
Radix Ophiopogonis	20 g
Radix Asparagi	20 g
Radix Angelicae Sinensis	12 g
Radix Polygoni Multiflori Praeparata	24 g
Radix Salviae Miltiorrhizae	21 g
stir-fried Ziziphi Spinosae	30 g
Semen Biotae	12 g
Poria cum Ligno Hospite	12 g
powder of Succinum (to be taken following its infusion)	1 g
powder of Cinnabaris (to be taken following its infusion)	1 g
Radix Pseudostellariae	10 g
Radix Scrophulariae	15 g

Decoct the above ingredients in water for oral dose.

Modification: In case of dysphoria with feverish sensation in the chest, palms and soles, add *Zhi Bai Di Huang Wan* 1 pill each time, twice a day.

7. Deficiency of *Yang* of both heart and kidney.

Symptoms and signs: Palpitation and shortness which become worse with exertion, pallor, avertion to cold, cold limbs, weakness and soreness of the lumbus and knees, oliguria, edema, pale tongue with white coating, and deep and forceless pulse.

Principle of treatment: Warming and nourishing the heart and kidney, calming the mind and arresting palpitation.

Recipe: *Shen Fu Tang* (29) plus *Ling Gui Zhu Gan Tang* (44) with modification.

Ingredients:

Radix Ginseng	10 g
Radix Aconiti Praeparata	10 g
Poria	24 g
Ramulus Cinnamomi	10 g
Rhizoma Atractylodis Macrocephalae	10 g
Radix Glycyrrhizae	3 g
Rhizoma Zingiberis	6 g
Os Draconis	15 g
Concha Ostreae	15 g

Decoct the above ingredients in water for oral dose.

Modification: In severe case characterized by palpitation and difficulty in lying flat, use Radix Ginseng and Radix Aconiti 30 g each.

Add 20 ml of *Shen Fu* Injection to 30 ml of 50% glucose solution for slow intravenous injection, 3 times a day.

2.7 Heart Failure

Heart failure corresponds to severe palpitation, asthma and cough, retention of phlegm-fluid and edema of traditional Chinese medicine.

Etiology and Pathogenesis

Prolonged illness causes injury of the five *Zang* organs in which hypofunction of the heart leads to palpitation, that of the lung leads to cough, that of the kidney leads to asthma and that of the spleen leads to edema.

Essentials of Diagnosis

1. Its main clinical manifestations are dyspnea, shortness of breath and cough, hemoptysis, cyanosis, abdominal distention and edema.

2. The patient usually has history of chronic diseases of the heart and lung.

Treatment Based on Differentiation of Syndromes

1. Deficiency of *Yang* of both heart and kidney leading to attack of the heart and lung by retained fluid.

Symptoms and signs: Palpitation, shortness of breath, stuffiness in the chest, cough with frothy expectoration, difficulty in lying flat, worsening with exertion, spontaneous sweating, lassitude, avertion to cold, cold limbs, spiritlessness dark pale tongue with white and slippery coating, and deep, thready and weak pulse.

Principle of treatment: Warming *Yang* to resolve the retained fluid, purging the lung and clearing the mind.

Recipe: *Shen Fu Tang*(29) plus modified *Ting Li Da Zao Xie Fei Tang*(51).

Ingredients:

Radix Ginseng	30 g
Radix Aconiti Praeparata	24 g
Ramulus Cinnamomi	15 g
Poria	30 g
Semen Lepidi seu Descurainiae	18 g

Rhizoma Acorus Calamus	12 g
Fuctus Ziziphi Jujubae	10 pieces

Decoct the above ingredients in water for oral dose.

Modification: In case of facial edema, add Semen Plantaginis 24 g (wrapped to be decocted).

In case of purple lips and tongue, add Radix Salviae Miltiorrhizae 24 g and Semen Persicae 12 g.

2. Upward perversion of turbid *Qi* due to stagnation of phlegm and stasis of blood.

Symptoms and signs: Palpitation, shortness of breath, stuffy chest with suffocation sensation, stuffiness and distention in epigastrium and abdomen, fullness and hardness in right hypochondriac region, nausea, edema of lower limbs, listlessness, aversion to cold, dislike of movement, scanty urine and loose stool, dark red tongue with white and sticky coating, and slow and hesitant pulse.

Principle of treatment: Dispersing stasis of blood, resolving phlegm, activating *Qi* circulation and invigorating the spleen.

Recipe: *Tiao Ying Yin*(52) with modification.

Ingredients:

tail of Radix Angelicae Sinensis	10 g
Rhizoma Ligustici Chuanxiong	10 g
Paeoniae Rubra	12 g
Rhizoma Zedoariae	12 g
Radix et Rhizoma Rhei	6 g
Cortex Magnoliae Officinalis	10 g
Poria	30 g
ginger juice—prepared Pinelliae	10 g
Semen Lepidii seu Descurainiae	12 g

Pericarpium Arecae	15 g
Semen Plantaginis	18 g
Ramulus Cinnamomi	6 g
Rhizoma Zingiberis	3 slices
Fructus Ziziphi Jujubae	10 pieces

Decoct the above ingredients in water for oral dose.

Modification: In case of obvious nause and vomiting, add pericarpium Citri Reticulatae 10 g and ginger juice–Prepared Caulis Bambusae in Taeniam 10 g.

In case of significant edema of lower limbs, add Radix Stephaniae Tetrandrae 12 g and Herba Lycopi 12 g.

3 Respiratory Diseases

Respiratory diseases in modern medicine correspond to those symptoms and signs appeared in respiratory system in traditional Chinese medicine.

3.1 Acute Bronchitis

This disease can be observed in *Gan Mao*(cold), *Ke Sou*(cough) and *Chuan Ni*(dyspnea) of traditional Chinese medicine.

Etiology and Pathogenesis

Impairment of purifying and descending function of the lung due to invasion of pathogenic factors gives rise to reversed flow of *Qi* of the lung, resulting in cough and dyspnea.

Essentials of Diagnosis

1. There are generally symptoms of infection of upper respiratory tract such as nasal obstruction, rhinorrhea, sorethroat, avertion to cold and fever.

2. Its main symptom is cough with little white expectoration at begining followed by yellow or blood—stained sputum.

3. Coarse respiration and dry or wet rale could be heard in the pulmonary region on auscultation.

4. In most cases, thickened lung marking could be observed during X—ray examination.

Treatment Based on Differentiation of Syndromes

1. Invasion of the lung by wind—cold

Symptoms and signs: Cough in deep sound, shortness of breath, itching throat, dilute and white expectoration, often accompanied by fever, chillness and headache, thin and white tongue coating, and superficial or superficial and tense pulse.

Principle of treatment: Dispersing the lung and arresting cough by dispelling wind and cold.

Recipe: *Ma Xing Tang*(53) with modification.

Ingredients:

Herba Ephedrae	9 g
Folium Perillae	9 g
Herba Schizonepetae	12 g
Semen Armeniacae	9 g
Radix Glycyrrhizae	6 g

Decoct the above ingredients in water twice to obtain about 300 ml of decoction to be taken half in the morning and half in the evening while it is warm.

Modification: In case of profuse sputum and white and sticky tongue coating, add Pericarpium Citri Reticulatae 9 g, Poria 12 g and Cortex Magnoliae Officinalis 9 g.

In case of accumulation of cold and fluid in the lung characterized by profuse sputum, bad cough, thick and sticky tongue coating and slippery pulse, use modified *Xiao Qing Long Tang*(54) Its ingredients are as follow:

Herba Ephedrae	20 g
Ramulus Cinnamomi	12 g
Rhizoma Zingiberis	20 g
Herba Asari	3 g
Fructus Schisandrae	3 g
Fuctus Ziziphi Jujubae	20 g

Radix Glycurrhizae	20 g
Rhizoma Pinelliae Praeparata	30 g
Gypsum Fibrosum	120 g

Decoct the above ingredients in water twice to obtain about 300 ml of decoction to be taken half in the morning and half in the evening while it is warm.

2. Invasion of the lung by wind—heat

Symptoms and signs: Frequent severe cough, coarse breathing, sorethroat, yellow, thick and sticky sputum, fever with thirst, yellow tongue coating, and superficial and rapid pulse.

Principle of treatment: Purifying the lung and arresting cough by dispelling wind—heat.

Recipe: *Yin Qiao San*(2) plus *Ma Xing Shi Gan Tang*(8) with modification.

Ingredients:

Flos Lonicerae	30 g
Fructus Forsythiae	15 g
Herba Ephedrae Praeparata	9 g
Gypsum Fibrosum	
(to be decocted first)	45 g
Semen Armeniacae	10 g
Radix Platycodi	12 g
Fructus Arctii	12 g
Rhizoma Phragmitis	30 g
Cortex Mori Radicis	12 g
Radix Scutellariae	15 g
Radix Glycyrrhizae	6 g

Decoct the above ingredients in water twice to obtain 300 ml of decoction to be taken half in the morning and half in the evening while it is warm.

Modification: In case of thirst with difficult expectoration, add Radix Adenophora 30 g and Radix Ophiopogonis 15 g.

In case of yellow and sticky sputum, add Herba Houttuyniae 30 g.

In case of high fever, add Rhizoma Anemarrhenae 12 g and Gypsum Fibrosum 60 g.

3. Accumulation of phlegm—heat in the lung.

Symptoms and signs: Severe cough and dyspnea with yellow and sticky sputum, high fever, yellow and sticky tongue coating, slippery pulse, thirst and dry stool.

Principle of treatment: Clearing heat, reducing fire, resolving phlegm and srresting cough.

Recipe: *Ma Xing Shi Gan Tang*(8) plus *Xiao Bao Xia Tang*(55) with modification.

Ingredients:

Herba Ephedrae Praeparata	12 g
Gypsum Bibrosum	30 g
Semen Ameniacae	9 g
Fructus Trichosanthis	30 g
Rhizoma Coptidis	9 g
Radix Scutellariae	12 g
Herba Houttuyniae	30 g
Rhizoma Pinelliae Praeparata	9 g
Radix Glycyrrhizae	6 g

Decoct the above ingredients in water twice to obtain about 300 ml of decoction to be taken half in the morning and half in the evening while it is warm.

Modification: In case of yellow and stinking sputum, add

Reed Stem 30 g, Semen Coicis 30 g, Semen Persicae 12 g, Semen Benincasae 30 g and Rhizoma Paridis 20 g to clear away toxic heat, resolve phlegm and evacuate pus.

In case of dry stool, add Rhizoma et Radix Rhei 9 g.

3.2 Lobar Pneumonia

Lobar pneumonia corresponds to *Ke Sou*(cough) in traditional Chinese medicine.

Etiology and Pathogenesis

Invasion of the lung by pathogenic wind—heat through mouth and nose leads to accumulation of phlegm—heat and dysfunction of the lung in purification and dispersion, giving rise to cough and chest pain.

Essentials of Diagnosis

1. This disease is usually caused by pathogenic wind(cold, heat and pestilential factors), overfatigue, mental injury or asthenia after illness.

2. It is characterized by such symptoms as chillness, fever, cough, chest pain, and yellow or rusty sputum.

3. Dry or wet rale can be heard on auscultation and dull resonance of the lung on percussion. The X—ray examination of the patient at the satge of consolidation shows patchy shadow with an even increase in density. With the blood test, it could be found that the total white blood cells is over $10,000/mm^3$, the value of neutrophilic granulocytes is over 70% with the nucleus shifting to the left.

4. In severe cases, there might appear semiconsciousness or even coma accompanied by a fall of blood pressure, cold sweating and pallor, which indicate peripheral circulatory failure.

Treatment Based on Differentiation of Syndromes

1. Dysfunction of the lung in dispersing due to invasion by pathogenic warm.

Symptoms and signs: Chillness, fever(with sudden rise of body temperature up to over 38 °C), headache, general ache, chest pain, bad cough with sticky expectoration, thin and white tongue coating or thin, yellow coating with less fluid, and superficial and rapid pulse.

Principle of treatment: Clearing heat, dispelling wind, dispersing the lung and arresting cough.

Recipe: *Yin Qiao San*(2) with modification.

Ingredients:

Flos Lonicerae	30 g
Fructus Forsythiae	12 g
Herba Houttuyniae	30 g
Herba Menthae	9 g
Semen Armeniacae	9 g
Radix Platycodi	9 g
Rhizoma Phragmitis	30 g
Herba Lophatheri	3 g
Fructus Arctii	12 g
Fructus Trichosanthis	24 g
Indigo Naturalis	6 g
(to be taken following its infusion)	6 g
Radix Glycyrrhizae	6 g

Decoct the above ingredients in water twice to obtain about 400 ml of decoction for the intake of 2 or 3 times.

Modification: In case of high fever, add Gypsum Fibrosum

30—45 g.

In case of high fever with coma, add 1.5 g of *Ling Yang Fen* (powder of Cornu Antelopis) to be taken following its infusion.

2. Preponderance of heat in the lung and stomach leading to intense heat in the *Ying* and blood system

Symptoms and signs: High fever (over 39 °C), severe cough, coarse breathing, burning pain in the chest, massive expectoration of rusty or blood—stained purulent sputum, polyhidrosis, thirst, red face, constipation, yellow and dry tongue coating, and full or full and rapid pulse.

Principle of treatment: Clearing away heat from the lung and stomach, and cooling toxic heat from the blood system.

Recipe: *Yin Qiao Bai Hu Tang* (56) with modification.

Ingredients:

Flos Lonicerae	30 g
Fructus Forsythiae	12 g
Folium Isatidis	30 g
Herba Patriniae	30 g
Herba Houttuyniae	30 g
Rhizoma Anemarrhenae	12 g
Gypsum Fibrosum	45 g
Cortex Mori Radicis	30 g
fresh Radix Rehmanniae	30 g
Semen Persicae	15 g
Semen Coicis	30 g

Decoct the above ingredients in water twice to obtain 300 ml of decoction to be taken half in the morning and half in the evening while it is warm.

Modification: In case of severe cough with chest pain, add Bulbus Fritillariae Thunbergii 12 g, Radix Curcumae 12 g and Pericarpium Trichosanthis 15 g.

In case of constipation with absence of bowel movement for days, add Rhizoma et radix Rhei 12 g, Natrii Sulfas 9—12 g(to be taken following its infusion) and Fructus Trichosanthis 30 g.

3. Impairment of *Yin* of the lung and stomach

Symptoms and signs: Gradual fall of high fever, or persistent low—grade fever, decrease of expectoration of blood—stained sputum, mental restlessness, thirst, red tongue with little coating, and thready and rapid pulse.

Principle of treatment: Nourishing *Yin* of the stomach, moistening the lung and clearing away the heat from the lung.

Recipe: *Zhu Ye Shi Gao Tang*(57) plus *Yang Yin Qing Fei Tang*(58).

Ingredients:

Gypsum Fibrosum	30 g
Radix Ophiopogonis	30 g
Radix Pseudostellariae	30 g
Herba Lophatheri	3 g
Cortex Moutan Radicis	9 g
Radix Rehmanniae	15 g
Radix Scrophulariae	15 g
Radix Paeoniae Alba	12 g
Bulbus Fritillariae Cirrhosae	12 g

Decoct the above ingredients in water twice to obtain 300 ml of decoction to be taken half in the morning and half in the evening while it is warm.

Modification: In case of constipation, add Rhizoma et Radix

Rhei 9 g and Fructus Cannabis 30 g.

4. Deteriorated syndromes

(1) Excessive heat in the lung and stomach leading to impairment of both *Qi* and *Yin*.

Symptoms and signs: High fever, cough, chest pain, coarse breathing, expectoration of bloody or rusty sputum, polyhidrosis, thirst, red tongue with yellow and dry coating hollow and rapid pulse.

Principle of treatment: Clearing away pathogenic heat from the lung and stomach, tonifying *Qi* and nourishing *Yin*.

Recipe: *Yin Qiao Bai Hu Tang*(56) with modification or *Sheng Mai San*(26) with modification.

Ingredients:

Flos Lonicerae	30 g
Fructus Forsythiae	15 g
Herba Houttuyniae	30 g
Gypsum Eibrosum	45 g
Rhizoma Anemarrhenae	12 g
Radix Pannacis Quinquefolii	15 g
Radix Pseudostellariae	30 g
Radix Ophiopogonis	30 g
Cortex Mori	15

Decoct the above ingredients twice in water to obtain decoction to be taken once in the morning and once in the evening.

(2) Collapse due to exhaustion of *Yang*

Symptoms and signs: Sudden drop of high fever(down to below 36 ° C), profuse cold sweating, pallor, sudden fall of blood pressure(with systolic pressure below 80 mmHg), oliguria, fading or rapid and faint pulse.

Principle of treatment: Restoring *Yang* and rescuing the patient from the collapse.

Recipe: ① *Du Shen Tang*(59).

Ingredients: Radix Ginseng　　　　　　　　　　　　30 g

Decoct the ingredient in water twice to obtain about 200 ml of decoction, Take it while it is warm.

②*Shen Fu Tang*(29) with modification.

Ingredients:
Radix Ginseng	15 g
Radix Aconiti Praeparata	30 g
Fructus Evodiae]	15 g
Semen Schisandrae	9 g

Decoct the above ingredients in water twice to obtain 200 ml of decoction. Take it while it is warm.

③ *Hui Yang Fan Ben Tang*(60).

Ingredients:
Radix Ginseng	12 g
Radix Ophiopogonis	24 g
Fructus Schisandrae	9 g
Radix Aconiti Praeparata	15 g
Rhizoma Zingiberis	12 g
Radix Glycyrrhizae Praeparata	6 g

Decoct the above ingredients in water for oral dose.

Besides, add 10 ml of *Shen Fu* Injection(61) or *Sheng Mai* Injection to 500 ml of 10% glucose solution for intravenous drip, 1–2 times daily.

3.3 Acute Aspiration Pulmonary Abscess

Acute aspiration pulmonary abscess corresponds to *Fei Yong* in traditional Chinese medicine.

Etiology and Pathogenesis

This disease is resulted from retention of toxic heat which is combined with stasis of blood to cause abscess in the lung.

Essentials of Diagnosis

1. This disease is characterized by persistent fever, cough, shortness of breath, chest pain, yellow and stinking expectoration, red tongue with yellow coating, full and slippery or rapid pulse.

2. Blood test shows distinct rise in the count of leucocytes and neutrophilic granulocytes.

3. Chest X-ray examination shows alveolar infiltrative infection at early stage and cavities full of fluid in the focus after pus is formed.

Treatment Based on Differentiation of Syndromes

1. Retention of toxic heat in the lung

Symptoms and signs: Persistent high fever, cough and dyspnea with chest pain, yellow, thick and sticky expectoration, dark red tongue with yellow and sticky coating, slippery and rapid or full and rapid pulse.

Principle of treatment: Clearing away toxic heat, resolving phlegm and dispersing stasis of blood.

Recipe: *Wei Jing Yu Xing Tang*(62) with modification.

Ingredients:

Reed Stem	30 g
Herba Houttuyniae	30 g

Radix Platycodi	9 g
Herba Patriniae	24 g
Semen Persicae	15 g
Semen Benincasae	30 g
Flos Lonicerae	30 g
Semen Coicis	30 g
Fructus Trichosanthis	18 g
Radix Glycyrrhizae	12 g

Decoct the above ingredients in water twice to obtain 300 ml of decoction. Take it once in the morning and once in the evening while it is warm.

Modification: In case of high fever, add Gypsum Fibrosum 45 g.

In case of constipation, add Rhizoma et Radix Rhei 6–10 g.

2. Internal ulceration with toxic pus

Symptoms and signs: Excessive yellow, sticky and stinking sputum, fever, chest pain, yellow and sticky tongue coating, and slippery and rapid pulse.

Principle of treatment: Removing toxic heat and promoting evacuation of pus.

Recipe: *Wei Jing Tang*(63) with modification

Ingredients:

Rhizoma Phragmitis	30 g
Semen Coicis	45 g
Semen Benincasae	30 g
Radix Platycodi	12 g
Houttuyniae	30 g
Bulbus Fritillariae Thunbergii	15 g
Flos Lonicerae	30 g

Nodus Nelumbinis Rhizomati   12 g
Semen Persicae   12 g
Pericarpium Citri Reticulatae   9 g

Decoct the above ingredients in water twice to obtain about 400 ml of decoction. Take it 3 times separately after breakfast, lunch and supper, one dose a day.

Modification: In case of high fever, add Gypsum Fibrosum 45 g and Rhizoma Anemarrhenae 9 g.

In case of blood-stained purulent expectoration, add powder of Radix Notoginseng 6 g to be taken with warm boiled water.

In case of constipation, add Rhizoma et Radix Rhei 6-10 g.

In case of high fever and coma, add *Zi Xue Dan* 2 g to be taken following its infusion twice a day, or *An Gong Niu Huang Wan*(20) 1 pill each time and 2 times a day, which is taken with warm boiled water.

4 Neurogenic and Psychogenic Diseases

4.1 Trigeminal Neuralgia

This disease is more or less included in *Tou Feng Tou Tong* (headache caused by attach of the head by pathogenic wind) and *Pian Tou Tong* (migraine) in traditional Chinese medicine.

Etiology and Pathogenesis

Attack of the head by pathogenic wind leads to stagnation of the blood and obtruction of the channel, giving rise to pain. Since pathogenic wind is characterized by constant movement and rapid change, the occurrence of this disease is abrupt with intermittent pain.

Essentials of Diagnosis

1. It is characterized by paroxysmal burning sensation on the face with severe flash pain accompanied with facial spasm and lacrimation which usually last for few seconds and then relieves spontaneously. The patient feels nothing during interval of attacks.

2. The pain is usually induced by muscular movement of the face and referred to the lips, ala nasi and jaw.

3. Generally, there is no positive symptoms of nerve system.

4. Clinically, it should be differentiated from toothache, sinusitis and glossopharyngeal neuralgia.

Treatment Based on Differentiation of Syndromes

1. Invasion of exogenous pathogenic wind—heat.

Symptoms and signs: Paroxysmal burning stabbing pain on one side of the face which is accompanied usually by fever, thirst, sorethroat, thin and yellow tongue coating, and wiry and rapid pulse.

Principle of treatment: Dispelling pathogenic wind and heat, removing obstruction from the channel and arresting pain.

Recipe: *Xiong Zhi Shi Gao Tang*(64) with modification.

Ingredients:

Rhizoma Ligustici Chuanxiong	20 g
Gypsum Fibrosum	30 g
Flos Chrysanthemi	9 g
Herba Menthae	9 g
Lumbricus	9 g
Fructus Arctii	12 g
Radix Angelicae Dahuricae	12 g
Rhizoma Phragmitis	30 g

Decoct the above ingredients in water for oral dose.

Application of acupuncture in combination with needling Point Tinggong(SI19), Xiaguan(ST7), Hegu(LI4), Sibai(ST2), Yuyao(EX-HN4), etc. will obtain better result.

2. Invasion of exogenous pathogenic wind-cold.

Symptoms and signs: Severe pain, aversion to cold, fever, thin and white tongue coating, and wiry and tense pulse.

Principle of treatment: Dispelling cold and Arresting Pain.

Recipe: *Chuan Xiong Cha Tiao San*(65) with modification.

Ingredients:

Rhizoma Ligustici Chuanxiong	30 g
Radix Angelicae Dahuricae	12 g
Rhizoma seu Radix Notopterygii	12 g

Radix Ledebouriellae	12 g
Herba Asari	6 g
Herba Schizonepetae	9 g
Herba Menthae	6 g
Radix Aconiti Kusnezoffii Praeparata (to be decocted 1 hour before)	9 g
Radix Glycyrrhizae	6 g

Decoct the above ingredients in water for oral dose.

Modification: In case of severe pain or no significant therpeutic effect after taking the above decoction, add Scolopendra 6 pieces and Scorpio 6 g which are charred by baking and then ground into fine powder to be taken with warm boiled water.

Application of acupuncture in combination with needling Point *Tou Wei*(ST8), Lieque(LU7), Jiache(ST6), Xiaguan(ST7), Hegu(LI4), etc.may obtain even better result.

3. Obstruction of the channel by stasis of blood.

Symptoms and signs: Repeated attack of fixed stabbing pain, tinnitus, deafness, dark red tongue with possible stagnant spots, and thready and hesitant pulse.

Principle of treatment: Activating blood circulation, dispersing blood stasis and arresting pain.

Recipe: *Tong Qiao Zhi Tong Tang*(66) with modification.

Ingredients:

Rhizoma Ligustici Chuanxiong	30 g
Radix Paeoniae Rubra	15 g
Semen Persicae	15 g
Flos Carthami	9 g
Rhizoma Zingiberis	3 slices

Bulbus Allii Fistulosi 1 decimetre

Mochus

   (to be taken following its

   infusion) 0.2 g

Decoct the above ingredients in water for oral dose.

4.2 Migraine

Migraine corresponds to *Nao Feng*(headache caused by pathogenic wind), *Shou Feng*(headache caused by pathogenic wind) and *Tou Feng Tou Tong*(headache caused by attack of the head by pathogenic wind) in trditional Chinese medicine.

Etiology and Pathogenesis

Congenital deficiency or lack of proper care after birth leads to injuries of *Zang-Fu* organs and imbalance of *Yin* and *Yang*, which are complicated with invasion of exogenous pathogenic facotrs or excessive emotional changes or overfatigue, resulting in reversed flow of *Qi* and blood and malnourishment of the brain, giving rise to abrupt attack of headache with intermittent pain.

Essentials of Diagnosis

1. This disease is usually related to inheritance, and may occur repeatedly with the first attack at childhood.

2. It is often induced by seasonal pathogenic factors, overfatigue, tension, excitement, poor sleep or menstrual period.

3. The patient may have such symptoms and signs before attacks as lethargy, spiritlessness or hyperhedonia, blurred vision, photophobia, and possiblly blind spot, hemianopsia, distention and pain of the eyeballs, or abnormal sensation of the limbs, or kinesic disturbance, etc.

4. There are repeated attacks of intolerable burning throb-

bing or boring pain on the forehead, temple and obit unilaterally in most cases and bilaterally in few cases. The pain usually lasts for few minutes or even 1—2 days. Sometimes it attacks several times a day. It may reattack in few months or few years.

5. It is often accompanied by nausea, vomiting, abdominal distention, diarrhea, polyhidrosis, lacrimation, pallor, bluish purple skin and edema, etc.

Treatment Based on Differentiation of Syndromes

1. Obstruction of the collaterals by stagnation of wind and phlegm and stasis of blood.

Symptoms and signs: Intermittent headache which is aggravate by wind and cold, no thirst, thin and white tongue coating, and superficial, wiry and tense pulse.

Principle of treatment: Dispelling wind, resolving phlegm, activating blood circulation and arresting pain.

Recipe: ① *San Pian Tang*(67) with modification.

Ingredients:

Rhizoma Ligustici Chuanxiong	30 g
Radix Angelicae Dahuricae	15 g
Radix Paeoniae Alba	15 g
Semen Sinapis Albae	9 g
Rhizoma Cyperi	6 g
Radix Bupleuri	3 g
Semen Pruni	3 g
Radix Glycyrrhizae	3 g

Decoct the above ingredients in water twice to obtain 300 ml of decoction to be taken half in the morning and half in the evening while it is warm, one dose a day.

② *Chuan Xiong Ding Tong Tang*(68) with modification.

Ingredients:

Rhizoma Ligustici Chuanxiong	30 g
Radix Paeoniae Rubra	15 g
Radix Salviae Miltiorrhizae	30 g
Radix Ledebouriellae	12 g
Herba Asari	6 g
Radix Aconiti	6 g
Semen Sinapis Albae	15 g
Semen Coicis	30 g
Fructus Amomi	9 g

Decoct the above ingredients in water for oral dose.

The following prescription could be used to prevent reattacks:

Rhizoma Ligustici Chuanxiong	15 g
Radix Angelicae Sinensis	9 g
Flos Carthami	9 g
Radix Angelicae Dahuricae	9 g
Fructus Tribuli	9 g
Flos Chrysanthemi	9 g
Ramulus Uncariae cum Uncis	6 g
Concha Margaritifera Usta	30 g

Decoct the above ingredients in water for oral dose.

2. Flaring-up of wind-fire

Symptoms and signs: Sudden attack of bursting or stabbing headache, red face with sweating, thirst, mental restlessness, red tongue with thin and yellow coating, and wiry and rapid pulse.

Principle of treatment: Dispelling wind-heat, removing obstruction from the channel and arresting pain.

Recipe: *Qing Shang Juan Tong Tang* (69).

Ingredients:

Radix Ophiopogonis	15 g
Radix Scutellariae	12 g
Fructus Viticis	12 g
Flos Chrysanthemi	12 g
Rhizoma seu Radix Notopterygii	9 g
Radix Ledebouriellae	9 g
Rhizoma Atractylodis	9 g
Radix Angelicae Sinensis	9 g
Radix Angelicae Dahuricae	9 g
Rhizoma Ligustici Chuanxiong	15 g
Herba Asari	6 g
Radix Glycyrrhizae	3 g

Decoct the above ingredients in water for oral dose.

Modification: In case of left migraine, add Flos Carthami 9 g, Radix Bupleuri 9 g, Radix Gentianae 12 g and Radix Rehmanniae 9 g.

In case of right migraine, add Rhizoma Gastrodiae 12 g Rhizoma Pinelliae 12 g, Fructus Crataegi 12 g and Fructus Citri Aurantii Immaturus 12 g.

In case of vertex headache, add Rhizoma Ligustici 9 g, Rhizoma et Radix Rhei 6 g and Herba Schizonepetae 9 g.

In case of headache caused by invasion of the brain by pathogenic wind complicated with nasal obstruction or turbid nasal discharge, add Fructus Xanthii, 12 g , Fructus Chaenomelis 9 g and Herba Schizonepetae 9 g.

In case of headache caused by deficiency of *Qi* and blood, add Radix Astragali seu Hedysari 15 g, Radix Ginseng 9 g or Radix Pseudostellariae 30 g, Radix Paeoniae Rubra 12 g, Radix

Paeoniae 12 g, *Sheng Di*(Radix Rehmanniae)12 g and *Shu Di*(Radix Rehmanniae)12 g;Or use *Quan Xie Gou Teng San*(70) with modification.

Ingredients:
Scorpio	6 pieces
Ramulus Uncariae cum Uncis	9 g
Radix Ginseng Rubra	6 g
Radix Angelicae Dahuricae	9 g
Radix Aconiti Praeparata	6 g

Administration: Bake the above ingredients together, grind them into fine powder which is divided into 9 equal portions.Put 1 portion of the powder into 250 ml of boiling water and cover it tightly for 20 minutes.Then, remove the residue and drink the juice.Take the above medicinal juice 3 times a day.

In case of attack during menstrual period, *Yang Jiao Chong Ji*(71) may be used.

Ingredients:
Cornu Naemorhedus goral Hardwicke	18 g
or Cornu Antelopis	3 g
Rhizoma Ligustici Chuanxiong	6 g
Radix Angelicae Dahuricae	9 g
Radix Aconiti Praeparata	6 g

Administration: Grind the above ingredients together into fine powder which is divided into 2 equal portions.Put 1 portion in a cup with 250 ml of boiling water, cover it tightly for 30 minutes, and then take away the residue and drink the juice while it is warm.Such administration is taken once in the morning and once in the evening and 10 days are considered as one course.

Besides, *Zhi Tong San*(72) may also be used.

Ingredients:

Borneolum Syntheticum	3 g
Natrii Sulfas	6 g
Moschus	0.5 g
Menthol	2 g

Administration: Grind the above ingredients together into fine powder which is put in a bottle tightly closed. During an attack of migraine, wrap 0.3 g of the powder in a small piece of gauze and then put it in the nostril of opposite side, that is, left nostril for right migraine and right nostril for left migriane. This method is indicated in various kinds of headache and can obtain immediate effect.

4.3 Acute Polyneuritis

Polyneuritis is also known as "peripheral neuritis" and "multiple peripheral neuritis". According to its clinical manifestations, it belongs to *Wei Zheng (Wei* syndrome) in traditional Chinese medicine.

Etiology and Pathogenesis

This disease is usually resulted from invasion of toxic heat which consumes blood and body fluid, deprives tendons and vessels of nourishment and gives rise to flaccid paralysis of the limbs.

Essentials of Diagnosis

1. The occurrence of the disease is abrupt.

2. Such paresthesia as prickling-like sensation, formication, burning sensation as well as tenderness of the muscle may appear at the distal end of the limbs symmetrically.

3. Symmetrically, myodynamia attenuates, myodystonia occurs and the tendon reflex becomes weak or even disappears at

the distal part of the limbs. Besides, there may appear muscular atrophy at the end of acute stage.

4. The skin at the distal end of the limbs is smooth with either hyperhidrosis or anhidrosis.

Treatment Based on Differentiation of Syndromes

1. Accumulation of toxic heat and malnourishment of tendons and muscles

Symptoms and signs: Fever, thirst, prickling or burning pain at the distal end of the limbs, yellow urine, thin and yellow tongue coating, and rapid pulse.

Principle of treatment: Clearing away toxic heat, activating blood circulation and removing obstruction from the channel.

Recipe: *Jie Du Tong Luo Yin*(73) with modification.

Ingredients:

Rhizoma Polygoni Cuspidati	15 g
Herba Bidentis	15 g
Radix Notoginseng	15 g
Radix Salviae Miltiorrhizae	15 g
Flos Lonicerae	60 g
Rhizoma Dryopteris	30 g

Decoct the above ingredients in water for oral dose.

Modification: In case of preponderance of damp-heat, add Rhizoma Smilacis Glabrae 30 g and Rhizoma Alismatis 15 g.

In case of significant pain, add Rhizoma Corydalis 15 g and Caulis Aristolochiae 6 g.

For mental restlessness, thirst and red tongue with little coating, add Semen Coicis 30 g, Radix Glehniae 30 g, Radix Trichosanthis 15 g and Radix Ophiopogonis 15 g.

2. Obstruction of the channel by stasis of blood.

Symptoms and signs: Severe numbness and pain at the distal end of the limbs which gets worse at night, dark red tongue, and wiry and hesitant pulse.

Principle of treatment: Activating blood circulation, removing obstruction from the channel and arresting pain.

Recipe: ① *Huo Xue Qu Tong Tang*(74) with modification.
Ingredients:

Radix Angelicae Sinensis	12 g
Radix Salviae Miltiorrhizae	20 g
Flos Carthami	12 g
Radix Achyranthis Bidentatae	12 g
Caulis Spatholobi	30 g
Radix Angelicae Dahuricae	9 g
Rhizoma Corydalis	9 g
Radix Astragali seu Hedysari	30 g
Radix Glycyrrhizae	12 g
Radix Stephaniae Tetrandrae	12 g
Rhizoma seu Radix Notopterygii	9 g
Resina Olibani	6 g
Myrrha	6 g
stir-fried Semen Ziziphi Spinosa	15 g

Decoct the above ingredients in water for oral dose.

② *Dang Gui Wei Ling Xian Tang*(75) with modification
Ingredients:

Radix Angelicae Sinensis	12 g
Radix Clematidis	12 g
Rhizoma Ligustici Chuanxiong	12 g
Radix Angelicae Dahuricae	9 g
Radix Stephaniae Tetrandrae	9 g

Rhizoma Atractylodis	6 g
Rhizoma seu Radix Notopterygii	6 g
Ramulus Cinnamomi	6 g
Rhizoma Zingiberis Recens	2 slices

Decoction the above ingredients in water for oral dose.

4.4 Acute Infective Polyneuroradiculitis

This disease is a special kind of polyneuritis. In general, it belongs to *Wei Zheng* (*Wei* Syndrome) in traditional Chinese medicine.

Etiology and Pathogenesis

Accumulation of toxic heat due to invasion by pathogenic wind-heat consumes blood and body fluid, deprives the tendons and muscles of nourishment, giving rise to flaccidity of the limbs.

Essentials of Diagnosis

1. The patient usually has infection of upper respiratory tract or diarrhea 2-3 weeks prior to the onset of the disease.

2. The disease occurs abruptly with numbness sensation of the limbs. Examination may suggest no significant symptoms. The patient may have hypoesthesia, anesthesia or hyperesthesia at the distal end of the limbs.

3. Myasthenia of the limbs and trunk is the main symptom which will lead to symmetrical paralysis of the limbs and reach its climax within one week. The proximal part of the limbs is worse than the distal part. In severe case, there may be dyspnea.

4. Hypomyotonia, tendon hyporefexia and significant muscular tenderness of the limbs may occur.

5. Laboratory examination of the cerebrospinal fluid will find that the count of lymphocytes is normal or slightly higher

while the protein content is significantly increased.

Treatment Based on Differentiation of Syndromes

1. Excessive accumulation of toxic heat and retardation of blood circulation.

Symptoms and signs: Sudden occurrence of numbness of the limbs, accompanied with fever, thirst, yellow urine, dry stool, red tongue with yellow coating, and rapid pulse.

Principle of treatment: Clearing away toxic heat, activating blood circulation and removing obstruction from the channel.

Recipe: *Jie Du Huo Xue Tong Luo Tang*(76) with modification.

Ingredients:

Radix Isatidis	30 g
Folium Isatidis	15 g
Radix Sophorae Subprostratae	15 g
Lasiosphaera seu Calvatia	9 g
Radix Platycodi	6 g
Ramulus Cinnamomi	6 g
Radix Salviae Miltiorrhizae	30 g
Radix Paeoniae Rubra	15 g
Radix Angelicae Sinensis	15 g
Radix Achyranthis Bidentatae	12 g
Retinervus Luffae Fructus	18 g
Radix Astragali seu Hedysari	15 g
Radix Glycyrrhizae	9 g

Decoct the above ingredients in water for oral dose.

Modification: In case of severe toxic heat, *Jie Du Tong Luo Yin*(73) could be used with modification.

Ingredients:

Rhizoma Polygoni Cuspidati	15 g
Herba Bidentis	15 g
Radix Notoginseng	15 g
Radix Salviae Miltiorrhizae	15 g
Caulis Lonicerae	60 g
Rhizoma Dryopteris	30 g

Decoct the above ingredients in water for oral dose.

Modification: In case of severe pain, add Rhizoma Corydalis 15 g and Radix Clematidis 15 g.

For dry mouth, mental restlessness, red tongue and rapid pulse, add Radix Glehniae 30 g, Radix Ophiopogonis 15 g and Radix Trichosanthis 12 g.

2. Deficiency of both *Qi* and blood and malnourishment of muscles and tendons.

Symtoms and signs: This syndrome mostly occurs at the Convalescent stage with such manifestations as flaccidity of the four limbs, general lassitude, normal or poor appetite, normal bowel movement and micturition, thready and forceless pulse, and thin and white tongue coating.

Principle of treatment: Tonifying *Qi*, nourishing blood and activating circulation in the channels and collaterals.

Recipe: ①Modified *Bu Qi Yang Xue Chu Wei Tang*(77).
Ingredients:

Radix Astragali seu hedysari	60 g
sugared Radix Ginseng	10 g
Rhizoma Polygonati	12 g
Radix Codonopsis Pilosulae	15 g
Ramulus Cinnamomi	6 g
stir-fried Fructus Hordei	

Germinatus	18 g
Radix Angelicae Sinensis	12 g
Radix Salviae Miltiorrhizae	18 g
Radix Achyranthis Bidentatae	15 g
Radix Ophiopogonis	15 g
Radix Dipsaci	12 g
Rhizoma Cimicifugae	6 g
Rhizoma Homalomenae	15 g
Herba Cistanchis	15 g
Caulis Spatholobi	18 g

Decoct the above ingredients in water for oral dose, one dose a day.

② *Du Huo Ji Sheng Tang*(78) with modification.

Ingredients:

Radix Angelicae Pubescentis	6 g
Ramulus Loranthi	12 g
Radix Gentianae Macrophyllae	6 g
Radix Ledebouriellae	6 g
Herba Asari	3 g
Radix Angelicae Sinensis	12 g
Radix Paeoniae Albae	12 g
Rhizoma Ligustici Chuanxiong	10 g
Radix Rehmanniae Praeparata	12 g
Cortex Eucommiae	9 g
Radix Achyranthis Bidentatae	9 g
Radix Ginseng Rubra	6 g

Decoct the above ingredients in water for oral dose.

Acupuncture and point-injection therapy: When acupuncture is applied, points Shousanli (LI10) and Hegu(LI4) are taken as the main points; and Jianyu(LI15), Jian liao(SJ14) and Quchi(LI11) as the secondary for the upper limbs; Points Shen shu(BL23), Dachangshu(BL25) and Huantiao(GB30) are taken as the main ones, and Zusanli(ST36) and Yanglingquan(GB34) as the secondary for lower limbs, Treatment should be given once every other day and ten treatments are taken as one course. Warm needling may obtain even better therapeutic effect.

In point-injection therapy, vitamin B_1, Vitamin B_{12} or *Dang Gui* injection could be used for the corresponding point injection, or use *Fu Fang Dan Shen* Injection for intramuscular injection, 2 ml each point and once a day.

4.5 Hypertensive Cerebral Hemorrhage

According to its clinical manifestations, hypertensive cerebral hemorrhage corresponds to apoplexy in which the *Zang* organs are attacked.

Etiology and Pathogenesis

Fundamentally, the patient has imbalance of *Yin* and *Yang* in which *Yin* is insufficient in the lower and *Yang* hyperactive in the upper due to congenital deficiency or improper care after birth.Besides, the patient is affected by invasion of exogenous pathogenic factors, excessive emotional changes, improper work and rest or irregular food-intake which lead to adverse movement of *Qi* and blood in ascending and descending, and extravasation of blood, giving rise to sudden attack of apoplexy.

Essentials of Diagnosis

1. This disease usually abruptly occurs during emotional excitement or physical exertion with sudden coma and fall.

2. It happens more often to those people over 40 years old.

3. Most of the patients have the history of headache, vertigo or hypertention which gets worse before the stroke with such signs as temperal aphasia, numbness of the limbs, and blurred vision. The attack is often induced by invasion of exogenous pathogenous factors, excitement, voracious eating and alcohol drinking, and overfatigue.

4. The main clinical manifestations are sudden fall and unconsciousness, hemiplegia, deviation of mouth and tongue, dysphasia or even aphasia.

5. Cerebrospinal fluid presents a bloody colour and the hemorrhagic focus can be found during a GT scan.

Treatment Based on Differentiation of Syndromes

1. Attack of the brain by wind—fire

Symptoms and signs: The patient often has headache, dizziness and vertigo. with excessive emotional change, voracious alcohol drinking or over strain and stress, the pathological condition may suddenly change, giving rise to such manifestations as trance, semiconsciousness, hemiplegia, rigidity or contracture of the limbs, dry stool or constipation, deep red tongue with yellow, sticky and dry coating, and wiry, slippery, full and rapid pulse.

Principle of treatment: Dispelling wind, reducing fire, resolving phlegm and opening the mind.

Recipe: ① *An Gong Niu Huang Wan*(20) to be dissolved in water for oral application or nasal feeding 1 pill each time and once every 6 hours during the first 3 days, and 1 pill each time and twice a day during the next 4 days.

② *An Gong Niu Huang San* to be taken following its infusion or by nasal feeding 1.6 g each time and once or twice a day.

③ Add 40–60 ml of *Qing Kai Ling* to 250 ml of 5% glucose solution for intravenous drip, once a day.

④ Experienced recipe:

Scorpio	3 g
Concretio Silicea Bambusae	5 g
powder of Cornu Antelopis	3 g
Margarita	0.5 g

Grind the above ingredients into fine powder which is divided into 3–4 portions to be taken following its infusion or by nasal feeding separatedly, and it is the dose for 1 day.

⑤ Experienced recipe:

powder of Cornu Antelopis]	
(to be taken following its infusion)	4.5 g
Concha Haliotidis	
(to be decocted first)	30 g
Plastrum Testudinis	
(to be decocted first)	30 g
Os Draconis	
(to be decocted first)	30 g
Radix Praeoniae Albae	30 g
Ramulus Uncariae cum Uncis	
(to be decocted later)	15 g
Radix Achyranthis Bidentatae	15 g
Radix Rehmanniae	30 g
Cortex Moutan Tadicis	9 g
Spica Prunellae	15 g

Decoct the above ingredients in water twice to obtain 300 ml of decoction to be taken separately in the morning, at noon and in the evening orally or by nasal feeding.

⑥ Take *Bai Yao* orally or by nasal feeding 0.5 g each time, once every 4 hours during the first 3–4 day, once every 6 hours during the 5–10th days, and 3 times a day during the 11–14th days.

2. Mist of the mind by phlegm–damp

Symptoms and signs: The patient has constitutional insufficiency of *Yang* with accumulation of phlegm–damp in the interior. After the attack of the disease, there may be coma, hemiplegia, flaccidity of the limbs, or even cold limbs, pale complexion, dark lips, preponderance of phlegm, dark pale tongue with white and sticky coating, and deep and slippery or deep and slow pulse.

Principle of treatment: Resolving phlegm and opening the mind.

Recipe: *Di Tan Tang*(79) with modification.

Ingredients:

Rhizoma Arisaematis Praeparata	12 g
Rhizoma Pinelliae Praeparata	12 g
Fructus Citri Aurantii Immaturus	9 g
Pericarpium Citri Reticulatae	9 g
Rhizoma Acori Graminei	6 g
Poria	9 g
Rhizoma Zingiberis	3 slices
Radix Ginseng	3 g
Radix Glycyrrhizae	3 g
Succus Bambosae	
(to be taken following its infusion)	30 g

Decoct the above ingredients in water twice to obtain 250 ml of decoction to be taken for 3 times through oral or nasal feeding.

1—2 doses a day.

Modification: In case of critical condition or if no good therapeutic result is obtained, *Su He Xiang Wan*(1 pill each time) or *Hou Zao San*(3 g each time) could be taken together with the above decoction 1—2 times a day.

3. Block of the mind by phlegm—heat.

Symptoms and signs: Abrupt onset with coma or mental confusion, snoring, rattle in the throat, hemiplegia, rigitidy and contracture of the limbs, stiffness of the neck, fever, restlessness, or even cold extremities, frequent convulsion, occasional hematemesis, deep red tongue with brownish yellow, dry and sticky coating, and wiry, slippery and rapid coating.

Principle of treatment: Clearing away heat, resolving phlegm, dispelling wind and opening the mind.

Recipe: *Chang Pu Yu Jin Tang*(33) with modification.

Ingredients:

Rhizoma Acori Graminei	12 g
Radix Curcumae	12 g
Fructus Gardeniae	9 g
Succus Bambosae	
(to be taken following its infusion)	30 ml
Cortex Moutan Radicis	12 g
Folium Bambosae	3 g
Fructus Firsythiae	9 g
to be taken following its infusion	3 g
Arisaema cum Bile	6 g
Concretio Silicea Bambusae	6 g
powder of Cornu Antelopis	
(to be taken following its infusion	1.5 g

Decoct the above ingredients in water twice to obtain 200 ml of decoction to be taken through oral or nasal feeding for several times, 1 dose a day.

Modification: In case of severe coma, add *Zhi Bao Dan*(80) or *An Gong Niu Huang Wan*(20) 1 pill each time and 1 or 2 times a day, or grind Scorpio 3 g, Concretio Silicea Bambosae 5 g, powder of Cornu Antelopis 3 g and Margarita 0.5 g together into fine powder to be taken through oral or nasal feeding for 3—4 times, once every 6—8 hours.

4. Exhaustion of *Yuan* (source) *Qi* and scatter of the mind.

Symptoms and signs: Sudden coma, mental confusion, flaccicity of the limbs, cold limbs with sweating, and in severe case, clammy body, incontenence of bowel movement and micturition, flaccid tong which is dark purple with white and sticky coating, and deep and slow or deep and weak pulse.

Principle of treatment: Tonifying *Qi*, restoring *Yang*, strengthening the anti—pathogenic *Qi* and saving the patient from collapse.

Recipe: *Shen Fu Tang*(29) with modification.

Ingredients:

Radix Ginseng	15 g
Radix Aconiti Praeparata	15 g
Radix Ophiopogonis	30 g

Decoct the above ingredients in water twice to obtain 150—200 ml of decoction to be taken through roal or nasal feeding for 3—4 times.

Modification: In case of persistent hiccup, add Haematitum 30 g and Rhizoma et Redix Rhei 6 g.

In case of persistent profuse sweating, add Os Draconis 20 g,

Concha Ostreae 20 g, Fructus Corni 12 g and Fructus Schisandrae 6 g. Decoct the above ingredients in water and take the decoction through nasal feeding for 3—4 times.

In case of hematemesis, add *Bai Yao* 0.5—1 g (to be taken following its infusion).

In case of convulsion, grind Scorpio 3 g and powder of Cornu Antelopis 1.5 g into fine powder to be taken twice following its infusion.

4.6 Arteriosclerotic Cerebral Infarction

This disease belongs to *Zhong Feng* (wind stroke) in traditional Chinese medicine.

Etiology and Pathogenesis

Congenital insufficiency or improper care after birth injures the *Zang—Fu* organs, consumes *Yin* essence, leading to imbalance of *Yin* and *Yang*. Besides, the patient is affected by certain pathogenic factors such as excessive drinking of alcohol, overfatigue and excessive emotional change, leading to sudden adverse movement of *Qi* and blood in ascending and descending and obstruction of the channel by phlegm, fire or stasis of blood, giving rise to wind stroke.

Essentials of Diagnosis

1. The onset of the disease is abrupt and it usually occurs at night when the patient is taking a rest.

2. People over 40 years old have a predilection for the disease.

3. The initial attack is often preceded by some signs such as dizziness, numbness of the limbs, deviation of mouth and stiff tongue, etc.

4. It is often induced by mental depression, overstrain and overstress, excessive alcohol drinking and voracious eating, etc.

5. The main symptoms are coma and mental confusion, hemiplegia, numbness of the affected side, deviation of mouth and tongue, dysphasia or aphasia.

6. The pulse of the affected side is more wiry and slippery and the tongue coating mostly white and sticky or yellow and dry.

Treatment Based on Differentiation of Syndromes

1. Sudden hyperactivity of liver *Yang* and stirring-up of wind-fire.

Symptoms and signs: Hemiplegia, deviation of mouth and tongue, dysphasia or aphasia, numbness of affected side, headache, dizziness, red face and eyes, bitter mouth, dry throat, mental restlessness, irritability, deep yellow urine, dry stool, red or deep red tongue with thin and yellow coating, and wiry and forceful pulse.

Principle of treatment: Soothing the liver *Yang*, reducing fire, activating blood circulation and removing obstruction from the channel.

Recipe: ① *Ping Gan Xie Huo Tang*(81)
Ingredients:

Ramulus Uncariae cum Uncis	30 g
Flos Chrysanthemi	10 g
Spica Prunellae	15 g
Cortex Moutan Radicis	15 g
Concha Margaritifera Usta	30 g
Radix Achyranthis Bidentatae	20 g
Radix Paeoniae Rubra	10 g

Decoct the ingredients in water for oral dose.

② *Dan Gou Liu Zhi Yin* (82) with modification.

Ingredients:

Radix Salviae Miltiorrhizae	30–60 g
Ramulus Uncariae cum Uncis (to be decocted later)	15–30 g
Herba Siegesbeckiae	12–24 g
Spica Prunellae	12–24 g
Lumbricus	9 g
Flos Carthami	6 g
Ramulus Mori	15 g
Ramulus Citri Reticulatae	15 g
Ramulus Pini	15 g
Ramulus Persicae	15 g
Ramulus Abies	15 g
Ramulus Bambosae	15 g
Radix Glycyrrhizae	3 g

Decoct the above ingredients in water for oral dose.

Modification: In case of preponderance of phlegm, add Fructus Trichosanthis 15 g and Semen Raphani 20 g.

In case of coma, add Radix Curcumae 9 g and Rhizoma Acori Graminei 9 g.

In case of persistent high blood pressure, add Ochra Haematitum 30 g and Achyranthis Bidentatae 20 g.

③ *Li Xiu Lin Shi Fan Wei Jian Zheng Fang* (83)

Ingredients:

Ochra Haematitum	30 g
Gypsum Fibrosum	30 g
Radix Paeoniae Alba	15 g

ginger juice-prepared Caulis Bambusae in Taeniam	20 g
Exocarpium Citri Grandis	9 g
Rhizoma Acori Graminei	9 g
Poria	30 g
Semen Amomi Cardamomi	3 g
Herba Eupatorii	20 g

Decoct the above ingredients in water for oral dose.

Modification: In case of hematemesis and hematochezia, **add** *San Qi Fen*(powder of Radix Notoginseng) 3-6 g(to be taken following its infusion), or *Yun Nan Bai Yao* to be taken following its infusion 0.5 g each time and 4 times a day.

④ *Ping Gan Huo Xue Tang*(84)

Ingredients:

powder of Cornu Antelopis (to be taken following its infusion)	1.5-3 g
Concha Haliotidis	30 g
Ramulus Uncariae cum Uncis (to be decocted later)	15-30 g
Radix Achyranthis Bidentatae	9-15 g
Rhizoma Ligustici Chuanxiong	9 g
Flos Carthami	9-15 g
Semen Persicae	9 g
Eupolyphaga seu Steleophaga	9 g

Decoct the above ingredients in water for oral dose.

Modification: In case of constipation with fullness in the abdomen, add Rhizoma et Radix Rhei 9 g and Fructus Trichosanthis 30 g.

In case of aphasia with thick and sticky tongue coating, add

Arisaema cum Bile 9 g and Rhizoma Acori Graminei 9 g.

In case of convulsion, add powder of Scorpio 3—6 g to be taken following its infusion.

In case of mental restlessness, add Calcitum 30 g.

2. Obstruction of the channel by wind—phlegm and blood stasis.

Symptoms and signs: Hemiplegia, deviation of mouth and tongue, dysphasia or aphasia, numbness of affected side, dizziness and vertigo, dark pale tongue with thin and white or white and sticky coating, and wiry and slippery pulse.

Principle of treatment: Dispelling wind, resolving phlegm, activating blood circulation and removing obstruction from the channel.

Recipe: ①*Hua Tan Tong Luo Yin*(85).

Ingredients:

Rhizoma Pinelliae Praeparata	10 g
Rhizoma Atractylodis Macrocephalae	10 g
Rhizoma Gastrodiae	10 g
Arisaema cum Bile	6 g
Radix Salviae Miltiorrhizae	30 g
Rhizoma Cyperi	15 g
Radix et Rhizoma Rhei Praeparata	5 g

Decoct the above ingredients in water for oral dose.

② *Xi Feng Hua Tan Tang*(86) with modification.

Ingredients:

Poria	21 g
Rhizoma Pinelliae	12 g
Exocarpium Citri Grandis	12 g
Caulis Bambusae in Taeniam	12 g

Rhizoma Acori Graminei	12 g
Rhizoma Arisaema cum Bile Praeparata	12 g
Bombyx Batryticatus	9 g
Ramulus Uncariae cum Uncis (to be decocted later)	20 g
Scorpio	9 g
Herba Siegesbeckiae	30 g
Semen Persicae	12 g
Flos Carthami	9 g

Decoct the above ingredients in water for oral dose.

3. Excess syndrome of *Fu* organ due to phlegm—heat complicated with upward disturbance of wind—phlegm.

Symptoms and signs: Hemiplegia, deviation of mouth and tongue, dysphasia or aphasia, numbness of the affected side, abdominal distention, dry stool, constipation, dizziness, vertigo, expectoration with possible profuse sputum, dark red or dark pale tongue with yellow or yellow and sticky coating, and wiry, slippery and full pulse.

Principle of treatment: Removing obstruction from the *Fu* organ and resolving phlegm.

Recipe: ① *Tong Fu Hua Tan Yin*(87).

Ingredients:

Radix et Rhizoma Rhei	10 g
Fructus Trichosanthis	30 g
Arisaema cum Bile	6 g
Natrii Sulfas (to be taken following its infusion)	10 g

Decoct the above ingredients in water for oral dose.

Radix Astragali seu Hedysari	15–30 g
Rhizoma Ligustici Chuanxiong	6–9 g
Radix Angelicae Sinensis	9–12 g
Radix Paeoniae Rubra	9–12 g
Lumbricus	9–12 g
Semen Persicae	9–12 g
Radix Achyranthis Bidentatae	15 g
Radix Salviae Miltiorrhizae	15–30 g

Decoct the above ingredients in water for oral dose.

③ *Huang Qi Wǔ Wù Tàng* (90).

Ingredients:

Radix Astragali seu Hedysari	60 g
Radix Paeoniae Albae	15 g
Ramulus Cinnamomi	24 g
Rhizoma Zingiberis Recens	9 g
Fructus Ziziphi Jujubae	5 pieces

Decoct the above ingredients in water for oral dose.

Modification: In case of hemiplegia on the left side, add Radix Angelicae Sinensis 30 g. For the paralysis of the lower limb, add Achyranthis Bidentatae 9 g. To treat contracture of the tendons, add Fructus Chaenomelis 15 g. To cure weakness of the legs, add Os Tigris 9 g. In case of slow and thready pulse, and *Fu Zi Pian* (Radix Aconiti Praeparata) 15 g.

5. Stirring-up of endogenous wind due to deficiency of *Yin*.

Symptoms and signs: Hemiplegia, deviation of mouth and tongue, dysphasia or aphasia, numbness of the affected side, mental restlessness, insomnia, vertigo, tinnitus, hotness sensation in the palms and soles, deep red or dark red tongue with no or little coating, and thready and wiry or thready, wiry and rapid pulse.

Principle of treatment: Nourishing *Yin* and dispelling wind.

Recipe: ① *Xi Xian Zhi Yin Tang*(91)

Ingredients:

Herba Siegesbeckiae Praeparata	50 g
Radix Rehmanniae	15 g
salt solution—prepared Rhizoma Anemarrhenae	20 g
Radix Angelicae Sinensis	15 g
Fructus Lycii	15 g
stir-fried Radix Paeoniae Rubra	25 g
Plastrum Testudinis	10 g
Radix Achyranthis Bidentatae	10 g
Flos Chrysanthemi	15 g
Radix Curcumae	15 g
Radix Salviae Miltiorrhizae	15 g
Cortex Phellodendri	5 g

Decoct the above ingredients in water for oral dose.

② *Yu Yin Xi Feng Tang*(92)

Ingredients:

Radix Rehmanniae	20 g
Radix Scrophulariae	15 g
Fructus Ligustri Lucidi	15 g
Ramulus Loranthi	30 g
Ramulus Uncariae cum Uncis (to be decocted later)	30 g
Radix Paeoniae Albae	20 g
Radix Salviae Miltiorrhizae	15 g

Decoct the above ingredients in water for oral dose.

6. Block of the mind by phlegm—heat.

Symptoms and signs: Abrupt onset of coma or mental confusion with snoring, rattle in the throat, hemiplegia, contracture of limbs, neck rigidity, fever, restlessness, convulsion of the limbs, deep red tongue with brownish yellow, dry and sticky coating, and wiry, slippery and rapid pulse.

Principle of treatment: Clearing away heat, resolving phlegm and opening the mind.

Recipe: ① Add 40—60 ml of *Qing Kai Ling* Injection to 300 ml of 5% glucose solution for intravenous drip, once a day.

② Add 20 ml of *Xing Nao Jing* Injection to 300 ml of 5% glucose solution for intravenous drip, once a day.

③ *Ling Jiao Gou Teng Tang*(93) with modification.

Ingredients:

powder of Cornu Antelopis (to be taken following its infusion)	4.5 g
Concha Haliotidis (to be decocted first)	30 g
Os Draconis (to be decocted first)	30 g
Concha Ostreae (to be decocted first)	30 g
Radix Paeoniae Albae	30 g
Ramulus Uncariae cum Uncis (to be decocted later)	15 g
Radix Achyranthis Bidentatae	15 g
Radix Rehmanniae	30 g
Spica Prunellae	15 g

Cortex Moutan Radicis 9 g

Decoct the above ingredients in water twice to obtain 300 ml of decoction to be taken for 3—4 times by oral or nasal feeding, one dose a day.

④ Take *Zhi Bao Dan*(80) or *An Gong Niu Huang Wan*(20) or *An Gong Niu Huang San* 2.5 g through oral or nasal feeding once every 6 hours and contineously for 3—5 days.

7. Mist of the mind by phlegm—damp

Symptoms and signs: The patient has constitutional deficiency of *Yang* with accumulation of phlegm—damp in the interior. During the attack, there are coma, hemiplegia, cold and flaccid limbs, pallor, dark lips, profuse phlegm, dark pale tongue with white and sticky tongue coating, and deep and slippery or deep and slow pulse.

Principle of treatment: Dispelling wind, resolving phlegm and opening the mind.

Recipe:

①Rhizoma Arisaematis Praeparata	12 g
Rhizoma Pinelliae Praeparata	12 g
Fructus Citri Aurantii Immaturus	9 g
Pericarpium Citri Reticulatae	9 g
Rhizoma Acori Graminei	6 g
Poria	9 g
Rhizoma Zingiberis	6 g
Radix Ginseng	3 g
Radix Glycyrrhizae	3 g
Succus Bambusae (to be taken following its infusion)	30 g

Decoct the above ingredients in water twice to obtain 300 ml of decoction to be taken together with 1 pill of *Su He Xiang Wan* through oral or nasal feeding, 1 or 2 doses daily.

② In severe case or no good therapeutic effect is obtained, take *Hou Zao*(Calculus Macacae) 0.3 g together with the above decoction.

8. Exhaution of *Yuan*(source) *Qi* and scatter of mind.

Symptoms and signs: Sudden coma or mental confusion, flaccidity of limbs with hands relaxed, cold limbs, profuse sweating, and in severe case, clammy body, incontenence of bowel movement and micturition, flaccid tongue which is dark purple with white and sticky coating, and deep and slow or deep and fading pulse.

Principle of treatment: Tonify *Yuan* (source) *Qi*, restoring *Yang* and saving the patient from the collapse.

Recipe: *Shen Fu Tang*(29).

Ingredients:

Radix Ginseng	30 g
Radix Aconiti Praeparata	9 g

Decoct the above ingredients in water twice to obtain 200 ml of decoction to be taken through frequent oral or nasal feeding.

5 Digestive Diseases

This chapter mainly introduces non-operative treatment of the disorders of the stomach, intestine, liver, gallbladder and pancreas, as well as the disorder of the acute abdomen. These diseases respectively correspond to *Ji Xing Fu Tong*(acute abdominal pain), *Chang Jie*(constipation), *Guan Ge*(dysuria and constipation with incessant vomiting), *Xie Tong*(hypochondriac pain), *Huang Dan*(jaundice), *Ou Tu*(nausea and vomiting) and *Chang Feng*(hemotochezia), etc. in TCM caused by accumulation of heat, stagnation of cold, stagnation of *Qi*, stasis of blood, retention of food and parasites. The principles "treating the acute symptoms for the emergency and the cause when the acute symptoms are relieved" and "keeping the six *Fu* organs unobstructe to make them function well" are taken as the basic principles of treatment in this chapter.

5.1 Acute Gastroenteritis

Acute gastroenteritis takes nausea, vomiting, diarrhea and epigastric and abdominal pain as the main manifestations. It corresponds to *Ou Tu*(nausea and vomiting), *Xie Xie*(diarrhea) and *Wei Wan Tong*(epigastric pain) in traditional Chinese medicine.

Etiology and Pathogenesis

This disease is usually caused by dysfunction of the spleen in transportation and transformation and that of the stomach in descending due to voracious eating and drinking, insanitary food-intake and invasion of pathogenic wind, cold, summer heat and damp. When excessive vomiting and diarrhea occur, which will severely consume *Qi* and body fluid, there may appear collapse.

Essentials of Diagnosis

1. This disease usually has the history of improper food-intake. It often abruptly occurs in summer and autumn with such symptoms as sudden vomiting, diarrhea and epigastric and abdominal pain complicated with exterior symptoms caused by exogenous pathogenic factors.

2. During severe vomiting and diarrhea, there may appear the symptoms of dehydration such as sunk eyes, flabby skin, mental restlessness, dry mouth and oliguria, etc.

3. There are tenderness on the upper abdomen and around umbilicus and borbory gmus with increased leucocytes and neutrophilic granulocytes.

4. This disease must be differentiated from cholera and bacillary dysentery.

Treatment Based on Differentiation of Syndromes

1. Invasion of exogenous pathogenic factors

Symptoms and signs: Sudden nausea and vomiting, fullness and stuffiness in the chest and epigastrium, abdominal pain with borborygmus, diarrhea, accompanied with chillness, fever, headache, general ache, white or white and sticky tongue coating, and soft pulse.

Principle of treatment: Relieving the exterior by dispelling pathogenic factors, and resolving the turbid by using aromatic drugs.

Recipe: *Huo Xiang Zheng Qi San* (3) with modification.

Ingredients:

Herba Agastachis　　　　　　　　　　　　　　　　9 g

Folium Perillae	9 g
Cortex Magnoliae Officinalis	9 g
Rhizoma Pinelliae	9 g
Pericarpium Citri Reticulatae	9 g
Poria	21 g
Pericarpium Arecae	9 g
stir-fried Rhizoma Atractylodis Macrocephalae	12 g
Radix Platycodi	6 g
Radix Angelicae Dahuricae	6 g
Radix Glycyrrhizae	3 g
Rhizoma Zingiberis(a proper amount)	
Fructus Ziziphi Jujubae (a proper amount)	

Decoct the above ingredients in water for oral dose.

Modification: In case of dysfunction in transportation and transformation due to invasion of the spleen by pathogenic damp-heat in late summer and early autumn,*Ge Gen Huang Qin Huang Lian Tang*(162) with modification may be used.

In case of high fever,add Fructus Forsythiae 15 g and Herba Euphorbiae Humifusae 9 g.

In case of preponderance of damp,add Rhizoma Atractylodis 9 g and Semen Amomi Cardamomi 9 g.

2. Retention of food

Symptoms and signs: Sour and foul vomiting,fullness and distention in epigastrium and abdomen,abdominal pain with borborygmus which could be alleviated by diarrhea with discharge of spoiled egg-like offensive feces,belching,poor appetite,turbid tongue coating,and slippery and excess or deep and wiry pulse.

Principle of treatment: Promoting digestion and removing food retention.

Recipe: *Bao He Wan*(94) with modification.

Ingredients:

Massa Fermentata Medicinalis	20 g
Carbonized Fructus Crataegi	21 g
Semen Raphani	12 g
Poria	5 g
Pericarpium Citri Reticulatae	12 g
Rhizoma Pinelliae	12 g
Fructus Forsythiae	15 g
Fructus Amomi	9 g
Radix Saussureae	9 g
Fructus Aurantii	10 g
stir-fried Hordei Germinatus	30 g

Decoct the above ingredients in water for oral dose.

Modification: In case of severe abdominal distention and pain with hesitant bowel movement, add Radix et Rhizoma Rhei 9 g, Fructus Aurantii Immaturus 12 g and Semen Arecae 9 g.

If the food retention has turned into heat, add Rhizoma Coptidis 9 g.

For nausea and vomiting with yellow tongue coating and rapid pulse, add Caulis Bambusae in Taeniam 9 g and Semen Amomi Cardamomi 9 g.

5.2 Gastroduodenal Ulcerative Bleeding

This disease is known as *Xue Zheng* (hemorrhage) in traditional Chinese medicine.

Etiology and Pathogenesis

It is usually caused by either accumulation of heat in the stomach,or stagnation of liver *Qi* turning into fire and affecting the stomach,which,injures the vessels and gives rise to bleeding.It may also be resulted from overstrain and overstress which injure the spleen and stomach,giving rise to deficiency of *Qi* in controlling blood.

Essentials of Diagnosis

1. Most of the patients have a history of typical gastroduodenal ulcer before bleeding.However,some patients may suffer from sudden hematemesis and melena (result of occult blood examination is positive) without any symptoms and the pain may alleviate or disappear after that.

2. Most of those patients with bleeding amount over 400 ml will have such manifestations in varying degrees as vertigo,white tongue coating,cold sweating,thready and rapid pulse,drop of blood pressure or even coma.

3. About half of the patients will have lower grade fever within 24 hours after bleeding,which usually lasts for one week or so.

4. There may be decrease of erythrocytes and leucocytes 6—12 or even 20—30 hours after massive bleeding.

5. After massive bleeding,the blood urea nitrogen may increase in few hours and reach its maximum in 24—48 hours.

Treatment Based on Differentiation

1. Burn of the vessels due to excessive stomach—heat

Symptoms and signs: Burning sensation or stuffiness and pain in epigastrium,hematemesis or vomiting of dark pauple indigested food,foul breathing,constipation or melena,red tongue with yellow and sticky coating,and slippery and rapid pulse.

Principle of treatment: Clearing away heat from the stomach, reducing fire, dispersing stasis of blood and stopping bleeding.

Recipe: *Xie Xin Tang*(95) plus *Shi Hui San*(96) with modification.

Ingredients:

Radix et Rhizoma Rhei	9 g
Rhizoma Coptidis	9 g
Radix Scutellariae	9 g
Carbonized Cacumen Biotae	12 g
Herba seu Radix Cirsii Japonici	15 g
Herba Cephalanoploris	15 g
Carbonized Radix Rubiae	15 g
Cortex Moutan Radicis	10 g

Decoct the above ingredients in water for oral dose.

Modification: For epigastric distention, add Pericarpium Citri Reticulatae 9 g. For Epigastric pain, add Rhizoma Cyperi 9 g. For nausea and vomiting, add Caulis Bambusae in Taeniam 9 g.

2. Invasion of the stomach by liver fire and extravasation of blood.

Symptoms and signs: Sudden hematemesis with fresh red or dark purple blood, mental restlessness, irritability, dry mouth and lips, red tongue with little coating, and wiry, thready and rapid pulse.

Principle of treatment: Reducing the liver fire, cooling the blood and stopping pain.

Recipe: *Long Dan Xie Gan Tang*(97) with modification.

Ingredients:

Radix Gentianae	12 g

Radix Scutellariae	12 g
stir-heated Fructus Gardeniae	9 g
Radix Rehmanniae	30 g
Cortex Moutan Radicis	12 g
Rhizoma Imperatae	30 g
Nodus Nelumbinis Rhizomatis	18 g
Herba Ecliptae	12 g
Radix Paeoniae Rubra	9 g

Decoct the above ingredients in water for oral dose.

Modification: In case persistent bleeding accompanied by stuffy epigastrium and thirst, add powder of Ophicalcitum 15-30g, *San Qi Fen*(powder of Radix Notoginseng)6-9 g powder of Rhizoma Bletillae 9-12 g and Folium Callicarpae Pedunculatae 18 g.

In case of sudden and severe hematemesis, take the decoction of *Xi Jiao Di Huang Tang*(25) together with *San Qi Fen*(powder of Radix Notoginseng)9 g and powder of Radix et Rhizoma Rhei 12 g.

In case of persistent hematemesis, add Cacumen Biotae 30 g and *San Qi Fen*(powder of Radix Notoginseng) 10 g to be taken following its infusion.

3. Deficiency of spleen *Qi* and dysfunction of the spleen in controlling blood.

Symptoms and signs: Prolonged hematemesis which goes up and down, palpitation, shortness of breath, pallor, lassitude, pale tongue with thin and white coating, and thready and forceless pulse.

Principle of treatment: Strengthening the spleen *Qi* and promoting the function of the spleen in controlling blood to stop bleeding.

Recipe: *Gui Pi Tang*(50) with modification.
Ingredients:

Radix Astragali seu Hedysari	15 g
Radix Codonopsis Pilosulae	15 g
Poria cum Ligno Hospite	15 g
Arillus Longan	20 g
Radix Saussureae	6 g
Radix Polygalae	9 g
Semen Ziziphi Spinosae	15 g
Radix Glycyrrhizae	6 g
Radix Angelicae Sinensis	12 g
Rhizoma Bletillae	12 g
Os Sepiella seu Sepiae	30 g
carbonized Radix Sanguisorba officinalis L.	15 g

Decoct the above ingredients in water for oral dose.

Modification: The astringent drugs mentioned above could be adjusted according to the clinical symptoms and signs.

Acupuncture: Select Point Zusanli(ST36), Neiguan(PC6), Neiting(ST44) and Gongsun(SP4), etc. When ear acupuncture is applied, puncture Point Subcortex, Heart, Adrenal combined with Liver, Spleen and Shenmen(HT7), etc.

Besides, take orally powder of *Xue Yu Tan* (Crinis Carbonisatus) or *Di Yu Tan* (Radix Sanguisorbae Carbonisatus) 3–9 g each time and 3 times a day, or *Yun Nan Bai Yao* 0.5 g each time and 3–4 times a day, or *Bai Ji Fen*(powder of Rhizoma Bletillae) 3 g and *Sheng Da Huang Fen*(powder of Radix et Rhizoma Rhei) 1.5 g each time and 2 times a day, and 4–6 times a day in severe case.

5.3 Acute Gastric Dilatation

Acute gastric dilatation belongs to *Su Shi*(dyspepsia) or *Shang Shi*(impairment of the stomach by overeating) in traditional Chinese medicine.

Etiology and Pathogenesis

This disease is usually resulted from voracious eating or drinking at the recovery stage of an internal injury or after operation,which injures the stomach and intestine,and leads to dysfunction of the spleen in transportation and transformation and that of the stomach in descending,giving rise to stagnation of *Qi*.

Essentials of Diagnosis

1. It often occurs after major chest and abdominal operation or at the recovery stage of a severe case. Besides, infection, traumatic injury and overeating are also the common causes.

2. There may be fluxional vomiting of brownish green substance which becomes coffee-like with the development of pathological condition.In severe cases,there may appear dehydrant alkalosis and coma.

3. Epigastric bulge can be observed.With palpation,enlarged stomach without peristalsis and tenderness could be found.During percussion,there could be excessive echo and splashing sound.Plain abdominal X-ray film may show fluid level in the dilated stomach.

4. There could be complication of acute gastric perforation and peritonitis.

Treatment Based on Differentiation of Syndromes

First of all, fasting, gastrointestinal decompression and intravenous drip should be given to restore water-electrolyte balance. Then herbal decoction should be taken orally or through gastrogavage.

1. Stagnation of *Qi* and dysfunction of the stomach in descending due to weakness of the spleen.

Symptoms and signs: Fullness, distention and solidity in epigastrium, vomiting of fetid substance, sticky tongue coating and wiry and slippery pulse.

Principle of treatment: Strengthening spleen, removing stagnation, regulating stomach and checking upward adverse flow of *Qi*.

Recipe: *Zhi Zhu Tang*(98) with modification.

Ingredients:

Fructus Aurantii Immaturus	30 g
Rhizoma Atractylodis Macrocephalae	15 g
Cortex Magnoliae Officinalis	10 g
stir-heated Semen Raphani	15 g
Radix Glycyrrhizae	9 g

Decoct the above ingredients in water for oral dose.

Modification: In case of accumulation of heat in the stomach and intestine, add Radix et Rhizoma Rhei 9-15 g(to be decocted later).

If the above method is ineffective, take orally refined powder of Semen Strychni 0.6 g each time and 2 times a day.

2. Dysfunction of the spleen in transportation and transformation due to food retention.

Symptoms and signs: Fullness in the chest and abdomen, sour and foul vomiting, belching, anorexia, constipation, yellow and sticky tongue coating, and deep and excess pulse.

Principle of treatment: Promoting digestion to remove food retention and strengthening the spleen to activate *Qi* circulation.

Recipe: *Mu Xiang Bing Lang Wan*(99) with modification.

Ingredients:

Radix Aucklandiae	6–9 g
Semen Arecae	9 g
Pericarpium Citri Reticulatae Viride	9 g
Pericarpium Citri Reticulatae	9 g
Rhizoma Zedoariae	6 g
Rhizoma Coptidis	6 g
Cortex Phellodendri	6 g
Radix et Rhizoma Rhei	6 g
Rhizoma Cyperi Praeparata	9 g
Fructus Aurantii	9 g
Semen Pharbitidis	9 g

Decoct the above ingredients in water for oral dose.

Modification: In case of abdominal fullness, distention and pain aggravated by pressure complicated with fullness, dryness and obstruction in the intestine, use *Da Cheng Qi Tang*(11).

In case of phlegm and food retention giving rise to fullness and hardness in the chest and epigastrium and restlessness, use *Gua Di San*(100).

If the disease is caused by cold, remove the drugs which are bitter, cold and purgative in flavour and property from the above mentioned prescription, and add *Liang Jiang*(Rhizoma Alpiniae Officinari) 6 g and *Wu Yao*(Radix Linderae)6 g.

5.4 Hepatic Coma

This disease is known as *Hun Bu Zhi Ren*(unconsciousness), *Hun Meng*(syncope), *Hun Kui*(mental confusion)and *Shen Hun*(coma)in traditional Chinese medicine.

Etiology and Pathofenesis

This disease is caused by preponderance of combined pathogenic water,fire,damp and heat which interiorly invade the pericardium and mist the mind,giving rise to a tense syndrome that might turn into a flaccid syndrome in case of exhaustion of *Qi* and *Yin*.

Essentials of Diagnosis

1. The patients usually have a history of primary hepatic diseases.Meanwhile,special attention should be paid to possible latent hepatocirrhosis in some cases.

Mostly, the attack of hepatic coma is induced by hemorrhage of the digestive tract, acute infection, large amount of ascitic fluid by abdominal paracentesis, long–time use of diuretic, or excessive intake of high protein–contained food and certain medicines which are harmful to the liver.

2. Typical manifestations:

(1) Prodromal stage: Gradual onset of the disease with such manifestations as poor appetite, nausea and vomiting, hiccup, diarrhea, oliguria, fetor hepaticus, flutter–fibrillation, followed by mental symptoms such as euphoria or abnormal silence, poor memory, anoesia, and changed character and behaviour.

(2) Hepatic precoma stage: Poor comprehension, excitement with restlessness, slurred speech, clumsiness in writing, disorientation, hallucination, lethargy, etc. Some patients may

have stuporous state, flutter-fibrillation, hyperreflexia, of the knee, and positive Babinski's sign.

(3) Coma stage: Lethargy, stuporous state, unconsciousness, followed by deep coma. In severe cases, there may be acidosis, deep breathing, convulsion and high fever, etc.

3. Obvious injury of the liver function, lowered combining power of kalium, Natrium, chlorine and carbon dioxide, increased urea nitrogen, blood nitrogen over 100% μg in case of nitrogenous hepatic coma and normal blood nitrogen in case of non-nitrogenous hepatic coma, From hepatic precoma stage to coma stage, there may appear abnormal EEG (It should be distinguished from other kinds of coma).

Treatment Based on Differentiation of Syndromes

1. Tense syndrome of coma.

(1) Mist of the heart by upward attack of turbid damp

Symptoms and signs: Dull expression, vague mind, lethargy, dysaphasia, hesitant bowel movement, or even coma, tympanites, jaundice, turbid and sticky tongue coating, and deep and slow or deep and loose pulse.

Principle of treatment: Warming *Yang*, resolving dampness, reducing the turbid and opening the mind.

Recipe: *Yin Chen Zhu Fu Tang*(101) plus *Chang Pu Yu Jin Tang*(33) with modification.

Ingredients:

Herba Artemisiae Scopariae	24 g
Rhizoma Atractylodis Macrocephalae	9 g
Radix Aconiti Praeparata	6 g
Rhizoma Zingiberis	6 g
Radix et Rhizoma Rhei	9 g

Rhizoma Acori Graminei	6 g
Radix Curcumae	10 g
Poria	6 g
Radix Polygalae	6 g

Decoct the above ingredients in water for oral dose.

Modification: In case of coma, add *Su He Xiang Wan*(102) 1 pill each time and 2 times a day.

In case of preminence of damp characterized by thick tongue coating, add Rhizoma Atractylodis 9 g and Cortex Magnoliae Officinalis 6 g.

For preponderance of turbid phlegm, add Arisaema cun Bile 6 g and Rhizoma Pinelliae 6 g.

(2) Block of the mind by phlegm—fire

Symptoms and signs: Tympanites, or jaundice, fever, mental restlessness, delerium, vague mind or even unconsciousness, dry stool, deep yellow urine, red tongue with yellow coating, and rapid pulse.

Principle of treatment: Clearing heat, reducing fire, removing obstruction from the *Fu* organ and opening the mind.

Recipe: *Qian Jin Xi Jiao San*(103) plus *Da Cheng Qi Tang*(11) with modification.

Ingredients:

Cornu Rhinoceri	3 g
Herba Artemisiae Scopariae	20 g
Rhizoma Coptidis	6 g
Fructus Gardeniae	6 g
Radix Arnebiae seu Lithospermi	20 g
Radix Isatidis	20 g
Herba Taraxaci	20 g

Rhizoma Acori Graminei	9 g
Radix et Rhizoma Rhei	9 g
Cortex Magnoliae Officinalis	6 g
Fructus Aurantii Immaturus	6 g
Natrii Sulfas (to be taken after infusion with decoction)	6 g

Decoct the above ingredients in water for oral dose.

Modification: In case of convulsion, add Ramulus Uncariae cum Uncis 12 g and Concha Haliotidis 20 g.

In case of red tongue with little coating, add Radix Glehniae 15 g and Radix Ophiopogonis 9 g.

In case of blood extravasation such as gingival bleeding, epistaxis and subcutanious bleeding, increase the dosage of the ingredients which have the function to clear away toxic heat and cool the blood. Meanwhile, take *Xi Jiao Fen*(powder or Cornu Rhinoceri)1.5 g (to be taken following its infusion), Cortex Moutan Radicis 9 g and Radix Paeoniae Rubra 9 g.

2. Flaccid syndrome of coma

(1) Collapse of *Yin*

Symptoms and signs: Coma, tremulous hands, feeble breathing, cold limbs with sweating, pale tongue and thready and rapid pulse.

Principle of treatment: Saving *Yin*, astringing *Yang*, tonifying *Qi* and rescuing the collapse.

Recipe: *Sheng Mai San*(26) with modification.

Ingredients:

Radix Ginseng	12 g
Radix Ophiopogonis	12 g

Fructus Schisandrae	6 g
Fructus Corni	9 g
Rhizoma Polygonati	9 g
calcined Os Draconis	15 g
calcined Concha Ostreae	15 g

Decoct the above ingredients in water for oral dose.

Modification: In case of red tongue with less coating, add Radix Rehmanniae 12 g, Plastrum Testudinis 15 g and Colla Corii Asini 6 g.

In case of cold extremities, add Radix Aconiti Praeparata 6 g.

(2) Collapse of *Yang*

Symptoms and signs: Coma, opened mouth with the eyes closed, cold limbs with hands relaxed, profuse sweating, incontinence of feces and urine, pale and moist tongue and fading pulse.

Principle of treatment: Restoring *Yang* to save collapse.

Recipe: *Shen Fu Tang*(29).

Ingredients:

Radix Ginseng	10 g
Radix Aconiti Praeparata	10 g

Decoct the above ingredients immediately for nasal feeding.

5.5 Biliary Ascariasis

Biliary ascariasis is known as *Hui Jue*(colic caused by ascariasis) in traditional Chinese medicine.

Etiology and Pathogenesis

According to the theory of TCM, ascarides in the intestine prefer warmth and dislike cold. Stagnation of cold in the intestine may drive them up into the stomach or biliary tract, giving rise to this disease.

Essentials of Diagnosis

1. Paroxymal colic pain in epigastrium and below xiphoid process, which may radiate to the shoulder in some cases. The patient often holds the abdomen, bends the knees, tosses and turns in bed, or even groans. The pain is usually accompanied by nausea and vomiting, chillness and fever, and jaundice. The patient returns to normal when the pain is relieved.

2. The patient has a history of vomiting and defecation of ascarided and ascaris eggs can be found in the stool routine test. This disease is often induced by hunger, cold, fever, pregnancy or improper ascaris-expelling treatment.

3. Biochemical blood tests should be carried out for differential diagnosis according to the pathological condition.

Treatment Based on Differentiation of Syndromes

1. Stagnation of cold in the intesting and accumulation of heat in the stomach.

Symptoms and signs: Sudden attack of intermittent abdominal pain, toss and turn in bed when the pain is severe, nausea and vomiting, or even vomiting of ascarides, cold limbs with sweating, thin and white tongue coating, and wiry and tense pulse.

Principle of treatment: Calming ascarides and arresting pain by using both the drugs cold and heat in property.

Recipe: *Wu Mei Wan*(104) with modification.

Ingredients:

Fructus Mume	30 g
Pericarpium Zanthoxyli	9 g
Rhizoma Zingiberis	6 g
Herba Asari	3 g

Rhizoma Coptidis	9 g
Rhizoma Pinelliae	12 g
Semen Arecae	10 g
Radix Glycyrrhizae	3 g

Decoct the above ingredients in water to be taken twice, or decoct the ingredients immediatedly in water for frequent multiple oral dose.

Modification: In case of deficiency of *Qi* with lassitude, add Radix Codonopsis Pilosusae 9 g and Radix Angelicae Sinensis 6 g.

If there are severe cold manifestations, add Ramulus Cinnamomi 6 g and Radix Aconiti Praeparata 6 g.

In case of prominent heat, add Cortex Phellodendri 9 g.

2. Accumulation of damp—heat in the gallbladder channel

Symptoms and signs: Chillness, fever, abcominal pain which is aggravated by pressure, bitter mouth, dry throat, vomiting of ascarides, or even red face, hotness sensation of the body, fidget, cold limbs, red tongue, and wiry and rapid pulse.

Principle of treatment: Eliminating damp—heat, promoting function of the gallbladder and expelling ascarides.

Recipe: *Qu Hui Li Dan Tang*(105) with modification.

Ingredients:

Herba Artemisiae Scopariae	30 g
Flos Lonicerae	40 g
Radix Scutellariae	15 g
Radix Bupleuri	9 g
Radix et Rhizoma Rhei	9 g
Fructus Meliae Toosendan	9 g
Fructus Quisqualis	15 g
Semen Arecae	9 g

Cortex Meliae	9 g
Radix Glycyrrhizae	6 g
powder of Natrii Sulfas (to be following its infusion)	9 g

Decoct the above ingredients in water for oral dose.

Modification: In case of high fever, add Fructus Forsythiae 15 g and Fructus Gardeniae 9 g.

In case of severe nausea and vomiting, add Caulis Bambusae in Taeniam 9 g and ginger juice—prepared Rhizoma Pinelliae 9 g.

In case of jaundice, add Herba Lysimachiae 30 g.

5.6 Acute Cholecystitis and Cholelithiasis

Acute cholecystitis and cholelithiasis are two different diseases, which often exist simultaneously, and cause and effect each other. They belong to *Fu Tong*(abdominal pain), *Xie Tong*(hypochondriac pain), *Huang Dan*(jaundice) and *Jie Xiong Fa Huang*(jaundice with accumulation of pathogens in the chest) in traditional Chinese medicine.

Etilogy and Pathogenesis

Traditional Chinese medicine holds that gallbladder is the *Fu* organ that contains refined fluid, and is favourable when its free—going condition is normal. Excessive emotional change, improper food—intake, invasion by cold and damp, and retention of ascarides may all lead to stagnation of *Qi* in the liver and gallbladder and dysfunction of the two organs in maintaining unrestrained and free—going condition, giving rise to accumulation of damp—heat in the liver and gallbladder and obstruction of the two channels, thus, pain and other symptoms occur.

Essentials of Diagnosis

1. This disease is characterized by pain, vomiting, fever and jaundice. The colic pain in the right upper abdomen and below xiphoid process is continuous or gets worse paroxysmally, and may radiate to the left or right shoulder, lumbus and back. It is often accompanied by nausea and vomiting, alternate chills and fever, and jaundice.

2. There is obvious tenderness in the gallbladder region. Deep inhalation may give rise to haphalgesia. In some cases, there may be palpable enlarged gallbladder. In severe cases, it could be associated with the manifestations of localized peritonitis such as muscular tension of the right upper abdomen and rebounding pain.

3. Exact diagnosis could be made with the assistance of abdominal B-type ultrasonography, plain abdominal X-ray film taking and cholecystography.

Treatment Based on Differentiation of Syndromes

1. Stagnation of *Qi* in the liver and gallbladder

Symptoms and signs: Distending or colic pain in the hypochondriac and epigastric regions, bitter mouth, dry throat, anorexia, nausea and vomiting, thin and white or slightly yellow coating, and wiry and tense pulse.

Principle of treatment: Promoting the function of the liver and gallbladder in maintaining unrestrained and free-going condition for flow of *Qi*, relieving spasm and arresting pain.

Recipe: *Chai Hu Shu Gan San*(106) with modification.

Ingredients:

Radix Bupleuri	9 g
Radix Paeoniae Albae	15 g
Fructus Aurantii	9 g

Rhizoma Cyperi	9 g
Rhizoma Pinelliae	9 g
Radix Curcumae	9 g
Rhizoma Corydalis	9 g
Radix Aucklandiae	9 g
Radix Glycyrrhizae	3 g
Herba Artemisiae Scopariae	12 g

Decoct the above ingredients in water for oral dose.

Modification: In case of gall stone that has existed for long, add Spora Lygodii 18 g, Endothelium Corneum Gigeriae Galli 9 g and Herba Lysimachiae 15 g.

2. Accumulation of damp-heat in the liver and gallbladder

Symptoms and signs: Abodominal fullness, distention and pain, alternate chills and fever, bitter mouth, dry throat, mental restlessness, poor appetite, jaundice, constipation, deep yellow urine, red tongue with yellow and sticky or thick coating, and wiry and rapid pulse.

Principle of treatment: Promoting the function of the liver and gallbladder in maintaining unrestrained and free-going condition for flow of *Qi*, clearing heat and eliminating damp by promoting diurisis.

Recipe: *Da Chai Hu Tang* (107) with modification.

Ingredients:

Radix Bupleuri	9 g
Radix et Rhizoma Rhei	
(to be decocted later)	9 g
Fructus Aurantii Immaturus	9 g
Radix Scutellariae	9 g
Rhizoma Pinelliae	9 g

Radix Paeoniae Albae	9 g
Herba Artemisiae Scopariae	15 g
Radix Curcumae	9 g
Radix Aucklandiae	6 g
Semen Plantaginis (wrapped to be decocted)	6 g
Herba Lysimachiae	30 g

Decoct the above ingredients in water for oral dose.

Modification: In case of prominence of pathogenic damp, add Semen Coicis 30 g and Semen Amomi Cardamomi 9 g.

In case of prominence of heat, add Radix Isatidis 15 g, Flos Lonicerae 15 g and Fructus Forsythiae 15 g.

In case of constipation, add Natrii Sufas 9 g (to be taken following its infusion) and Cortex Magnoliae Officinalis 12 g.

3. Retention of excess fire in the liver and gallbladder

Symptoms and signs: Hypochondriac and epigastric distention and pain which are aggravated by pressure, abdominal fullness and distention, alternate chills and fever, bitter mouth, dry throat, jaundice, yellow and turbid urine or deep yellow and difficult urination, constipation, red or deep red tongue with yellow and dry or thorny coating, and wiry, slippery and rapid or full and rapid pulse.

Principle of treatment: Clearing heat and reducing fire from the liver and gallbladder.

Recipe: *Long Dan Xie Gan Tang*(97) plus *Yin Chen Hao Tang*(35) with modification.

Ingredients:

Radix Gentianae	6 g
Radix Scutellariae	9 g

Radix Bupleuri	9 g
Fructus Gardeniae	6 g
Radix Rehmanniae	6 g
Radix Angelicae Sinensis	3 g
Rhizoma Alismatis	6 g
Radix Curcumae	9 g
Radix et Rhizoma Rhei (to be decocted later)	9 g
Natrii Sulfas	9 g
Radix Aucklandiae	6 g
Herba Artemisiae Scopariae	10 g

Decoct the above ingredients in water for oral dose.

Modification: In case of domination of toxic heat, add Radix Scutellariae 9 g, Herba Oldenlandiae Diffusae 18 g and Herba Scutellariae Barbatae 9 g.

5.7 Acute Pancreatitis

Acute pancreatitis is involved in *Wei Wan Tong* (epigastric pain), *Ge Tong* (pain of diaphragm), *Fu Tong* (abdominal pain), *Pi Xin Tong* (pain of spleen) and *Jie Xiong* (accumulation of pathogens in the chest) in traditional Chinese medicine.

Etiology and Pathogenesis

This condition is often resulted from improper food-intake, excessive emotional changes or parasites, which leads to dysfunction of the liver, gallbladder, spleen and stomach, causes stagnation of *Qi* and blood, produces damp and heat, giving rise to the symptoms of accumulation of excess heat. When the anti-pathogenic *Qi* is too weak to dispel pathogenic factors, there may appear syncope and collapse.

Essentials of Diagnosis

1. The abdominal pain usually occurs 1–2 hours after excessive food-intake or alcohol drinking. It is severe and continuous, and accompanied with lancinating pain which gets worse paroxysmally. The pain is mostly located in the upper abdomen or around umbilicus, and radiates to the left shoulder, lumbus and back without tenderness.

2. It is accompanied by nausea and vomiting at begining, which alleviates gradually later. Stop of vomiting will not help relieve the pain, but is followed by fever, abdominal distention, jaundice and even shock.

3. Chemical examination may find increase of blood amylase, urine amylase and ascites amylase.

Treatment Based on Differentiation of Syndromes

1. Accumulation of heat due to stagnation of *Qi*

Symptoms and signs: Sudden attack of epigastric and abdominal pain which might get worse from time to time and radiate to the hypochondriac region, nausea and vomiting, bitter mouth, dry throat, alternate chills and fever, abdominal distention, constipation, thin and yellow tongue coating, and wiry and tense pulse.

Principle of treatment: Dispersing stagnation of the liver, reducing heat and removing obstruction from the *Fu* organ.

Recipe: *Da Chai Hu Tang*(107) with modification.

Ingredients:

Radix Bupleuri	12 g
Radix et Rhizoma Rhei	
(to be decocted later)	15 g
Radix Scutellariae	15 g

Fructus Aurantii Immaturus 6 g
Radix Paeoniae Albae 15 g
Rhizoma Pinelliae 9 g

Decoct the above ingredients in water for oral dose.

If the patient can not take the decoction because of nausea and vomiting, help take the decoction by enema, and give nasal feeding after gastrointestinal decompression when necessary.

Modification: In case of severe abdominal pain, add Rhizoma Corydalis 9 g and Fructus Meliae Toosendan 9 g.

In case constipation, add Natrii Sulfas 6 g (to be taken following its infusion).

In case of abdominal distention, add Cortex Magnoliae Officinalis 9 g and Radix Aucklandiae 6 g.

2. Excess syndrome of *Fu* organs due to domination of heat

Symptoms and signs: Abdominal pain which is aggravated by pressure, fullness, distention and hardness in the abdomen, persistent fever, thirst, red tongue with yellow, thick and sticky or dry coating, and full and rapid pulse.

Principle of treatment: Clearing away toxic heat and removing obstruction from the *Fu* organ by purgation.

Recipe: *Da Cheng Qi Tang*(11) with modification.

Ingredients:
Radix et Rhizoma Rhei 12 g
Natrii Sulfas 12 g
Fructus Aurantii Immaturus 9 g
Cortex Magnoliae Officinalis 15 g
Radix Scutellariae 15 g
Fructus Gardeniae 9 g
Fructus Forsythiae 15 g

Herba Taraxaci	15 g
Caulis Sargentodoxae	15 g
Herba Patriniae	15 g

Decoct the above ingredients in water, take the decoction 1–2 times a day, or by enema when necessary.

Modification: In case of heat complicated with retention of water (hydroperitoneum and hydrothorax), use *Da Xian Xiong Tang* (108): Decoct Radix et Rhizoma Rhei first, then add in the powder of Natrii Sulfas and Radix Euphorbiae Kansui. Drink the first half of the decoction first, and the next half if no stool discharge.

In case of blood stasis, add Radix salviae Miltiorrhizae 9 g, Flos Carthami 9 g, Radix Paeoniae Rubra 9 g, Cortex Moutan Radicis 9 g and Semen Persicae 9 g.

In case of collapse with cold limbs, add *Shen Fu Tang* (29): Radix Ginseng 9 g and Radix Coptidis Praeparata 9 g. Decoct the ingredients immediately for oral use.

5.8 Acute Appendicitis

Acute appendicitis corrensponds to *Chang Yong* (intestinal abscess) and *Su Jiao Chang Yong* (intestinal abscess with flexed legs).

Etiology and Pathogenesis

This disease is mainly resulted from overfatigue or excessive mental irritation which leads to dysfunction of *Zang—Fu* organs, as well as weather changes, improper food—intake, etc. giving rise to stagnation of *Qi* and blood and accumulation of damp—heat in the intestine.

Essentials of Diagnosis

1. Attack of acute appendicitis usually begins with the pain

in epigastrium or around umbilicus, which later shifts to the right lower abdomen. The pain is persistent and gets worse paroxysmally. It is often accompanied by nausea, vomiting, fever, yellow urine, etc. When standing or walking, the patient often bends over with the two hands placed on the right lower abdomen. Besides, the patient often flexes the hip joint when lying flat.

2. There is a marked fixed tenderness with rebounding pain in severe case over the M'Burney's point on the right lower abdomen. The Psoas muscle irritation sign indicates retroposition of the appendix. In case of pregnancy, the tenderness may move to the right upper abdomen according to the size of the uterus. If a infantile patient has obvious general symtoms with spasm of the right lower abdominal muscle, acute appendicitis should be considered first. With a history of appendicitis, a palpable painful mass often suggests appendiceal abscess.

3. Routine blood test will find increase of white blood cells and neutrophilic granulocytes.

Treatment Based on Differentiation of Syndromes

1. Accumulation of heat and stagnation of *Qi*

Symptoms and signs: Abdominal pain wandering around the umbilicus and moving down to localize in the right lower abdomen finally, accompanied by nausea, vomiting, fever, or constipation, deep yellow urine, thin and white tongue coating, and wiry or slightly rapid pulse.

Principle of treatment: Activating circulation of *Qi* and blood and clearing away toxic heat.

Recipe: *Lan Wei Hua Yu Tang*(109).

Ingredients:

Fructus Meliae Toosendan   15 g

Rhizoma Corydalis	9 g
Cortex Moutan Radicis	9 g
Semen Persicae	9 g
Radix Aucklandiae	9 g
Flos Lonicerae	15 g
Radix et Rhizoma Rhei	
(to be decocted later)	9 g

Decoct the above ingredients in water for oral dose.

Modification: In case of severe heat with blood stasis, add Caulis Sargentodoxae 30 g and Radix Salviae Miltiorrhizae 9 g.

2. Stasis of toxic heat in the intestine

Symptoms: Right lower abdominal pain which is aggravated by pressure and right leg stretching, dry mouth, fever, constipation, yellow urine, red tongue with yellow and sticky or yellow and dry coating, and wiry and rapid or slippery and rapid pulse.

Principle of treatment: Clearing away toxic heat, eliminating swelling and resolving mass.

Recipe: *Lan Wei Qing Hua Tang*(110).

Ingredients:

Flos Lonicerae	30 g
Herba Taraxaci	30 g
Cortex Moutan Radicis	15 g
Radis et Rhizoma Rhei	
(to be decocted later)	15 g
Fructus Meliae Toosendan	9 g
Semen Persicae	9 g
Radix Paeoniae Rubra	12 g
Radix Glycyrrhizae	9 g

Decoct the above ingredients in water for oral dose.

Modification: In case of prominence of damp-heat, add Rhizoma Coptidis 12 g and Radix Scutellariae 12 g.

In case of prominence of damp, add Herba Eupatorii 12 g, Semen Amomi Cardamomi 15 g and Petiolus Agastachi 10 g.

In case of constipation, *Da Huang Mu Dan Pi Tang*(111) can be used.

3. Flaming of Toxic heat

Symptoms and signs: Right lower abdominal pain which is aggravated by pressure and accompanied by high fever, red face and eyes, dry mouth and lips, constipation, deep yellow urine, red tongue with yellow and dry or yellow and sticky coating, and wiry and rapid pulse.

Principle: Clearing away toxic heat and removing stagnation.

Recipe: *Lan Wei Jie Du Tang* (112).

Ingredients:

Caulis Sargentodoxae	30 g
Herba Patriniae	30 g
Flos Lonicerae	30 g
Herba Taraxaci	30 g
Semen Benineasae	15 g
Radix Paeoniae Rubra	12 g
Radix et Rhizoma Rhei (to be decocted later)	12 g
Radix Aucklandiae	9 g
Radix Scutellariae	9 g
Semen Persicae	9 g
Fructus Meliae Toosendan	9 g

Decoct the above ingredients in water for oral dose.

Modification: In case of severe abdominal distention, add

Cortex Magnoliae 12 g and Fructus Aurantii Immaturus 12 g.

In case of severe abdominal pain, add Rhizoma Corydalis 12 g.

In case of nausea and vomiting, add Caulis Bambusae in Taeniam 9 g, Rhizoma Pinelliae 9 g.

Common medicines for external use:

1. *Xiao Yan San.*

It is used in combination when there is peritonitis with appendiceal abscess.

Ingredients:

Folium Jibisci	30 g
Radix et Rhizoma Rhei	30 g
Radix Scutellariae	240 g
Rhizoma Coptidis	240 g
Cortex Phellodendri	240 g
Folium Lycopi	240 g
Borneolum Syntheticum	9 g

Grind the above ingredients into fine powder to be mixed and heated with Chinese rice wine or onion Chinese rice wine for external use, twice a day.

2. *Liu Huang San.*

Mix powder of Radix et Rhizoma Rhei and Natrii Sulfas with certain amount of vinegar into a paste for external application.

5.9 Gastroduodenal Ulcer and Perforation

This disease is known as *Jue Xin Tong* (precordial pain with cold limbs), *Xin Fu Tong* (cardiac and abdominal pain), *Wei Wan Tong* (epigastric pain) and *Jue Ni* (epigastric pain with cold limbs) in traditional Chinese medicine.

Etiology and Pathogenesis

It is mostly resulted from voracious eating, hunger, excitement and overfatigue, etc. In case of large perforation with severe peritonitis in which the pathogenic factors are dominating and anti-pathogenic *Qi* is weak, there may appear obstruction and rejection, and collapse with cold limbs.

Essentials of Diagnosis

1. Sudden attack of epigastric lancinating pain which is persistent and rapidly radiates to the right lower abdomen and the whole abdomen. Part of the patients may have the pain radiating to the right shoulder, lumbus and back. The abdominal pain is not so obvious in some aged or asthenic patients. Some cases may have shock due to severe pain at early stage and bacterial peritonitis at late stage. Half of the patients have nausea and vomiting, and abdominal distention and constipation due to enteroparalysis at late stage.

2. At early stage, the abdominal wall is flat. At late stage, there is obvious abdominal distention which leads to limited or even disappeared abdominal breathing, tenderness and rebounding pain on the whole abdomen, especially the upper and right lower abdomen, where the muscle of abdominal wall is tense and rigid like a piece of board. There is diminution or absence of hepatic dullness and borborygmus. X-ray examination will find signs of pneumoperitoneum and slight increase of white blood cells.

Treatment Based on Differentiation of Syndromes

1. Stagnation of *Qi* and stasis of blood

Symptoms and signs: Sudden onset of severe lancinating abdominal pain which is aggravated by pressure, possibly accompanied with nausea and vomiting, white tongue coating and wiry pulse. This syndrome usually occurs 1–2 days after formation of perforation, the period known as occlusive perforation stage.

Principle of treatment: Relieving spasm, arresting pain and promoting healing of perforation.

Treatment should be given mainly with acupuncture by needling point *Shangwan*(RN13), *Zhongwan*(RN12), *Liangmen*(ST21)(bilateral), *Tianshu*(ST25)(bilateral), *Neiguan*(PC6)(bilateral) and *Zusanli*(ST36)(bilateral). Strong stimulation should be given with the needles retained in the points for 30–60 minutes after the needling sensation arrives. During retention of the needles, manipulate the needles once every 15 minutes or apply electro-stimulation. Treatment should be given once every 6 hours with the combination of fasting, gastrointestinal decompression and adjustment of waterelectrolyte balance.

2. Stagnation of *Qi* and blood producing heat

Symptoms and signs: localized abdominal pain, fever, dry stool, deep yellow urine, yellow tongue coating, and rapid pulse. This syndrome has occlusion of perforation complicated with abdominal hydrops and infection, corresponding to the resolution stage of inflammation.

Principle of treatment: Clearing away heat with purgative drugs, regulating *Qi* and removing obstruction from *Fu* organs.

Recipe: *Liang Ge San*(9) with modification.

Ingredients:

Radix et Rhizoma Rhei	9 g
Natrii Sulfas (to be taken following its infusion)	9 g
Radix Glycyrrhizae	6 g

Fructus Gardeniae	9 g
Herba Menthae	6 g
Radix Scutellariae	12 g
Fructus Forsythiae	15 g
Herba Lophatheri	9 g
Fructus Aurantii	9 g
Radix Aucklandiae	6 g

Decoct the above ingredients in water for oral dose.

Modification: In case of prominent stasis of blood, add Semen Persicae 9 g, Flos Carthami 9 g, Pollen Typhae 9 g and Rhizoma Ligustici Chuanxiong 9 g.

In case of severe abdominal infection, add Flos Lonicerae 30 g, Herba Taraxaci 30 g and Caulis Sargentodoxae 15 g.

In case of persistent abdominal pain, add Rhizoma Corydalis 9 g, Fructus Meliae Toosendan 9 g and RadixPaeoniae Albae 12 g.

3. Deficiency of the spleen and stomach

Symptoms and signs: Sallowish face, emaciation, stuffiness in the chest, abdominal distention, diarrhea or constipation, indigestion, belching, sour regurgitation, pale tongue with thin and white coating, and deep, thready and weak pulse. This syndrome corresponds to reparative stage of ulcer.

Principle of treatment: Strengthening spleen and regulating stomach.

Recipe: *Xiang Sha Liu Jun Zi Tang*(113) with modification.

Ingredients:

Radix Codonopsis Pilosulae	15 g
Rhizoma Atractylodis Macrocephalae	9 g
Poria	12 g
Pericarpium Citri Reticulatae	9 g

Rhizoma Pinelliae	9 g
Radix Aucklandiae	9 g
Fructus Amomi	9 g
Radix Glycyrrhizae	6 g
Os Sepiellae seu Sepiae	30 g
Membrana Follicularis Ovi	9 g

Decoct the above ingredients in water for oral dose.

Modification: In case of hiccup due to cold in the stomach, add Fructus Evodiae 9 g, Rhizoma Zingiberis 9 g and Radix Linderae 9 g.

In case of stomachache, add Rhizoma Corydalis 9 g, Fructus Meliae Toosendan 9 g and Radix Paeoniae Albae 12 g.

In case of stagnation of the liver *Qi*, add Rhizoma Cyperi 9 g, Fructus Citri Sarcodactylis 9 g and Semen Arecae 9 g.

5.10 Acute Peritonitis

Acute purulent peritonitis is included in *Ji Xing Fu Tong* (acute abdominal pain) in traditional Chinese medicine.

Etiology and Pathogenesis

This disease is generally caused by weakness of antipathogenic *Qi* and invasion of the interior by toxic pathogens which lead to accumulation of damp-heat, stagnation of *Qi* and blood, stasis of blood and heat in the interior and obstruction and abscess in the intestine.

Essentials of Diagnosis

1. There appears persistent severe abdominal pain which is more obvious at the primary focus, and is accompanied by nausea, vomiting and signs of general poisoning.

2. The patient has weakened abdominal breathing, tender-

ness, rebounding pain and muscular tension of the abdomen, decreased borborygmus, tympany during percussion, and movable dullness when there is much fluid in the abdominal cavity.

3. Rectoabdominal examination, abdominal paracentesis, laboratory tests and X-ray examination are helpful for confirmation of the diagnosis.

Treatment Based on Differentiation

1. Stagnation of heat in the three *Jiao*

Symtoms and signs: Abdominal distention and pain aggravated by pressure, high fever, thirst, dry mouth and throat, deep yellow urine, yellow and dry tongue coating, and wiry and rapid pulse. This syndrome is often observed in acute diffusional infection of abdominal viscera.

Principle of treatment: Clearing away toxic heat.

Recipe: *Huang Lian Jie Du Tang* (114) with modification.

Ingredients:

Rhizoma Coptidis	6 g
Radix Scutellariae	9 g
Fructus Gardeniae	9 g
Cortex phellodendri	6 g
Flos Lonicerae	30 g
Fructus Forsythiae	15 g
Herba Taraxaci	30 g
Herba Violae	30 g
Radix Paeoniae Albae	18 g
Radix Glycyrrhizae	6 g

Decoct the above ingredients in water for oral dose.

Modification: In case of jaundice, add Herba Artemisiae Scopariae 30 g and Semen Phaseoli 30 g.

In case of hydroperitoneum, add Poria 30 g, Polyporus Umbellatus 30 g, Rhizoma Alismatis 15 g and Exocarpium Benincasae 30 g.

In case of nausea and vomiting, add Caulis Bambusae in Taeniam 12 g and Rhizoma Pinelliae 9 g.

2. Stagnation of *Qi* and blood producing heat

Symptoms and signs: Abdominal distention and pain aggravated by pressure, fever, dry stool, deep yellow urine, yellow tongue coating, and rapid pulse. This syndrome corresponds to infection of abdominal cavity after occlusion of gastrointestinal perforation.

Principle of treatment: Clearing away toxic heat, resolving blood stasis and removing obstruction from the *Fu* organs.

Recipe: *Da Chai Hu Tang*(107).

Ingredients:

Radix Bupleuri	9 g
Radix Scutellariae	9 g
Rhizoma Pinelliae	9 g
Fructus Aurantii Immaturus	9 g
Radix Paeoniae Albae	12 g
Radix et Rhizoma Rhei (to be decocted later)	9 g
Flos Lonicerae	30 g
Semen Benincasae	15 g

Decoct the above ingredients in water for oral dose.

Modification: If it is complicated with bitter mouth and hypochondriac pain, add Radix Gentianae 6 g, Radix Curcumae 12 g and Rhizoma Cyperi 9 g.

In case of excessive heat, add Gypsum Fibrosum 30 g,

Rhizoma Anemarrhenae 12 g and Herba Taraxaci 30 g.

In case of severe blood stasis, add Semen Persicae 9 g, Radix Paeoniae Rubra 12 g, Spina Gleditsiae 9 g and Squama Manitis 9 g.

5.11　Pseudomembranous Enteritis

Pseudomembranous Enteritis in involved in *Xie Xie*(diarrhea) in traditional Chinese medicine.

Etiology and Pathogenesis

Pathogenic cold, damp and heat invade the spleen and stomach when the anti-pathogenic *Qi* is weak due to prolonged illness, lead to obstruction of *Qi* in the two organs, giving rise to dysfunction of them in transformation and transportation.

Essentials of Diagnosis

1. If diarrhea suddenly occurs during or shortly after application of broad-spectrum antibiotics, or the pathological condition after abdominal operation is deteriorated with diarrhea, this disease should be put into consideration.

2. Stool bacterial cultures under special condition will show the unidentified growing clostridia. Besides, feces examination, colonoscope and X-ray examination are also necessary for diagnosis.

Treatment Based on Differentiation of Syndromes

1. Heat in the upper and cold in the lower

Symptoms and signs: Fulminant watery diarrhea, abdominal distention and pain, nausea and vomiting, fever, thirst, fullness in the chest with mental restlessness, thin and yellow tongue coating, and deep and rapid pulse.

Principle of treatment: Clearing away heat from the upper, warming the lower, and regulating the spleen and stomach.

Recipe: *Huang Lian Tang*(115) with modification.

Ingredients:

Rhizoma Coptidis	4.5 g
Rhizoma Zingiberis	6 g
Cortex Cinnamomi	4.5 g
Rhizoma Pinelliae	6 g
Radix Codonopsis Pilosulae	9 g
Radix Glycyrrhizae Praeparata	3 g
Fructus Ziziphi Jujubae	4 pieces

Decoct the above ingredients in water for oral dose.

Modification: In case of severe abdominal pain, add Radix Paeoniae Albae 9 g and Rhizoma Corydalis 9 g.

In case of oliguria, add Rhizoma Alismatis 9 g and Herba Plantaginis 30 g.

In case of severe nausea and vomiting, add Semen Amomi Cardamomi 9 g, Caulis Bambusae in Taenium 9 g and Folium Perillae 9 g.

2. Deficiency of the spleen and stomach with remained damp—heat

Symptoms and signs: Poor appetite, lassitude, emaciation, low fever, loose stool with peudomembranous pieces, pale tongue with white coating, and deficient and rapid pulse.

Principle of treatment: Strengthening spleen, regulating stomach and eliminating damp—heat.

Recipe: *Shen Ling Bai Zhu San*(116) with modification.

Ingredients:

Radix Codonopsis Pilosulae	9 g

Rhizoma Atractylodis Macrocephalae	9 g
stir-heated Semen Dolichoris Album	9 g
Semen Nelumbinis	9 g
Poria	15 g
stir-heated Rhizoma Dioscoreae	15 g
Semen Coicis	15 g
Pericarpium Citri Reticulatae	6 g
Radix Platycodi	6 g
Fructus Amomi	9 g
Rhizoma Coptidis	4 g

Decoct the above ingredients in water for oral dose.

Modification: If there appears collapse syndrome, immediately decoct Radix Ginseng 9 g and Radix Aconiti Praeparata 9 g for oral dose.

Experienced precription: Pound 2-3 heads of garlic (Bulbus Allii) into a paste to be taken with warm boiled water twice a day until the symptoms disappear.

6 Diseases of the Urinary System

The emergency diseases of the urinary system belong respectively to *Guan Ge*(obstruction and rejection), *Long Bi*(urine retention), *Niao Xue*(hematuria) and *Lin Zheng*(Lin syndrome) in traditional Chinese medicine. They are often resulted from congenital insufficiency, seven emotional factors, indulgence in sex, insanitary food-intake or invasion by exogenic pathogenic factors. The principle of treatment to be introduced in this chapter is " Treating the excess with purgation and the deficiency with tonification."

6.1 Acute Renal Failure

Acute renal failure belongs to *Long Bi*(urine retention), *Guan Ge*(obstruction and rejection) and *Shui Zhong*(edema) in traditional Chinese medicine.

Etiology and Pathogenesis

This disease is resulted from pathogenic heat, toxin, stasis of blood, damp or traumatic injury due to extruding, which lead to consumption of kidney *Qi*, giving rise to dysfunction of the kidney in opening and closing, and dysfunction of *Qi* in ascending the clear and descending the turbid, resulting in accumulation of pathogenic factors in the three *Jiao*.

Essentials of Diagnosis

1. Oliguria stage

(1) With inducing factors, shock in perticular, the patient's

urinary output is reduced to less than 17 ml per hour or 400 ml per day after the shock relieved and sufficient fluid infusion given.

(2) Oliguria is accompanied by nausea, vomiting, poor appetite, headache, lethargy, trance, mental restlessness, or even coma, convulsion, raise of blood pressure, or hemorrhagic tendency.

(3) Results of routine urine examination vary with the causes of the disease. There may appear aciduria and the specific gravity of urine is kept around 1.012. The blood creatinine and urea nitrogen level rises day by day, so does that of blood kalium and blood magnesium.

2. Polyuria stage

Urine putput is over 3 000 ml per day. Improper management may lead to dehydration, hypokalemia, hyponatremia, weakness, nausea, vomiting, or even paralysis and heart failure.

3. At early stage, there may appear dehydration and shock due to primary diseases, it must be distinguished from functional oliguria stage by test of mannitol intravenous injection.

4. No evidence of chronic renal failure is found before the onset of the disease.

Treatment Based on Differentiation of Syndromes

1. Oliguria Stage

(1) Stagnation of toxic heat

Symptoms and signs: High fever, thirst, oliguria of anuria, dysphoria, ecchymosis or hematemesis, epistaxis, hemoptysis, hematuria, deep red tongue with dry and yellow coating, and slippery and rapid or thready and rapid pulse.

Principle of treatment: Clearing away heat, cooling blood, dispersing stasis of blood and promote urination.

Recipe: *Qing Wen Bai Du Yin*(117) with modification.

Ingredients:

Gypsum Fibrosum (to be decocted first)	30 g
Radix et Rhizoma Rhei	15 g
powder of Cornu Rhinoceri (to be taken following its infusion)	3 g
Radix Scutellariae	9 g
Fructus Gardeniae	9 g
Rhizoma Anemarrhenae	9 g
Radix Paeoniae Rubra	9 g
Radix Scrophulariae	9 g
Fructus Forsythiae	9 g
Cortex Moutan Radicis	9 g
Radix Coptidis	6 g
Radix Platycodi	6 g
Herba Lophatheri	6 g
Radix Glycyrrhizae	3 g
Radix et Rhizoma Rhei (to be decocted later)	9 g
Rhizoma Imperatae	30 g

Decoct the above ingredients in water to obtain 200 ml of decoction which is taken frequently with small dose.

Modification: In case of constipation, add Natrii Sulfas 9 g(to be taken following its infusion).

In case of massive bleeding, add *San Qi Fen*(powder of Radix Notoginseng) 6 g (to be taken following its infusion).

In case of dribbling urination with scanty and deep yellow urine, add Polyporus Umbellatus 15 g, Herba Lycopi 9 g and Herba Leonuri 12 g.

(2) Stagnation and stasis due to deficiency of *Qi*

Symptoms and signs: Oliguria or anuria, edema of the whole body, lassitude, poor appetite, nausea and vomiting, palpitation, shortness of breath, pallor, pale tongue with white coating, and deep and thready pulse.

Principle of treatment: Warming *Yang* to promote diuresis, and replenishing *Qi* to disperse blood stasis.

Recipe: *Zhen Wu Tang*(45) with modification.

Ingredients:

Radix Aconiti Praeparata	6 g
Rhizoma Atractylodis Macrocephalae	6 g
Rhizoma Zingiberis	6 g
Poria	15 g
Radix Paeoniae Albae	12 g
Radix Ginseng	12 g
Radix Salviae Miltiorrhizae	12 g
Herba Leonuri	15 g
Radix et Rhizoma Rhei	6 g
Rhizoma Alismatis	15 g

Decoct the above ingredients in water to obtain 150–200 ml of decoction for oral dose.

Modification: Enema of Chinese herbal decoction *Jie Chang Guan Zhu Ye* No. 1 could be used in combination.

Ingredients:

Radix et Rhizoma Rhei	12 g
Radix Astragali seu Hedysari	30 g
Radix Salviae Miltiorrhizae	30 g
Flos Carthami	15 g

Decoct the above ingredients in water to obtain 150 ml of

decoction for enema, once a day.

In case of severe stasis of blood, add Semen Persicae 9 g, Rhizoma Ligustici Chuanxiong 9 g, Flos Carthami 12 g and Hirudo 6 g.

For deficiency of kidney *Yang*, add Rhizoma Curculiginis 9 g and Epimedii 9 g.

2. Polyuria stage: Deficiency of kidney *Yin*

Symptoms and signs: Low fever, mental restlessness, increased urination, malar flush, thirst, dizziness, tinnitus, soreness and weakness of lumbus and knees, red tongue with little coating, and thready and rapid pulse.

Principle of treatment: Strengthening the kidney to promote its astringent function.

Recipe: *Liu Wei Di Huang Wan*(118) plus *Suo Quan Wan*(119).

Ingredients:

Radix Rehmanniae Praeparata	24 g
Fructus Corni	12 g
stir-heated Rhizoma Dioscoreae	12 g
Cortex Moutan Radicis	9 g
Poria	9 g
Rhizoma Alismatis	9 g
Radix Ophiopogonis	12 g
Fructus Schisandrae	6 g
Fructus Alpiniae Oxyphyllae	15 g
Radix Linderae	9 g

Decoct the above ingredients in water for oral dose.

3. Restoration stage: Deficiency of both *Qi* and *Yin*

Symptoms and signs: Asthenia, shortness of breath, dislike

of speaking, general lassitude, normal urination, red tongue with little coating, or pale tongue, and thready and forceless pulse.

Principle of treatment: Replenishing *Qi*, nourishing *Yin*, invigorating the spleen and strengthening the kidney.

Recipe: *Sheng Mai San*(26) with modification.

Ingredients:

Radix Pseudostellariae	30 g
Radix Ophiopogonis	12 g
Fructus Schisandrae	6 g
Herba Dendrobii	12 g
Rhizoma Dioscoreae	30 g
Fructus Amomi	9 g
Massa Fermentata Medicinalis (stir-heated)	12 g
Fructus Oryzae Germinatus (stir-heated)	12 g
Fructus Hordei Germinatus (stir- heated)	12 g

Decoct the above ingredients in water for oral dose.

Modification: If it is accompanied with deficiency of both spleen and lung, the patient may take *Du Qi Wan* (120) 1 pill each time and twice a day.

6.2 Uremia

Uremia due to chronic renal insufficiency is usually a secondary affection of the illness in the urinary system. It corresponds to *Xu Sun*(consumptive disease), *Long Bi*(urine retention) and *Guan Ge*(obstruction and rejection) in traditional Chinese medicine.

Etiology and Pathogenesis

This disease is resulted from prolonged renal illness with deficiency of anti-pathogenic *Qi* which lead to accumulation of pathogenic damp and excessive fluid, stasis of blood and damp-heat. Deficiency of kidney *Yang* causes upward disturbance of the pathogenic damp and deficiency of kidney *Yin* results in hyperactivity of liver *Yang*, or even sinking of the pathogenic factors into the heart and liver, stirring up the endogenous wind due to heat in the blood.

Essentials of Diagnosis

1. The patient has a history of primary nephrosis and certain kinds of inducing factors of uremia.

2. During the early stage, there are such manifestations as emaciation, anemia, listlessness, lethargy, poor appetite, nausea and vomiting, elevation of blood pressure, abdominal pain, diarrhea, skin pruritus and bleeding. During the late stage, there may appear haziness, dysphoria, delirium, convulsion, hyperpnea with ammonia odor, more serious coma, and pericardial friction sound can be heard.

3. At the early stage of uremia, creatinine clearance is 10-25%, blood creatinine 2.5-5% mg. At the late stage, creatinine clearance is less than 10%, blood creatinine more than 5%mg. Other examinations such as biochemical check-up to the bloody urine or nephrogram are also helpful for diagnosis.

Treatment Based on Differentiation of Syndromes

1. Stagnation of cold due to weakness of *Yang*, and retention of turbid *Yin* in the interior.

Symptoms and signs: Oliguria, anuresis, poor appetite, lassitude, dark complexion, aversion to cold, nausea and vomiting, constipation, thin and sticky tongue coating, and deep and wiry pulse.

Principle of treatment: Warming the kidney, dispelling cold, removing obstruction from the *Fu* organ to excrete the turbid *Yin*.

Recipe: *Da Huang Fu Zi Tang*(121) with modification.
Ingredients:

Radix et Rhizoma Rhei	9 g
Radix Aconiti Praeparata	12 g
Herba Asari	3 g
Fructus Aurantii Immaturus	9 g
Radix Astragali seu Hedysari	18 g
Radix Codonopsis Pilosulae	24 g
Rhizoma Atractylodis Macrocephalae	10 g
Pericarpium Citri Reticulatae	9 g
Rhizoma Pinelliae	9 g
Radix Glycyrrhizae	9 g

Decoct the above ingredients in water for oral dose.

Modification: In case of severe nausea and vomiting, add Evodiae 6 g and Rhizoma Zingiberis 3 g.

For weakness of *Yang*, add Herba cistanchis 9 g and Herba Epimedii 9 g.

If it is accompanied with deficiency of both *Yin* and *Yang*, add *Dong Chong Xia Cao*(Cordyceps)9 g.

2. Deficiency of both *Qi* and *Yin*, and accumulation of turbid materials combined with heat

Symptoms and signs: Sallowish face, spiritlessness, lassitude, regurgitation and nausea, poor appetite, urine odor in mouth, tidal fever, mental restlessness, heartburn sensation in the chest, dizziness and tinnitus, red tongue with yellow and sticky coating, and thready and rapid pulse.

Principle of treatment: Tonifying *Qi*, nourishing *Yin*, Clearing heat and excreting the turbid.

Recipe: *Feng Sui Dan*(122) with modification.

Ingredients:

Radix Ginseng	9 g
Radix Asparagi	18 g
Radix Rehmanniae Praeparata	9 g
Cortex Phellodendri	9 g
Fructus Amomi	9 g
Rhizoma Coptidis	9 g
Radix Scrophulariae	18 g
Rhizoma Imperatae	30 g
Radix et Rhizoma Rhei	6 g

Decoct the above ingredients in water for oral dose.

Modification: In case of stasis of blood in the interior, add Radix Salviae Miltiorrhizae 30 g, Semen Persicae 9 g and Herba Leonuri 30 g.

In case of constipation, add Natrii Sulfas 9 g (to be taken following its infusion) and Cortex Magnoliae Officinalis 9 g.

3. Invasion of the heart and liver by pathogens, and stirring-up of the endogenous wind due to heat in the blood.

Symptoms and signs: Oliguria or anuresis, coma, delirium, carphologia, gingival atrophy, epistaxis, clonic convulsion, red tongue with yellow coating, and thready, wiry and rapid pulse.

Principle of treatment: Promoting resuscitation by cooling heat in the blood and dispelling wind.

Recipe: *Xi Jiao Di Huang Tang*(25) plus *Ling Yang Gou Teng*

Tang(22) with modification.

Ingredients:

Cornu Rhinoceri (to be ground into powder and taken following its infusion)	1 g
Cornu Antelopis (to be ground into powder and taken following its infusion)	1 g
Radix Rehmanniae	30 g
Radix Paeoniae Rubra	12 g
Radix Paeoniae Albae	12 g
Cortex Moutan Radicis	9 g
Ramulus Uncariae cum Uncis	12 g
Flos Chrysanthemi	9 g
Concretio Silicea Bambusae	9 g
Radix Glycyrrhizae	3 g

Decoct the above ingredients in water for oral dose.

Modification: In case of mist of the brain by turbid phlegm, add Rhizoma Acori Graminei 12 g and Radix Curcumae 12 g.

In case of coma, use *Qing Kai Ling* Injection by means of either intravenous injection or intravenous drip.

6.3 Acute Urinary Tract Infection

Acute infection of urinary tract corresponds to *Lin Zheng* (*Lin* syndrome) in traditional Chinese medicine.

Etiology and Pathogenesis

This disease is resulted from invasion of exogenous damp—heat which accumulates in the lower *Jiao*, or excessive sweet and greasy food—intake which leads to dysfunction of the spleen in transportation and transformation and gives rise to accumulation of damp that produces heat, resulting in downward infusion of damp—heat in the lower *Jiao*, or overstrain or sexual intemperance which causes deficiency of the kidney in astringency.

Essentials of Diagnosis

1. This disease is characterized by such manifestations as frequency, urgency and pain of micturition, fever and chills.

2. There may appear increase of white blood cells and neutrophilic granulocytes, hematuria, and result of qualitative test of urinary protein is negative or positive with ++. More than 5 white cells could be observed in high power field, and there are great number of pus cells and white blood cell casts.

3. Fresh midstream urine culture and its count, or uropsammus counting are needed.

Treatment Based on Differentiation of Syndromes

1. Accumulation of damp—heat in the lower *Jiao*

Symptoms and signs: Frequency, urgency and pain of micturition, burning sensation in the urethra, fever, lumbago, deep yellow urine, dry mouth with desire of cold drinking, yellow and sticky tongue coating, and soft and rapid pulse.

Principle of treatment: Clearing heat, reducing fire, promoting diuresis to eliminate *Lin* syndrome.

Recipe: *Ba Zheng San* (15) with modification.

Ingredients:

Herba Plantaginis	
(to be wrapped for decocting)	30 g
Caulis Akebiae	9 g
Herba Dianthi	12 g

Herba Polygoni Avicularis	12 g
Fructus Gardeniae	6 g
Talcum	30 g
Radix et Rhizoma Rhei	12 g
Medulla Junci	3 g
tip of Radix Glycyrrhizae	6 g
Fructus Forsythiae	15 g

Decoct the above ingredients in water for oral dose.

Modification: In case of hematuria, add Radix Rehmanniae 15 g, Rhizoma Imperatae 30 g and Folium Pyrrosiae 20 g.

In case of pyuria, add Radix Scutellariae 9 g, Radix Sphorae Flavescentis 9 g and Rhizoma Smilacis Glabrae 12 g.

2. Accumulation of damp-heat due to deficiency of *Yin*

Symptoms and signs: Frequency and urgency of micturition, scanty and deep yellow urine, low fever, mental restlessness, soreness and pain of lumbus and back, weakness of legs, general lassitude, red tongue with little coating, and thready and rapid pulse.

Principle of treatment: Nourishing *Yin*, clearing heat, dispersing blood stasis and removing *Yin* syndrome.

Recipe: *Zhi Bai Di Huang Tang*(123) with modification.

Ingredients:

Rhizoma Anemarrhenae	9 g
Cortex Phellodendri	12 g
Radix Rehmanniae	12 g
Rhizoma Dioscoreae	9 g
Rhizoma Alismatis	15 g
Poria	24 g
Cortex Moutan Radicis	12 g

Herba Plantaginis (to be wrapped for decocting)	30 g
Herba Leonuri	30 g
Rhizoma Imperatae	30 g
Radix Cyathulae	9 g
Folium Pyrrosiae	30 g
Herba Lophatheri	9 g

Decoct the above ingredients in water for oral dose.

Modification: In case of high fever, add Flos Lonicerae 30 g and Fructus Forsythiae 15 g.

If it is accompanied by swelling and pain in the throat, add Fructus Arctii 10 g and Radix Isatidis 15 g.

6.4 Acute Urine Retention

Acute urine retention corresponds to *Long Bi* (urine retention) in traditional Chinese medicine.

Etiology and Pathogenesis

This disease is usually resulted from accumulation of damp-heat in the lower *Jiao* which leads to weakness of kidney *Qi* and dysfunction of *Qi* of the urinary bladder in transportation, giving rise to formation of turbid essential substance, stasis of blood and stone in the interior, obstructing the urinary bladder. Besides, stagnation of liver *Qi* may also lead to dysfunction of *Qi* of the bladder, giving rise to this disease.

Essentials of Diagnosis

1. The patient often has the chief complaint of urodialysis, and in severe case, lower abdominal distending pain and restlessness.

2. During examination, palpable distended urinary bladder

above pubis and percussion dullness could be found.

Treatment Based on Differentiation of Syndromes

1. Accumulation of damp-heat in the urinary bladder

Symptoms and signs: Dribbling urination with burning sensation and deep yellow urine, or anuresis with lower abdominal distention, dry mouth with no desire of drinking, or constipation, red tongue with yellow coating at its root, and rapid pulse.

Principle of treatment: Clearing heat and promoting urination.

Recipe: *Ba Zheng San*(15) plus *Shi Wei San*(124) with modification.

Ingredients:

Caulis Akebiae	9 g
Semen Plantaginis	
(to be wrapped for decocting)	20 g
Herba Dianthi	12 g
Herba Polygoni Avicularis	12 g
Talcum	30 g
Radix et Rhizoma Rhei	9 g
Fructus Gardeniae	6 g
Medulla Junci	3 g
Folium Pyrrosiae	30 g
Poria Rubra	20 g
tip of Radix Glycyrrhizae	6 g

Decoct the above ingredients in water for oral dose.

Modification: When the pathogenic heat consumes *Yin* and gives rise to dry mouth with little fluid, add Radix Rehmanniae 15 g and Radix Ophiopogonis 12 g.

In case of anuresis with lower abdominal distention, add

Rhizoma Anemarrhenae 9 g, Cortex Phellodendri 12 g and Cortex Cinnamomi 3 g.

When the pathogenic heat consumes the body fluid of the lung and leads to excessive thirst and rapid breathing, eliminate Radix et Rhizoma Rhei from the above prescription and add Radix Scutellariae 12 g, Cortex Mori Radicis 12 g, Rhizoma Phragmitis 30 g and Radix Glehniae 12 g.

2. Obstruction of the bladder by turbid essential substance and stasis of blood

Symptoms and signs: Dribbling or obstructed urination, lower abdominal distention and pain, purple tongue with stagnant spots, and uneven or thready and rapid pulse.

Principle of treatment: Removing blood stasis and dispersing masses to dredge the water passage.

Recipe: *Dai Di Dang Wan*(125) with modification.

Ingredients:

Radix et Rhizoma Rhei	9 g
tail of Radix Angelicae Sinensis	12 g
Squama Manitis	9 g
Semen Persicae	9 g
Radix Salviae Miltiorrhizae	30 g
Radix Cyathulae	9 g
Talcum	30 g
Medulla Tetrapanacis	6 g

Decoct the above ingredients in water for oral dose.

Modification: In case of deficiency of *Qi*, add Radix Astragali seu Hedysari 15 g and Radix Codonopsis Pilosulae 15 g.

In case of renal stone, add Herba Lysimachiae 18 g, Spora

Lygodii 18 g and Fructus Malvae Vertillatae 9 g.

For lower abdominal distention and pain, add Cortex Cinnamomi 6 g and Lignum Aquilariae Resinatum 2 g.

3. Stagnation of liver *Qi*

Symptoms and signs: Mental depression, or mental restlessness and irritability, urine retention or hesitant urination, hypochondriac and abdominal distention and pain, red tongue with thin and white or thin and yellow coating, and wiry and rapid pulse.

Principle of treatment: Dispersing stagnation of liver *Qi* and removing obstruction to promote urination.

Recipe: *Chai Hu Shu Gan San*(106) with modification.

Ingredients:

Radix Bupleuri	9 g
Fructus Aurantii	12 g
Radix Paeoniae	15 g
Rhizoma Cyperi	9 g
Pericarpium Citri Reticulatae Viride	9 g
Radix Curcumae	12 g
Radix Linderae	9 g
Fructus Foeniculi	9 g
tip of Radix Glycyrrhizae	3 g
Semen Plantaginis (to be wrapped for decocting)	30 g

Decoct the above ingredients in water for oral dose.

Modification: *Chen Xiang Fen*(powder of Lignum Aquilariae Resinatum) 3 g, *Hu Po Fen*(powder of Succinum) 3 g and *Xi Shuai*(criket) 2 pieces (to be ground into powder) can be mixed and divided into two portions to be taken with boiled water.

4. Declining of vital fire

Symptoms and signs: Urine retention or dribbling of urination, spiritlessness, aversion to cold, pallor, weakness of lumbus and knees, pale tongue, and deep and thready pulse with weakness on the *Chi* region.

Principle of treatment: Tonifying the kidney, warming *Yang*, replenishing *Qi* and removing obstruction from the orifice.

Recipe: *Ji Sheng Shen Qi Wan*(126) with modification.

Ingredients:

Radix Aconiti Praeparata	9 g
Cortex Cinnamomi	9 g
Radix Rehmanniae Praeparata	12 g
Fructus Corni	9 g
Poria	15 g
Rhizoma Alismatis	12 g
Radix Cyathulae	9 g
Semen Plantaginis	
(to be wrapped for decocting)	30 g
Herba Epimedii	30 g
Rhizoma Curculiginis	9 g
powder of Lignum Aquilariae Resinatum	
(to be taken following its infusion)	3 g

Decoct the above ingredients in water for oral dose.

Modification: In case of severe deficiency of *Yuan*(source) *Qi* and weakness of the kidney in governing due to senility, add Radix Ginseng Rubra 9 g and *Lu Jiao Pian* (Cornu Cervi) 15 g.

Acupuncture: Puncture Point Zusanli(ST36), Zhongji(RN3), Sanyinjiao(SP6) and Yinlingquan (SP9) with strong stimulation through repeated lifting, thrusting and rotation. In case of asthenia, give moxibustion to Guanyuan(RN4) and Qihai(RN6), and massage to the urinary area. Besides, extracting sneeze, inducing vomiting or external compress are also available.

7 Diseases of the Endocrine and Metabolic System

Acute diseases of the endocrine and metabolic system correspond to *Ying Bing*(goiter), *Xiao Ke*(diabetes), *Tuo Zheng*(prostration syndrome) and *Jue Zheng*(syncope) in traditional Chinese medicine. They are usually resulted from congenital insufficiency, lack of proper care after birth, invasion by exogenous pathogenic factors, internal injury due to excessive emotional changes, improper food-intake, overstrain and overstress, etc. Diseases of this system are often treated with the principle "Strengthening the body resistance and eliminating pathogenic factors".

7.1 Acute Suppurative Thyroiditis (Including Thyroid Crisis)

On the basis of goiter, acute suppurative thyroiditis often occurs due to hematogenous spread or local infection. It corresponds to *Jing Yong*(cervical carbuncle), *Suo Hou Yong*(throat-blocking phlegmon) and *Ying Bing*(goiter) in traditional Chinese medicine.

Etiology and Pathogenesis

Improper diet accumulates damp and produces phlegm. Stagnation of liver *Qi* turns into fire. When the exogenous pathogenic toxin invades the body, it combines itself with phlegm-heat and gives rise to the onset of the disease.

Essentials of Diagnosis

1. Most of such cases are of suppurative bacterial infection.

It mostly occurs in female aged from 20 to 40 years old.

2. It is characterized by severe pain and enlargement of thyroid, hotness sensation, local fluctuation sensation, red skin and high fever.

3. Increase of white blood cells.

Treatment Based on Differentiation of Syndromes

1. Upward attack of toxic heat

Symptoms and signs: Unilateral or bilateral swelling, distention, pain and hotness sensation of the neck accompanied by fever, aphagia, hoarse voice, yellow and sticky tongue coating, and wiry and slippery pulse.

Principle of treatment: Clearing away heat, eliminating toxin, resolving phlegm and subducing swelling.

Recipe: *Niu Bang Jie Ji Tang*(127) with modification.

Ingredients:

Fructus Arctii	12 g
Herba Menthae	9 g
Fructus Forsythiae	15 g
Fructus Gardeniae	9 g
Spica Prunellae	30 g
Cortex Moutan Radicis	9 g
Radix Scrophulariae	20 g
Herba Dendrobii	12 g
Herba Taraxaci	30 g
Herba Violae	15 g
Flos Lonicerae	30 g
Concha Ostreae	30 g
Lophatheri	9 g

Decoct the above ingredients in water for oral dose.

Modification: At early stage, *Jin Huang San*(128) could be used externally in combination. When there is pus formed, operation is needed for drainage.

2. Consumption of *Yin* due to pathogenic heat

Symptoms and signs: Soft or nodulose mass in the neck, palpitation, dysphoria with hotness sensation, dry mouth, deep yellow urine, red tongue with thin and yellow coating, and thready and rapid pulse.

Principle of treatment: Nourishing *Yin*, clearing away heat, dispersing stagnation and eliminating mass.

Recipe: *Yi Guan Jian*(129) with modification.

Ingredients:

Radix Rehmanniae	18 g
Radix Adenophorae	12 g
Radix Ophiopogonis	12 g
Radix Angelicae Sinensis	9 g
Fructus Meliae Toosendan	9 g
Radix Scrophulariae	20 g
Cortex Moutan Radicis	12 g
Radix Paeoniae Rubra	12 g
Fructus Forsythiae	30 g
Concha Ostreae	30 g
Spica Prunellae	12 g
Rhizoma Dioscoreae Bulbiferae	9 g

Decoct the above ingredients in water for oral dose.

7.2 Acute Hypoadrenocorticism

This disease corresponds to *Tuo Zheng* (prostration syndrome) in traditional Chinese medicine.

Etiology and Pathogenesis

This disease is usually caused by invasion of exogenous pathogenic factors, hemorrhge, vomiting, diarrhea, trauma or improper use of glucocorticoid, which leads to severe deficiency of *Yin* and *Yang* and prostration of *Yuan*(source) *Qi*, or exhaustion of *Yin* essence and floating of *Yang* that is deficient in the upper, or preponderance of *Yin* and cold in the interior, deficiency of *Qi* and collapse of *Yang*.

Essentials of Diagnosis

1. It is characterized by such manifestations as headache, diarrhea, anorexia, nausea and vomiting, spiritlessness, lassitude, shortness of breath, cyanosis, high fever or normal body temperature, disturbance of consciousness, or extensive submucosal and subcutaneous bleeding and shock.

2. With laboratory examination, it is found that the value of cortical hormone of plasma drops, acidocyte is uaually above $50/mm^3$, the level of blood sodium, chloride and blood sugar drop, the rate between sodium and kalium is less than 30, and the level of blood kalium, packed red blood cell volume, creatinine and blood acidity rise.

Treatment Based on Differentiation of Syndromes

1. Collapse of *Yin*

Symptoms and signs: High fever, headache, persistent profuse sweating, shortness of breath, nausea and vomiting, poor appetite, thirst with desire of cold drinking, or even coma, red and dry tongue, and rapid and forceless pulse.

Principle of treatment: Tonifying *Qi*, nourishing *Yin*, and promoting body fluid.

Recipe: *Di Huang Yin Zi*(130) with modification.

Ingredients:

Radix Ginseng	9 g
Radix Rehmanniae	30 g
Radix Rehmanniae Praeparata	30 g
Radix Astragali seu Hedysari	30 g
Radix Asparagi	20 g
Radix Ophiopogonis	20 g
Rhizoma Alismatis	9 g
Herba Dendrobii	12 g
Folium Eriobotryae	9 g
Fructus Aurantii	9 g
Radix Glycyrrhizae	9 g

Decoct the above ingredients in water for oral dose.

Modification: In case of coma, add 1 pill of *An Gong Niu Huang Wan* (20) by nasal or anorectal feeding.

In case of injury of vessels due to heat, add 1–3 g of *Xi Jiao Fen* (powder of Cornu Rhinoceri) to be taken following its infusion.

For persistent profuse sweating, add calcined Os Draconis 30 g, calcined Concha Ostreae 30 g, Fructus Schisandrae 6 g and Fructus Corni 9 g.

2. Collapse of *Yang*

Symptoms and signs: Temperature failing to rise, feeble breathing with snore, cold limbs, thirst with desire of hot drinking, or cyanosis, coma, pale tongue with white and moist coating, and fading pulse.

Principle of treatment: Tonifying *Qi*, restoring *Yang*, Strengthening body resistance to save collapse.

Recipe: *Shen Fu Tang* (29) with modification.

Ingredients:

Radix Ginseng	9 g
Radix Aconiti Praeparata	9 g
Rhizoma Zingiberis	9 g
Cortex Cinnamomi	9 g
Radix Glycyrrhizae Praeparata	6 g

Decoction the above ingredients in water for oral dose.

Modification: In case of coma, add *Su He Xiang Wan* to be taken after mixing with the decoction. Besides, acupuncture could be used in combination by needling point Shuigou(DU26) and Shixuan(EX−UE11) with moderate stimulation.

7.3 Spontaneous Hypoglycemia

Spontaneous hypoglycemia corresponds to *Jue Zheng*(syncope) and *Xu Lao*(consumptive disease) in traditional Chinese medicine.

Etiology and Pathogenesis

This disease is usually resulted from asthenia after long illness, congenital insufficiency, improper care after birth or indulgence in sex which leads to deficiency of the liver and kidney, causing consumption of essence and blood and poor nourishment of the brain, or dysfunction of the spleen in transportation and transformation, causing accumulation of damp and phlegm, misting the clear *Yang*, giving rise to this disease.

Essentials of Diagnosis

1. The patient often has paroxysmal attacks which usually stops ten minutes later or after eating. It may also last for long.

2. If the onset is abrupt, there may appear such manifestations as sweating, mental tension, pallor, palpitation, rapid pulse, tremor of limbs, and hypertention. If the onset is gradual, there

may appear mental symptoms such as muscular spasm, blurred vision and coma, etc.

3. During the attack, the blood sugar is usually below 60%mg. In case of dysfunction of vegetative nerve, the fall of blood sugar is not remarkable. At the same time, starvation test and tolbutamide test are needed for diagnosis.

Treatment Based on Differentiation of Syndromes

1. Deficiency of *Yin* of the liver and kidney

Symptoms and signs: Dizziness, halovision, palpitation, sweating, sallowish face, malar flush, soreness and weakness of lumbus and knees, seminal emission, irregular menstruation or leukorrhea, red tongue with little coating, and thready and rapid pulse.

Principle of treatment: Nourishing the liver and kidney.

Recipe: *Yi Guan Jian*(129) with modification.

Ingredients:

Radix Adenophorae	8 g
Radix Ophiopogonis	15 g
Radix Angelicae Sinensis	12 g
Radix Rehmanniae	15 g
Fructus Lycii	18 g
Fructus Meliae Toosendan	9 g
Radix Paeoniae Alba	18 g
stir-heated Semen Ziziphi Spinosae	12 g
Radix Glycyrrhizae	3 g

Decoct the above ingredients in water for oral dose.

Modification: In case of *Qi* and blood deficiency, modified *Shi Quan Da Bu Tang* should be used.

In case of deficiency of both heart and spleen, modified *Gui*

Pi Tang(50) should be applied.

2. Stagnation of *Qi* and phlegm

Symptoms and signs: Sudden coma, rattle in the throat, vomiting of phlegmy fluid, tremor of the four limbs, white and sticky tongue coating, and deep and slippery pulse.

Principle of treatment: Activating *Qi* circulation and resolving phlegm.

Recipe: *Wen Dan Tang*(131) with modification.

Ingredients:

Caulis Bambusae in Taeniam	9 g
Fructus Aurantii Immaturus	12 g
Rhizoma Pinelliae	9 g
Exocarpium Citri Grandis	9 g
Poria	30 g
Rhizoma Acori Graminei	9 g
Radix Curcumae	9 g
Radix Polygalae	9 g
Radix Glycyrrhizae Praeparata	6 g

Decoct the above ingredients in water for oral dose.

Modification: In case of blurred vision, add Semen Cassiae 30 g. Besides, acupuncture could be used in combination by neeling point Shuigou(DU26), Shixuan(EX-UE11), Hegu(LI4), etc. with moderate or strong stimulation. Then medicines should be given according to differentiation after resuscitation.

8 Diseases of Hematopoietic System

This chapter will mainly introduce anemia and purpura which, in traditional Chinese medicine, are known as *Tuo Xue*(collapse due to massive hemorrhage), *Xue Ku*(blood depletion), *Xu Lao*(consumptive disease), etc. The principle of treatment for the diseases to be introduced in this chapter is "treating the symptoms of the acute and the cause of the chronic."

8.1 Aplastic Anemia

Aplastic anemia corresponds to *Xu Lao*(consumptive disease) and *Xue Zheng*(blood trouble) in traditional Chinese medicine.

Etiology and Pathogenesis

The occurrence, development and change of this disease are mainly related to the heart, liver, spleen and kidney, especially to the deficiency of the spleen and kidney.

Essentials of Diagnosis

1. The disease is more commonly seen in the youth and the middle-aged, most of whom are males.

2. In the course of typical attack, there often appear such three major symptoms as progressive anema, extensive hemorrhage and repeated infection. In general, the liver, spleen and lymph node are not enlarged.

3. Laboratory examination

(1) Hemogram: The level of erythrocytes, leucocytes and platelets decreases, showing microcyte, hypochromic anemia, decreased absolute value of lymphocytes, and reticulocytopenia.

(2) Myelogram: The medullary hematopoietic tissue obviously decreases, while adipose tissue, lymphocytes, plasmacytes, reticulocytes and tissue basophile cells increase.

Treatment Based on Differentiation of Syndromes

1. Deficiency of both *Qi* and blood

Symptoms and signs: Pale and yellow complexion, vertigo, blurred vision, lassitude, palpitation, pale lips and nails, or menorrhagia, pale tongue with thin and white coating, and forceless or rapid pulse.

Principle of treatment: Tonifying *Qi* and blood.

Recipe: *Ba Zhen Tang*(132) with modification.

Ingredients:

Radix Codonopsis Pilosulae	15 g
Rhizoma Atractylodis Macrocephalae	12 g
Poria	9 g
Radix Glycyrrhizae Praeparata	6 g
Radix Rehmanniae Praeparata	12 g
Radix Angelicae Sinensis	9 g
Radix Paeoniae Alba	12 g
Rhizoma Ligustici Chuanxiong	6 g
Radix Astragali seu Hedysari	18 g

Decoct the above ingredients in water for oral dose.

2. Deficiency of *Yang* of the spleen and kidney

Symptoms and signs: Pale or dim complexion, aversion to cold, cold limbs, soreness and weakness of the lumbus and knees, frequency of micturition with profuse urination at night, edema, impotence, seminal emission, irregular menstruation, shortness of breathing, dislike of speaking, poor appetite, pale and swollen tongue with thin and white coating, and deep and weak of deep and slow pulse.

Principle of treatment: Strengthening the spleen and warming the kidney.

Recipe: *Si Jun Zi Tang*(133) plus *You Gui Wan*(134) with modification.

Ingredients:
Radix Codonopsis Pilosulae	15 g
Rhizoma Atractylodis Macrocephalae	12 g
Poria	9 g
Radix Glycyrrhizae	6 g
Radix Rehmanniae Praeparata	18 g
Rhizoma Dioscoreae	12 g
Colla Cornu Cervi	10 g
Cortex Cinnamomi	3 g
Radix Aconiti Praeparata	9 g
Cortex Eucommiae	12 g
Radix Angelicae Seninsis	12 g

Modification: If there appears deficiency of both heart and spleen, Modified *Gui Pi Tang* should be used instead. Besides, Cornu Cerri Pantotrichum, Gecko and Placenta Hominis may also be added, properly.

3. Deficiency of *Yin* of the liver and kidney

Symptoms and signs: Sallowish face, dizziness and vertigo, malar flush, soreness and weakness of the lumbus and knees, dull pain in hypochondriac region, menorrhagia, or uterine bleeding, epistaxis, hematochezia, pale lips and nails, afternoon fever, night sweating, mental restless with hotness sensation in the chest, palms and soles, dry tongue with no saliva and little coating, thready and rapid or wiry and rapid pulse.

Principle of treatment: Tonifying the liver and kidney

Recipe: *Gui Shao Di Huang Tang*(135) with modification.

Ingredients:

Radix Rehmanniae Praeparata	18 g
Rhizoma Dioscoreae	12 g
Fructus Corni	12 g
Poria	9 g
Rhizoma Alismatis	9 g
Cortex Moutan Radicis	9 g
Radix Paeoniae Alba	12 g
Radix Angelicae Sinensis	15 g
Fructus Schisandrae	9 g

Decoct the above ingredients in water for oral dose.

4. Deficiency of both *Yin* and *Yang*

Symptoms and signs: Pallor, soreness and weakness of lumbus and knees, lassitude, afternoon fever, night sweating, seminal emission, spermatorrhoea, insomnia or dream-disturbed sleep, or coldness sensation in the abdomen, loose stool, and deep, thready and forceless pulse.

Principle of treatment: Tonifying both *Yin* and *Yang*.

Recipe: *Jin Kui Shen Qi Wan*(136) with modification.

Ingredients:

Ramulus Cinnamomi	6 g
Radix Aconiti Praeparata	10 g
Radix Rehmanniae Praeparata	24 g
Fructus Corni	12 g

Rhizoma Dioscoreae	12 g
Cortex Moutan Radicis	9 g
Rhizoma Alismatis	9 g
Poria	9 g

Decoct the above ingredients in water for oral dose.

8.2 Acute Hemorrhagic Anemia

Acute hemorrhagic anemia corresponds to *Tuo Xue* (collapse due to hemorrhage), *Wang Xue* (hemorrhage) and *Xue Xu* (deficiency of blood) in traditional Chinese medicine.

Etiology and Pathogenesis

Acute hemorrhagic anemia due to various bleeding may affect the heart, spleen, liver and kidney, lead to dysfunction of the heart in dominating blood, that of the liver in storing blood, that of the spleen in controlling blood and that of the kidney in reception of *Qi*, giving rise to deficiency of *Qi*, blood, *Yin* and *Yang*.

Essentials of Diagnosis

First of all, it is necessary to find the cause of acute hemorrhagic anemia which could be the following diseases:

1. The diseases of digestive system such as gastroduodenal ulcer, carcinoma of stomach, rupture and varices of vein of the esophagus and fundus of the stomach, hookworm, hemorrhoid, etc.

2. The diseases of blood system such as hemophilia, thrombocytopenic purpura, acute leukemia, aplastic anemia, serious dysfunction of blood coagulation, etc.

3. The diseases of obstetrics and gynecology such as exfetation, Placenta previa and massive hemorrhage due to other causes.

Besides, it can also be resulted from bleeding due to various traumatic injuries.

Treatment Based on Differentiation of Syndromes

1. Deficiency of heart blood

Symptoms and signs: Palpitation, dizziness, vertigo, insomnia, dream-disturbed sleep, sallowish face, or pallor, pale tongue, thready and forceless pulse.

Principle of treatment: Nourishing blood and calming the mind.

Recipe: *Yang Xin Tang*(137) with modification.

Ingredients:

Radix Astragali seu Hedysari	15 g
Poria	9 g
Radix Angelicae Sinensis	12 g
Rhizoma Ligustici Chuanxiong	9 g
Radix Glycyrrhizae Praeparata	6 g
fermented Rhizoma Pinelliae	9 g
Semen Biotae	6 g
Fructus Schisandrae	6 g
Radix Ginseng	6 g
Cortex Cinnamomi	3 g
Fructus Ziziphi Jujubae	3 pieces

Decoct the above ingredients in water for oral dose.

2. Deficiency of spleen *Qi*

Symptoms and signs: Poor appetite, lassitude, dizziness, blurred vision, deficiency of *Qi*, dislike of speaking, general weakness, spontaneous sweating, loose stool, pale tongue, and deficient and forceless pulse.

Principle of treatment: Strengthening the spleen and tonifying *Qi*.

Recipe: *Shen Ling Bai Zhu San*(116) with modification.
Ingredients:

Radix Ginseng	9 g
Poria	10 g
Rhizoma Atractylodis Macrocephalae	12 g
Radix Platycodi	9 g
Rhizoma Dioscoreae	12 g
Semen Dolichoris Album	10 g
Semen Nelumbinis	10 g
Semen Coicis	15 g
Fructus Amomi	3 g
Radix Glycyrrhizae	6 g

Decoct the above ingredients in water for oral dose.

3. Deficiency of the liver blood

Symptoms and signs: Dizziness, blurred vision, tinnitus, hypochondriac pain, susceptability to panic, numbness of hands and feet, or even convulsion, scanty menstruation, pale tongue and wiry and thready pulse.

Principle of treatment: Nourishing the liver blood, activating blood circulation to disperse stagnation.

Recipe: *Si Wu Tang*(138) with modification.
Ingredients:

Radix Angelicae Sinensis	12 g
Radix Paeoniae Alba	15 g
Radix Ligustici Chuanxiong	9 g
Radix Rehmanniae Praeparata	12 g

Decoct the above ingredients in water for oral dose.

Modification: In case of dizziness, vertigo and tinnitus, add Fructus Ligustri Lucidi, Magnetitum and Concha Ostreae 30 g each to nourish *Yin* and pacify *Yang*.

In case of hypochondriac pain, add Radix Bupleuri 12 g, Radix Curcumae 12 g and Rhizoma Cyperi 12 g.

4. Deficiency of the kidney *Yin*

Symptoms and signs: Soreness of the lumbus, deafness, sorethroat, malar flush, weakness of the feet, blurred vision, deep red tongue with little fluid, and deep, thready and weak pulse.

Principle of treatment: Tonifying the kidney, nourishing blood, replenishing *Yin* and reducing fire.

Recipe: *Zhi Bai Di Huang Wan*(123) with modification.

Ingredients:

Rhizoma Anemarrhenae	9 g
Cortex Phellodendri	9 g
Radix Rehmanniae Praeparata	21 g
Rhizoma Dioscoreae	12 g
Cortex Moutan Radicis	9 g
Rhizoma Alismatis	9 g
Poria	9 g

Decoct the above ingredients in water for oral dose.

8.3 Allergic Purpura

Allergic purpura is also known as hemorrhagic capillary toxicosis. It corresponds to *Fa Ban*(eruption) and *Xue Zheng* (blood troubles) in traditional Chinese medicine.

Etiology and Pathogenesis

This disease is usually caused by invasion of the body by the exogenous pathogenic wind-dampness which mixes with *Qi* and blood. The heat injures the blood vessels, resulting in deranged blood flow in the blood vessels and extravasation of blood in the muscles and skin. In fact, it is the heat that causes bleeding. When the heat consumes blood, allergic purpura occurs.

Essentials of Diagnosis

1. This disease is more common in children and youth. The premonitory symptoms are lassitude, dizziness, poor appetite, and low fever. There are following types in general:

(1) Cutaneous allergic purpura: Skin eruption surrounded by a pale region, or urticaria-like rashes which are sporadic or coalescent into patches with itching, and distributed symmetrically on the lower limbs and buttocks in batches.

(2) Abdominal allergic purpura: The purpura is accompanied by the symptoms of digestive system such as abdominal distention and diarrhea.

(3) Arthro-allergic purpura: The purpura is accompanied by redness, swelling, hotness and pain of the joints, which are symmetrical and wandering, and mostly occur in the knee, ankle and wrist joints.

(4) Nephro-allergic purpura: The purpura is accompanied by facial edema, oliguria, proteinuria, hematuria and Cylindruria.

(5) Mixed type: The mixture of the above types is called a mixed type. Eruptions of each type may occur.

2. Laboratory examination: Laboratory examination shows no distinct abnormality but positiveness in tourniquet test.

Treatment Based on Differentiation of Syndromes

1. Eruption due to pathogenic wind-heat

Symptoms and signs: Fever, slight aversion to wind and cold, pantalgia and soreness of the bone, red eruption occurring one after another, red tongue with thin and yellow coating, and superficial and rapid pulse. This type mostly occurs at the early stage of the disease.

Principle of treatment: Relieving the exterior by clearing away the heat.

Recipe: *Yin Qiao San*(2) with modification.

Ingredients:

Flos Lonicerae	18 g
Fructus Forsythiae	18 g
Semen Sojae Praeparata	9 g
Fructus Arctii	9 g
Herba Menthae	6 g
Spica Schizonepetae	9 g
Radix Platycodi	9 g
Herba Lophatheri	6 g
Radix Ledebouriellae	9 g
fresh Rhizoma Phragmitis	18 g
Radix Glycyrrhizae	6 g

Decoct the above ingredients in water for oral dose.

2. Blood extravasation due to pathogenic heat

Symptoms and signs: High fever, red face, thirst, prominent purplish red eruption in patches, or accompanied by epistaxis, hematuria, dysphoria, or accompanied by redness, swelling, hotness and pain of joints, red tongue with possible stagnant spots and yellow dry coating, and slippery and rapid pulse.

This type is mostly seen at the middle stage of the disease.

Principle of treatment: Clearing away the toxic heat, cooling the blood and activating the blood circulation.

Recipe: *Xi Jiao Di Huang Tang*(25) with modification.

Ingredients:

Cornu Rhinoceri	3 g

Radix Rehmanniae	30 g
Radix Moutan Radicis	9 g
Radix Paeoniae Rubra	12 g
Rhizoma Anemarrhenae	12 g

Decoct the above ingredients in water for oral dose.

3. Hyperactivity of fire due to deficiency of *Yin*

Symptoms and signs: Persistent low fever, lassitude, dysphoria with hotness sensation in the chest, palms and soles, malar flush, dry lips, insomnia, night sweating, flat skin eruption with bright red colour, red tongue with little coating, and thready and rapid pulse.

Principle of treatment: Nourishing *Yin*, reducing fire, cooling the blood and activating blood circulation.

Recipe: *Zi Yin Jiang Huo Tang*(139) or *Zhi Bai Di Huang Wan*(123) plus *Xiao Ji Yin Zi*(140) with modification.

Ingredients:

Radix Paeoniae Alba	12 g
Radix Angelicae Sinensis	9 g
Radix Rehmanniae Praeparata	6 g
Radix Ophiopogonis	6 g
Rhizoma Atractylodis Macrocephalae	6 g
rice wine-prepared Radix Rehmanniae	12 g
Pericarpium Citri Reticulatae	3 g
Rhizoma Anemarrhenae	9 g
Cortex Phellodendri	6 g
Rhizoma Zingiberis	2 g
Fructus Ziziphi Jujubae	3 pieces

Decoct the above ingredients in water for oral dose.

4. Deficiency of the spleen and kidney

Symptoms and signs: Repeated attack of purpura, pallor, soreness and pain of the lumbus and back, dizziness, lassitude, poor appetite, frequency of micturition, menorrhagia, puffy tongue with tooth marks, and soft pulse.

This syndrome usually appears at the late stage of the disease.

Principle of treatment: Strengthening the spleen Qi, and tonifying the kidney to promote its astringent function.

Recipe: *Bu Zhong Yi Qi Tang*(141) plus *Wu Bi Shan Yao Wan*(142) with modification.

Ingredients:

Radix Astragali seu Hedysari	15 g
Radix Codonopsis Pilosulae	12 g
Rhizoma Atractylodis Macrocephalae	9 g
Radix Angelicae sinensis	12 g
Pericarpium Citri Reticulatae	9 g
Radix Glycyrrhizae Praeparata	6 g
Radix Bupleuri	6 g
Rhizoma Cimicifugae	9 g
Rhizoma Dioscoreae	12 g
Fructus Schisandrae	9 g
Cortex Eucommiae	12 g
Radix Morindae Officinalis	30 g
Rhizoma Alismatis	15 g

Decoct the above ingredients in water for oral dose.

Treatment of disease by acupuncture may obtain good therapeutic result. The points to be selected are Zusanli(ST36), Xuehai(SP10), Quchi(LI11), etc.

9 Diseases of the Connective Tissues and Allergic Reactions

In this chapter, systemic lupus erythematosus and allergic subacute septicemia are briefly introduced. There are no equivalent names to these diseases in traditional Chinese medicine and they are known as *Bi Zheng*(*Bi*-syndrome), *Fa Ban*(eruption), *Yao Tong*(lumbago), *Fa Re*(fever) and so on. The principles of treatment for these diseases are clearing up the *Ying* system, eliminating toxin, dispelling wind, removing heat, cooling the blood and protecting *Yin*.

9.1 Systemic Lupus Erythematosus

Systemic lupus erythematosus is a kind of autoimmune disease. It corresponds to *Bi Zheng* (*Bi*-syndrome) and *Fa Ban*(eruption) in traditional Chinese medicine.

Etiology and Pathogenesis

Invasion of *Zang Fu*, skin and muscles by pathogenic wind, cold and damp causes stagnation of *Qi* and blood. When stagnation of *Qi* and blood turns into fire, eruption is resulted. When it leads to obstruction, pain of the joints, skin and muscles occurs.

Essentials of Diagnosis

1. The disease is mostly seen in young and middle-aged females.

2. High or low fever, sensitivity to glucocorticoid hormone.

3. Manifestations in joints: Over 90% of the patients have wandering polyarticular soreness and pain and there may be redness, swelling, hotness and pain during the acute attacks.

4. Eruption: Erythematous eruption is most commonly seen. It appears symmetrically on the face, neck and limbs, especially on the finger tips, back, palms or around the elbow joints.

5. Over 50% of the patients suffer from lupus nephritis.

6. The examination of the blood and bone marrow smear shows lupus erythematosus cells. The positive rates in repeated examinations amount to 80% with no specificity.

Treatment Based on Differentiation of Syndromes

1. Preponderance of pathogenic toxic heat

Symptoms and signs: Sudden attack of the disease with high fever, butterfly-like red eruption on the cheeks, soreness and pain of the joints and muscles, and purpura on the skin, or even dysphoria, thirst, coma and delirium, deep red tongue with yellow and sticky coating, and full and rapid or wiry and rapid pulse.

Principle of treatment: Clearing away toxic heat from the *Ying* system, cooling the blood and protecting *Yin*.

Recipe: *Qing Wen Bai Du Yin*(117) with modification.

Ingredients:

powder of Cornu Budali	
(to be taken following its infusion)	6 g
Cortex Moutan Radicis	10 g
Radix Rehmanniae	30 g
Radix Paeoniae Rubra	10 g
Radix Scrophulariae	10 g
Flos Lonicerae	15 g

Radix Ophiopogonis	15 g
Gypsum Fibrosum	30 g
Rhizoma Anemarrhenae	10 g

Decoct the above ingredients in water for oral use.

2. *Bi*-syndrome caused by pathogenic wind, damp and heat

Symptoms and signs: Swelling, distention, soreness and pain of the joints, myalgia, or accompanied by low fever, red tongue with yellow and rough coating, and slippery and rapid or thready and rapid pulse.

Principle of treatment: Dispelling wind, removing obstruction from the channel, clearing away heat and regulating *Ying* system.

Recipe: *Du Huo Ji Sheng Tang* (78) with modification.

Ingredients:

Radix Angelicae Pubescentis	15 g
Ramulus Loranthi	10 g
Radix Gentianae Macrophyllae	10 g
Radix Ledebouriellae	10 g
Radix Rehmanniae	15 g
Radix Paeoniae Rubra	15 g
Rhizoma Ligustici Chuanxiong	6 g
Cortex Eucommiae	10 g
Radix Achyranthis Bidentatae	10 g
Radix Angelicae Sinensis	10 g
Caulis Lonicerae	10 g
Radix Salviae Miltiorrhizae	15 g

Decoct the above ingredients in water for oral dose.

3. Endogenous heat due to deficiency of *Yin*

Symptoms and signs: Low fever, mental restlessness with

hotness sensation in the palms and soles, dark red eruption, spontaneous sweating, night sweating, soreness and pain of the joints, alopecia, red tongue with peeled coating, and thready and rapid pulse.

Principle of treatment: Nourishing *Yin* and clearing away pathogenic heat.

Recipe: *Qing Hao Bie Jia Tang*(143) with modification.
Ingredients:

Herba Artemisiae Chinghao	10 g
Radix Stellariae	10 g
Cortex Lycii Radicis	10 g
Rhizoma Picrorhizae	10 g
Radix Rehmanniae	30 g
Radix Scrophulariae	10 g
Radix Ophiopogonis	10 g
Herba Dendrobii	10 g
Radix Cynanchi Atrati	10 g
Fructus Ligustri Lucici	15 g

Decoct the above ingredients in water for oral dose.

9.2 Allergic Subacute Septicemia

Allergic subacute septicemia corresponds to *Fa Re*(fever), *Yao Tong*(lumbago), etc. in traditional Chinese medicine. It is rarely seen in the clinic.

Etiology and Pathogenesis

This disease is resulted from invasion of the body by pathogenic wind, cold and damp because of deficiency of antipathogenic *Qi*. The pathogenic factors wander in the channels, block the joints, affect normal circulation of *Qi* and blood and cause blood stasis which turns into fire, giving rise to fever, joint pain or even pantalgia.

Essentials of Diagnosis

1. Long-term fever with intermittent remission, changeable patterns of fever which are mainly remittent and continuous.

2. Repeated appearance of transient skin rashes and joint pain.

3. Increased number of leucocytes with nucleus shifting to the left, speeded blood sedimentation.

4. No definite positive result in repeated blood cultures.

5. Ineffectiveness to antibiotic therapy.

6. Symptoms alleviated by hormonotherapy.

Treatment Based on Differentiation of Syndromes

1. Preponderance of pathogenic heat

Symptoms and signs: Sudden attack with high fever, dysphoria, thirst, constipation, scattered skin rashes on the chest and abdomen, joint pain, excessive sweating, deep yellow urine, yellow and dry tongue coating, and rapid pulse.

Principle of treatment: Clearing away toxic heat, cooling the blood and protecting *Yin*.

Recipe: *Xi Jiao Di Huang Tang*(25) with modification.

Ingredients:

powder of Cornu Rhinoceri (to be taken following its infusion)	6 g
Radix Rehmanniae	30 g
Radix Paeoniae Rubra	10 g
Flos Lonicerae	15 g
Fructus Forsythiae	12 g
Radix Ophiopogonis	15 g

Radix Scutellariae	12 g
Radix Trichosanthis	10 g
Cortex Moutan Radicis	9 g
Gypsum Fibrosum	30 g
Rhizoma Anemarrhenae	9 g
Fructus Gardeniae	10 g
Ramulus Mori	9 g

Decoct the above the ingredients in water for oral dose.

2. Accumulation of damp-heat in the interior

Symptoms and signs: Intermittent high fever, redness, swelling and pain of the joints, distending pain in the head with tightness sensation, dry mouth with no desire of drink, excessive sweating, yellow and sticky tongue coating, and soft and rapid pulse.

Principle of treatment: Clearing away heat and eliminating damp.

Recipe: *Du Huo Ji Sheng Tang*(78) with modification.

Ingredients:

Radix Angelicae Pubescentis	15 g
Rhizoma seu Radix Notopterygii	12 g
Flos Lonicerae	15 g
Ramulus Loranthi	15 g
Radix Gentianae Macrophyllae	9 g
Radix Ledebouriellae	9 g
Ramulus Cinnamomi	3 g
Herba Asari	3 g
Radix Angelicae Sinensis	12 g
Radix Paeoniae Rubra	12 g
Rhizoma Ligustici Chuanxiong	9 g

Radix Rehmanniae	15 g
Cortex Eucommiae	9 g
Radix Codonopsis Pilosulae	12 g
Poria	9 g
Radix Glycyrrhizae	6 g

Decoct the above ingredients in water for oral dose.

10 Infectious Diseases

This chapter will mainly deal with such infectious diseases as influenza, epidemic meningitis and encephalitis B, etc. which correspond to *Shi Xing Gan Mao*(influenza), *Chun Wen*(spring—warm syndrome), *Shu Shi*(summer—heat and damp) etc. respectively in traditional Chinese medicine. The corresponding principle of treatment is "treating the symptoms for the acute" but prevention is also important.

10.1 Influenza

This disease is known as *Shi Xing Gan Mao*(flu) in traditional Chinese medicine.

Etiology and Pathogenesis

This disease is caused by exogenous pathogenic factors when the function of defensive system is weakened and the emergency function is reduced due to changeable weather.

Essentials of Diagnosis

1. This disease is characterized by abrupt onset with symptoms of general poisoning such as high fever, general soreness and pain, chillness, headache, lassitude, etc. During the epidemic period, there may appear cough, expectoration and chest pain, or nausea, vomiting and diarrhea.

2. In the epidemic season, pay attention to differentiate influenza from other early acute epidemic diseases such as epidemic meningitis and measles.

Treatment Based on Differentiation of Syndromes

1. Invasion of the exterior by wind—cold

Symptoms and signs: Chills, fever, anhidrosis, headache, nasal obstruction, clear nasal discharge, sneezing, itching throat, slight cough with or without little dilute sputum, thin and white tongue coating, and superficial pulse.

Principle of treatment: Relieving the exterior by using the drugs pungent in flavour and warm in property.

Recipe: *Jing Fang Bai Du San*(144) with modification.

Ingredients:

Herba Schizonepetae	9 g
Radix Ledebouriellae	6 g
Folium Perillae	6 g
Radix Peucedani	9 g
Radix Platycodi	6 g
Semen Sojae Praeparata	9 g

Decoct the above ingredients in water for oral dose.

Modification: In case of soreness and pain of limbs and severe headache, add Rhizoma seu Radix Notopterygii 6 g and Radix Angelicae Pubescentis 6 g.

2. Invasion of the exterior by wind—heat

Symptoms and signs: Fever, slight aversion to wind and cold, little sweating, headache, cough with scanty yellow and sticky sputum, redness, swelling and pain of the throat, slight thirst, red tongue with thin and white or slightly yellow coating, and superficial and rapid pulse.

Principle of treatment: Relieving the exterior with drugs pungent in flavour and cool in property.

Recipe: *Yin Qiao San*(2) with modification.

Ingredients:

Fructus Arctii	9 g
Herba Menthae	3 g
Fructus Forsythiae	9 g
Flos Lonicerae	15 g
Semen Armeniacae Amarum	9 g
Folium Mori	9 g

Decoct the above ingredients in water for oral dose.

Modification: In case of redness and swelling of the tonsil, add Rhizoma Belamcandae 9 g and Radix Sophorae Subprostratae 15 g.

In case of epistaxis, add Fructus Gardeniae 9 g.

For persistent high fever, add Radix Bupleuri 9 g and Radix Scutellariae 9 g.

3. Invasion of the exterior by summer-heat and damp

Symptoms and signs: Invasion of summer-heat and damp gives rise to such manifestations as headache, lassitude or soreness and pain of the limbs, fever, sweating or little sweating, mental restlessness, thirst, slight cough, stuffiness and fullness in the chest and epigastrium, yellow urine, loose stool, thin, yellow and sticky tongue coating, and soft and rapid pulse.

Principle of treatment: Relieving the exterior by eliminating summer-heat and resolving damp.

Recipe: *Xin Jia Xiang Ru Yin*(6) with modification.

Ingredients:

Herba Elsholtziae seu Moslae	9 g
Semen Sojae Germinatum	9 g
Fructus Forsythiae	9 g
Flos Lonicerae	9 g

Cortex Magnoliae Officinalis   6 g
Herba Agastachis   9 g
Herba Eupatorii   9 g
Folium Nelumbinis   3 g

Decoct the above ingredients in water for oral dose.

Modification: In case of excessive fever with mental restlessness, add Rhizoma Coptidis 6 g.

If there is food retention, add Fructus Crataegi 9 g.

10.2 Epidemic Cerebrospinal Meningitis

Epidemic Cerebrospinal meningitis belongs to *Chun Wen*(spring-warm syndrome) in traditional Chinese medicine.

Etiology and Pathogenesis

This disease is caused by invasion of seasonal pestilence when the anti-pathogenic *Qi* is deficient and defensive function is weakened.

Essentials of Diagnosis

1. This disease is characterized by such manifestations as abrupt onset, high fever, aversion to cold, severe headache, projectile vomiting, dysphoria, convulsion, coma, neck rigidity, positive Brudzinski's sign and Kernig's sign, prominence of frontanel in infants, and opisthotonos, etc.

2. It mostly occurs in spring and winter, affecting preschool children. At its initial stage, there may appear purple spots or patches which are 2 cm in diametre on the skin, and herpes simplex in front of nostril and around the mouth 3-5 days after the attack.

3. Laboratory examination:

(1) Hydrocrania: Distinct elevation of cerebrospinal pressure

with cloudy appearance like rice water or pus, increase of cells and especially the neutrophil polykaryocytes, increase of protein, decrease of sugar, and the accordant bacteria found by smear examination or culture.

(2) Result of blood culture is positive. In blood routin, distinct increase of white blood cells, and neutrophil polykaryocytes in perticular.

Treatment Based on Differentiation of Syndromes

1. Stagnation of pathogenic factors in the defensive system

Symptoms and signs: Sudden onset with low fever, chillness, headache, nausea and vomiting, dry throat, thirst, neck rigidity (not typical), apathy, occasional bleeding spots on the skin, thin and white or light yellow tongue coating, and slippery and rapid pulse.

Principle of treatment: Clearing away toxic heat and secondarily relieving the exterior.

Recipe: *Yin Qiao San*(2) with modification.

Ingredients:

Flos Lonicerae	18 g
Fructus Forsythiae	15 g
Folium Isatidis	30 g
Radix Scutellariae	12 g
Radix Gentianae	9 g
Radix Puerariae	15 g
Herba Menthae	3 g
Succus Bambosae	10 g
Rhizoma Pinelliae	9 g

Decoct the above ingredients in water for oral dose.

2. Intense heat in both *Qi* and *Ying* system

Symptoms and signs: Persistent high fever, severe headache, neck rigidity, projectile vomiting, dysphoria or lethargy, paroxysmal convulsion, increasing of stagnant spots on the skin, red tongue with yellow and sticky coating, and slippery and rapid pulse.

Principle of treatment: Clearing away toxic heat and cooling the *Ying* system.

Recipe: *Qing Wen Bai Du Yin*(117) with modification.

Ingredients:

Flos Lonicerae	15 g
Folium Isatidis	30 g
Rhizoma Coptidis	3 g
Radix Scrophulariae	12 g
Cortex Moutan Radicis	9 g
Fructus Forsythiae	12 g
Radix Gentianae	6 g
Gypsum Fibrosum	15 g
fresh Radix Rehmanniae	15 g
Concha Haliotidis	30 g

Decoct the above ingredients in water for oral dose.

Modification: In case of high fever with convulsion and yellow tongue coating, add Ramulus Uncariae cum Uncis 15 g and Scorpio 9 g.

In case of high fever with coma and dysphoria, add *An Gong Niu Huang Wan* 1 pill each time and twice a day (for adult).

In case of consumption of *Yin* fluid, add Radix Ophiopogonis 15 g and Herba Dendrobii 15 g.

3. Collapse of anti-pathogenic *Qi* due to sinking of pathogenic factors into the interior

Symptoms and signs: Pale and dim complexion, sweating, cold limbs, feeble or irregular breathing, stagnant spots on the skin rapidly spreading into patches, drop of blood pressure, pale tongue, and faint thready pulse.

Principle of treatment: Tonifying *Qi*, restoring *Yang* and saving the collapse.

Recipe: *Shen Fu Tang*(29) with modification.

Ingredients:

Radix Ginseng Rubra	12 g
Radix Aconiti Praeparata	18 g
Radix Glycyrrhizae	15 g

Decoct the above ingredients in water for oral dose.

Modification: In case of red tongue and thready rapid pulse, add Radix Ophiopogonis 12 g and Fructus Schisandrae 9 g.

10.3 Epidemic Encephalitis B

Epidemic encephalitis B corresponds to *Shu Wen*(summer heat), *Shu Feng*(summer-heat spasm), *Shu Jue*(syncope due to summer-heat), etc. in traditional Chinese medicine.

Etiology and Pathogenesis

This disease is resulted from pestilential summer-heat in case of weakness of anti-pathogenic *Qi* due to heat in the summer.

Essentials of Diagnosis

1. Inquiry about the preventive case history.

2. In case of sudden attack with such symptoms as headache, nausea and vomiting, lethargy or dysphoria which becomes worse in 2-3 days during the epidemic season, this disease should be firstly considered. In severe case, there may appear prompt coma, convulsion or respiratory failure. In infants, there could be star-

ing of the eyes and startle.

3. During the inetial stage, no positive sign can be observed. 2—3 days later, there is usually meningeal irritation, and prominence of fontanelle in infants. Some patients may have disappearance of abdominal reflex, positive Babinski's sign and increased muscular tension, etc.

4. Laboratory examination:

(1) Blood test: White blood cells, mainly neutrophil polykaryocytes, increased.

(2) Cerebrospinal fluid test: The cells, mainly polykaryocytes increased at the initial stage, and lymphocytes at the late stage. The sugar is normal or slightly higher, protein slightly increased and chloride normal.

(3) The result of complement fixation test is positive.

Treatment Based on Differentiation of Syndromes

1. Invasion of the defensive system by pathogenic factors

Symptoms and signs: Fever, or slight chillness, sweating, headache, consciousness, or dysphoria, sethargy, slight rigidity of the neck, occasional slight convulsion, and thin and white tongue coating.

Principle of treatment: Relieving the exterior by clearing away heat and eliminating toxin.

Recipe: *Yin Qiao San*(2) with modification.

Ingredients:

Flos Lonicerae	15 g
Fructus Forsythiae	12 g
Gypsum Fibrosum	30 g
Herba Menthae	3 g
Herba Lophatheri	3 g

Radix Scutellariae	12 g
Folium Isatidis	30 g

Decoct the above ingredients in water for oral dose.

Modification: In case of preponderance of pathogenic damp characterized by nausea, vomiting, lethargy and white and sticky tongue coating, remove Gypsum Fibrosum, add Herba Agastachis 9 g, Herba Eupatorii 9 g and Cortex Magnoliae Officinalis 6 g.

2. Intense heat in both *Qi* and *Ying* system

Symptoms and signs: High fever, headache, neck rigidity, coma or semiconsciousness, delirium, convulsion, different pupils in size, irregular shallow breathing, red or deep red tongue with yellow and sticky coating, and full and rapid pulse.

Principle of treatment: Clearing away toxic heat and cooling the *Ying* system.

Recipe: *Bai Hu Tang*(10) plus *Qing Ying Tang*(18) with modification.

Ingredients:

Gypsum Fibrosum	45 g
Rhizoma Anemarrhenae	15 g
Flos Lonicerae	15 g
Fructus Forsythiae	9 g
Radix Isatidis	30 g
Rhizoma Coptidis	3 g
Rhizoma Paridis	18 g
Radix Rehmanniae	15 g
Radix Scrophulariae	15 g

Decoct the above ingredients in water for oral dose.

Modification: In case of high fever with constipation, add

Radix et Rhizoma Rhei 9 g.

In case of continuous convulsion, add Ramulus Uncariae cum Uncis 15 g (to be decocted later) and Scorpio 9 g.

For severe coma, add Rhizoma Acori Graminei 6 g and Herba Houttuyuniae Praeparata 3 g accordingly.

3. Weakness of anti-pathogenic *Qi* with lingering of pathogenic factors

Symptoms and signs: Persistent low fever which goes slightly higher in the afternoon, red face, dysphoria, red tongue, and thready and rapid pulse.

This syndrome usually occurs at the convalescent stage.

Principle of treatment: Nourishing *Yin* and clearing away heat.

Recipe: *Jia Jian Fu Mai Tang*(145) with modification.

Ingredients:

Radix Rehmanniae	15 g
Radix Ophiopogonis	12 g
Radix Glehniae	15 g
Radix Scrophulariae	12 g
Radix Paeoniae Alba	12 g
Herba Artemisiae Chinghao	9 g
Radix Cynanchi Atrati	12 g
Rhizoma Anemarrhenae	9 g
Cortex Moutan Radicis	12 g

Decoct the above ingredients in water for oral dose.

Modification: In case of dysphoria, insanity, and thin, yellow and sticky tongue coating, add Rhizoma Pinelliae 9 g, Arisaema cum Bile Praeparata 3 g and Rhizoma Acori Graminei 9 g.

In case of consumption of *Yin* fluid and stirring-up of

endogenous wind with such symptoms as tremor of hands and feet, and red tongue with little fluid, remove Herba Artemisiae Chinghao and add Plastrum Testudinis 15 g and Carapax Trionycis 15 g.

4. Obstruction of the channels by phlegm and stasis of blood

Symptoms and signs: Indifferent expression with no speaking, paralysis of limbs, pale complexion, and sticky tongue coating.

This syndrome usually appears at the convalescent stage of sequela.

Principle of treatment: Tonifying *Qi*, resolving phlegm and dispersing the stasis of blood.

Recipe: *Bu Yang Huan Wu Tang*(40) plus *Di Tan Tang*(79) with modification.

Ingredients:

Radix Astragali seu Hedysari	15 g
Radix Angelicae Sinensis	12 g
Radix Paeoniae Rubra	9 g
Flos Carthami	9 g
Semen Persicae	9 g
Rhizoma Acori Graminei	9 g
Rhizoma Pinelliae Praeparata	12 g
Rhizoma Typhonii	6 g
Lumbricus	9 g
Pericarpium Citri Reticulatae	6 g
Radix Glycyrrhizae	3 g

Decoct the above ingredients in water for oral dose.

5. Acupuncture treatment: In case of sequela of encephalitis B like deafness, dumbness and paralysis of limbs, acupuncture

treatment may obtain good therapeutic results. The commonly used points are Yamen(DU15), Tiantu(RN22), Chize(LU5), Weizhong(UB40), Taichong(LR3), Quchi(LI11), Huantiao(GB30), Yanglingquan(GB34), etc.

10.4 Epidemic Hemorrhagic Fever

Epidemic hemorrhagic fever is a kind of natural focal disease. It is known as *Dong Wen Shi Yi*(seasonal disease caused by winter-warm), *Yin Ban*(epidemic eruption), etc. in traditional Chinese medicine.

Etiology and Pathogenesis

When the pathogenic warm-heat invades the lung, it is quickly transmitted into the *Ying* system, giving rise to intense heat in both *Qi* and *Ying* systems and resulting in persistent high fever. When the intense heat consumes both *Qi* and *Yin* and caused collapse of *Yang*, there may appear the manifestations of dissociation of *Yin* and *Yang*. When it consumes the *Yin* of the kidney, there may be obstruction and rejection. When the pathogenic factors gradually subside, the anti-pathogenic *Qi* is insufficient and kidney weak, and there may appear polyuria.

Essentials of Diagnosis

1. Epidemiological deta.
2. Symptoms of general poisoning.
3. Capillary toxic signs.
4. Damage of the kidney.

Treatment Based on Differentiation of Syndromes

1. Pyrogenetic stage:

(1) Invasion of the *Wei*(defensive) system by pathogenic warm

Symptoms and signs: Chillness, fever, headache, lumbago, orbital pain, flushed neck and chest, redness of the tip and border of the tongue, white, thin and sticky tongue coating, and superficial and rapid pulse.

Principle of treatment: Relieving the exterior by clearing away heat.

Recipe: *Yin Qiao San*(2) with modification.

Ingredients:

Flos Lonicerae	15 g
Fructus Forsythiae	10 g
Herba Menthae	
(to be decocted later)	3 g
Radix Platycodi	5 g
fresh Rhizoma Phragmitis	
(with its nodes removed)	30 g
Radix Glycyrrhizae	6 g
Rhizoma Imperatae	30 g
Cortex Moutan Radicis	10 g
Radix Salviae Miltiorrhizae	15 g
Radix Scutellariae	10 g

Decoct the above ingredients in water for oral dose.

(2) Preponderance of heat in *Yang Ming* Channel

Symptoms and signs: High fever with no chillness, drunklike appearance, thirst, dysphoria, red tongue with yellow coating, and wiry and rapid pulse.

Principle of treatment: Clearing away heat and eliminating toxin.

Recipe: *Bai Hu Tang*(10) with modification.

Ingredients:

Gypsum Fibrosum	
(to be decocted first)	30 g
Rhizoma Anemarrhenae	10 g
Radix Rehmanniae	10 g
Herba Lophatheri	10 g
Radix Sophorae Subprostratae	10 g
Radix Isatidis	15 g
Semen Oryzae Glutinosae	30 g
Radix Glycyrrhizae	6 g
Radix Scrophulariae	10 g

Decoct the above ingredients in water for oral dose.

(3) Intense heat in both *Qi* and *Ying* systems

Symptoms and signs: High fever with thirst, blurred vision, dysphoria, eruption, hematemesis, deep red tongue with yellow and dry coating, and wiry and rapid or thready and rapid pulse.

Principle of treatment: Clearing away heat from the *Qi* and *Ying* systems, eliminating toxin and protecting *Yin*.

Recipe: *Qing Wen Bai Du Yin*(117) with modification.

Ingredients:

Gypsum Fibrosum	
(to be decocted first)	60 g
Rhizoma Anemarrhenae	10 g
Radix Rehmanniae	10 g
Radix Scutellariae	15 g
Fructus Gardeniae	10 g
Cornu Rhinoceri	5 g
Cortex Moutan Radicis	10 g
Herba Lophatheri	10 g
Radix et Rhizoma Rhei	6 g

Radix Scrophulariae	15 g
Flos Lonicerae	15 g
Radix Glycyrrhizae	6 g

Decoct the above ingredients in water for oral dose.

2. Hypotention and shock stage:

(1) Cold limbs due to excess of heat

Symptoms and signs: Cold limbs, burning sensation in the chest and abdomen, red face, mental restlessness, shortness of breath, red tongue with yellow, thick and dry coating, and slippery and rapid or deep and rapid pulse.

Principle of treatment: Clearing away toxic heat, tonifying *Qi* and promoting body fluid.

Recipe: *Bai Hu Tang*(10) plus *Sheng Mai San*(26) with modification.

Ingredients:

Gypsum Fibrosum (to be decocted first)	30 g
Rhizoma Anemarrhenae	10 g
Radix Isatidis	15 g
Radix Ginseng	3 g
Radix Ophiopogonis	15 g
Fructus Schisandrae	15 g
Rhizoma Cimicifugae	10 g
Radix Paeoniae Alba	10 g

Decoct the above ingredients in water for oral dose.

(2) Cold limbs due to deficiency of *Yang* and excess of *Yin*

Symptoms and signs: Aversion to cold, cold limbs, sleeping with shrunk legs, no thirst, feeble breathing, lassitude, pale complexion, cyanotic lips, and deep, slow and thready pulse of fading pulse.

Principle of treatment: Restoring *Yang* and saving collapse.

Recipe: *Shen Fu Tang* (29) with modification.

Ingredients:

Radix Ginseng	10 g
Radix Aconiti Praeparata (to be decocted first)	10 g
Fructus Schisandrae	10 g
Radix Rehmanniae Praeparata	18 g
Radix Ophiopogonis	10 g
Radix Salviae Miltiorrhizae	15 g
Radix Glycyrrhizae Praeparata	10 g
Os Draconis	30 g
Concha Ostreae	30 g

Decoct the above ingredients in water for oral dose.

3. Oliguria stage:

(1) Fire of deficiency type in the interior due to deficiency of kidney *Yin*

Symptoms and signs: Extreme exhaustion, listlessness, lethargy, soreness of the lumbus, uneven and scanty urination, dry mouth and throat, mental restlessness, insomnia, red tongue with dry coating, and thready and rapid pulse.

Principle of treatment: Nourishing the kidney and promoting body fluid.

Recipe: *Zhi Bai Di Huang Tang*(123) with modification.

Ingredients:

Rhizoma Anemarrhenae	10 g
Cortex Phellodendri	10 g
Radix Rehmanniae	10 g

Rhizoma Dioscoreae	15 g
Rhizoma Imperatae	30 g
Cortex Moutan Radicis	10 g
Radix Ophiopogonis	10 g
Fructus Corni	10 g
Rhizoma Alismatis	10 g

Decoct the above ingredients in water for oral dose.

(2) Attack of the pericardium by pathogenic heat and stirring-up of the liver wind

Symptoms and signs: Oliguria, urine retention, headache, nausea and vomiting, coma, delirium, convulsion, deep red tongue with dry coating, and wiry, thready and rapid pulse.

Principle of treatment: Clearing away toxic heat from the heart, dispelling wind and relieving convulsion.

Recipe: *Xi Jiao Di Huang Tang*(25) plus *Ling Yang Gou Teng Tang*(22) with modification.

Ingredients:

Cornu Rhinoceri	6 g
Radix Rehmanniae	15 g
Cortex Moutan Radicis	10 g
Ramulus Uncariae cum Uncis	12 g
Flos Chrysanthemi	10 g
Radix Paeoniae Rubra	12 g
Radix Paeoniae Alba	12 g
Caulis Bambusae in Taeniam	10 g
Semen Plantaginis	10 g
Rhizoma Imperatae	30 g

Decoct the above ingredients in water for oral dose.

(3) Retention of pathogenic fluid in the lung

Symptoms and signs: Fullness in the chest, asthmatic breathing, preponderance of phlegm, dysphoria, puffy tongue with white, thick and sticky coating, and wiry and rapid or slippery and rapid pulse.

Principle of treatment: Purging the lung and soothing asthma.

Recipe: *Ting Li Da Zao Xie Fei Tang*(146) with modification.
Ingredients:

Semen Lepidii seu Descurainiae	15 g
Fructus Ziziphi Jujubae	10 pieces
Fructus Plantaginis	
(wrapped in a piece of cloth for decocting)	10 g
Radix et Rhizoma Rhei	
(to be decocted later)	10 g
Rhizoma Imperatae	30 g
Poria	15 g

Decoct the above ingredients in water for oral dose.

4. Polyuria stage:

(1) Deficiency of kidney *Qi*

Symptoms and signs: Lassitude, dislike of speaking, thirst with much drinking, polyuria day and night, soreness and weakness of the lumbus and knees, light red tongue with little and dry coating, and deficient and large pulse.

Principle of treatment: Tonifying the kidney and promoting its function in controlling, replenishing *Qi* and producing body fluid.

Recipe: *Ba Xian Chang Shou Wan*(147) plus *Suo Quan Wan*(119) with modification.

Ingredients:

Radix Ophiopogonis	10 g
Radix Codonopsis Pilosulae	15 g
Radix Rehmanniae Praeparata	24 g
Rhizoma Dioscoreae	15 g
Fructus Rubi	10 g
Fructus Alpiniae Oxyphyllae	10 g
Fructus Schisandrae	5 g

Decoct the above ingredients in water for oral dose.

(2) Preponderance of heat in the lung and stomach

Symptoms and signs: Dry mouth and tongue, excessive thirst with desire or drinking, dry cough with little sputum, easy hunger with polyphagia, frequent and much urination, red tongue with yellow coating, and deep and rapid pulse.

Principle of treatment: Clearing away heat from the lung and stomach, nourishing *Yin* and promoting body fluid.

Recipe: *Sha Shen Mai Dong Tang*(5) with modification.

Ingredients:

Radix Glehniae	10 g
Radix Ophiopogonis	10 g
Folium Mori	10 g
Radix Trichosanthis	12 g
Rhizoma Polygonati Odorati	10 g
Gypsum Fibrosum	30 g
Herba Lophatheri	10 g
Rhizoma Dioscoreae	10 g
Fructus Alpiniae Oxyphyllae	10 g
Radix Glycyrrhizae	3 g

Decoct the above ingredients in water for oral dose.

5. Convalescent stage:

In case of deficiency of kidney *Yin*, use *Liu Wei Di Huang Wan*(118) with modification.

In case of weakness of spleen *Yang*, use *Shen Ling Bai Zhu San*(116) with modification.

In case of stomach *Yin* in recovery, use *Yi Wei Tang*(173) with modification.

10.5 Acute Fatal Hepatitis

Acute fatal hepatitis refers to acute live necrosis and subacute liver necrosis, taking up 0.2 per cent of all the cases of hepatitis. It corresponds to *Huang Dan*(jaundice), *Wen Huang*(epidemic jaundice) and *Xie Tong*(hypochondriac pain) in traditional Chinese medicine.

Etiology and Pathogenesis

It takes weakness of spleen and stomach or improper food-intake such as indulgence in alcohol drinking, which casues dysfunction of the spleen and stomach, plus invasion of seasonal pathogenic factors to cause this disease.

Essentials of Diagnosis

1. The condition at the initial stage

The condition of the onset at begining is similar to that of acute icteric hepatitis. However, the pathological condition develops quickly with the presence of fever, jaundice as well as distinct shrink of the liver, and hepatic coma in few days.

2. Characteristics of acute liver necrosis

In addition to those initial symptoms, the outstanding clinical manifestations are letharge or dysphoria, delirium, mental confusion, flutter-fibrillation, coma, convulsion and hemorrhage such as epistaxis, hematemesis, and mucocutaneous petechiae.

Besides, there may rapidly appear hydroperitoneum and hepatorenal syndrome. The patients often have distinct fector hepaticus.

3. Characteristics of subacute liver necrosis

The duration of subacute liver necrosis is longer than that (about 2 weeks) of the acute, and its clinical manifestations are milder. The main symptoms are jaundice, shrink of the liver, hydroperitoneum and disturbance of central nerve system. The jaundice is usually persistent and gets worse progressively.

Treatment Based on Differentiation of Syndromes

1. Steaming and fumigation of damp-heat

Symptoms and signs: Sallow body and face like orange colour, dysphoria with hotness sensation, stuffiness in epigastrium, poor appetite, nausea and vomiting, constipation or loose stool, yellow and sticky tongue coating, and wiry, slippery and rapid pulse or soft pulse.

Principle of treatment: Eliminating Damp-heat.

Recipe: *Yin Chen Hao Tang*(35) with modification.

Ingredients:

Herba Artemisiae Scopariae	30 g
Fructus Gardeniae	10 g
Radix et Rhizoma Rhei	6 g
Fructus Forsythiae	10 g
Radix Isatidis	15 g
Rhizoma Imperatae	30 g

Decoct the above ingredients in water for oral dose.

2. Stagnation of liver *Qi*

Symptoms and signs: Distention and pain in hypochondriac and intercostal region, fullness and distention in epigastrium, nausea, belching, poor appetite, light red tongue with thin coating, and wiry pulse.

Principle of treatment: Dispersing the stagnation of liver *Qi*.

Recipe: *Chai Hu Shu Gan San*(106) with modification.

Ingredients:

Radix Bupleuri	10 g
Fructus Aurantii	6 g
Radix Paeoniae Alba	10 g
Radix Curcumae	10 g
Radix Glycyrrhizae	3 g
Rhizoma Ligustici Chuanxiong	6 g
Rhizoma Cyperi	10 g
Fructus Citri Sarcodactylis	10 g

Decoct the ingredients in water for oral dose.

3. Invasion of the spleen by pathogenic damp

Symptoms and signs: Hypochondriac pain, fullness and distention in epigastrium and abdomen, nausea and vomiting, poor appetite, tastelessness with no desire of drinking, general heaviness sensation, loose stool, and yellow and sticky tongue coating.

Principle of treatment: Strengthening the spleen and resolving damp.

Recipe: *Wei Ling Tang*(148) with modification.

Ingredients:

Rhizoma Atractylodis	10 g
Cortex Magnoliae Officinalis	10 g
Pericarpium Citri Reticulatae	6 g
Fructus Alpiniae Oxyphyllae	12 g
Rhizoma Atractylodis Macrocephalae	10 g

Poria	10 g
Polyporus Umbellatus	6 g
Semen Plantaginis	
(wrapped for decocting)	15 g
Rhizoma Alismatis	6 g
Ramulus Cinnamomi	6 g
Radix Glycyrrhizae	3 g

Decoct the above ingredients in water for oral dose.

4. Deficiency of liver *Yin*

Symptoms and signs: Dull pain in hypochondriac region, low fever, soreness of the lumbus, dry and bitter mouth, feverish sensation in the palms and soles, red tongue with little or no coating, red prickles on the tip of the tongue, and wiry, thready and rapid pulse.

Principle of treatment: Tonifying *Yin* and nourishing the liver.

Recipe: *Yi Guan Jian*(129) with modification.

Ingredients:

Radix Glehniae	10 g
Radix Ophiopogonis	10 g
Radix Angelicae Sinensis	10 g
Radix Rehmanniae	12 g
Fructus Lycii	10 g
Fructus Meliae Toosendan	6 g
Herba Dendrobii	12 g

Decoct the above ingredients in water for oral dose.

5. Preponderance of toxic heat

Symptoms and signs: High fever, thirst, dysphoria, jaundice rapidly deepened, abdominal fullness and distention, constipation, deep yellow urine, or even coma, delirium, convulsion, or hematochezia, hematuria, deep red tongue with yellow and sticky or yellow and dry coating, and slippery and rapid pulse.

Principle of treatment: Clearing away toxic heat, cooling blood and saving *Yin*.

Recipe: *Xi Jiao Di Huang Tang*(25) with modification.

Ingredients:

Cornu Rhinoceri (to be ground into powder and taken with decoction)	3 g
Rhizoma Coptidis	6 g
Fructus Gardeniae	10 g
Radix Rehmanniae	15 g
Radix Isatidis	30 g
Herba Artemisiae Scopariae	30 g
Cortex Moutan Radicis	10 g
Herba Dendrobii	10 g

Decoct the above ingredients in water for oral dose.

11 Poisoning and Physicochemical Traumatic Diseases

Acute poisoning is one of commonly encountered diseases characterized by its abrupt onset and multiple changes. Without timely emergency treatment, it will bring on severe results or even fatal danger. Immediate diagnosis and treatment with combination of both Chinese and modern medicines are therefore vital.

Poisoning, in traditional Chinese medicine, is classified into five types: food poisoning, drug poisoning, alcoholism, poisonous insect and animal bite and poisoning caused by pestilential factors. This chapter will mainly deals with alcoholism, aconitine poisoning, dog button poisoning, etc. as well as physiocochemical traumatic diseases such as heliosis.

11.1 Alcoholism

Alcoholism generally refers to acute alcoholism caused by overdrinking of alcohol or accidental intake of certain liquid that contains alcohol.

Etiology and Pathogenesis

Alcohol is characterized by fieriness. When it goes into the stomach, it gives rise to distention of the stomach and upward perversion of *Qi*. When the rebellious *Qi* gose up to the chest, affecting the liver and gallbladder interiorly, it may cause stuffy chest and mental confusion, or even coma and dissociation of *Yin* and *Yang*, endangering one's life.

Essentials of Diagnosis

Alcoholic breathing, flushed face of pallor, accelerated pulse beating, palpitation, sweating, platycoria or miosis. At the initial stage, the patient may have drunk expression, nausea and vomiting, dysphoria, and in severe case, incontinence of micturition and bowel movement, convulsion, and coma, When the toxin goes into the lung affecting the canopy(the lung has the upmost location among the internal organs like a canopy), it will lead to suffocation and cause death.

The pulse is thready and rapid or full and rapid, the tongue is red, or deep red, or dark red with thin and yellow coating.

Treatment Based on Differentiation of Syndromes

1. Clearing the stomach by using emetics

(1) Recipe: *Gua Di San*(100) with modification.

Ingredients:

Pedicel of Cucumis Melo	15 g
Radix Glycyrrhizae	12 g
Radix Scrophulariae	30 g
Radix Sanguisorbae officinalis	15 g

Decoct the above ingredients in water for oral dose.

(2) Use 1: 2000 sodium bicarbonate solution for repeated gatric lavage.

(3) Automatic gastric lavage machine.

2. Eliminating the poisonous substance

(1)Removing the poisonous substance with disphoretics.

Recipe: *Cong Chi Tang*(174) with modification.

Ingredients:

Semen Sojae Praeparatum	30 g
Bulbus Allii Fistulosumi	15 g

Radix Glycyrrhizae　　　　　　　　　　　　　　　　15 g

Decoct the above ingredients in water for oral dose when it is warm.

(2) Removing the poisonous substance with purgatives.

Take *Da Huang Fen*(powder of Radix et Rhizoma Rhei) 12 g or *Yuan Ming Fen*(Natrii Sulfas) following its infusion.

(3) Removing the poisonous substance with diuretics.

① *Che Qian Cao*(Herba Plantaginis) 30 g and *Bai Mao Gen*(Rhizoma Imperatae) 30 g. Decoct the 2 above ingredients in water for oral dose.

② *Lu Dou Gan Cao Jie Du Tang*(149).

Ingredients:

Semen Phaseoli Radiati	90 g
Radix Glycyrrhizae	30 g
Radix Salviae Miltiorrhizae	15 g
Fructus Forsythiae	12 g
Herba Dendrobii	30 g
Radix et Rhizoma Rhei	9 g

Decoct the above ingredients in water for oral dose.

3. Regulating the middle *Jiao* and eliminating the poisonous substance.

Ingredients:

Radix Puerariae	30 g
Semen Phaseoli Radiati	30 g
Radix Scutellariae	15 g

Decoct the above ingredients in water for oral dose.

4. Other emergency managements:

(1) Care to preserve warmth for the patient and frequently change patient's posture.

(2) In case of dysphoria or over excitation, use librium or diazepam for muscular injection.

(3) In case of severe inhibition or coma, use respiratory stimulant, such as *Ma Qian Zi*(Semen Strychni) 0.5 fen (about 0.15 g) which is decocted in water for oral use. It has even better effect that ritalin. In case dyspnea, give oxygen therapy.

(4) In case of hypotention, use blood pressure-elevating drug. For dehydration, use *Sheng Mai San*(26). Besides, transfusion with combined Chinese and western medicines could be considered when necessary.

11.2 Aconite Poisoning

Aconite poisoning is aconitine poisoning. Aconitine refers to alkaloid of aconites (Radix Aconiti, Radix Aconiti Kusnezoffii, Radix Aconiti Praeparata). They are very poisonous. The clinical cases are usually resulted from accidental poisoning or overtaking of the drugs in treatment.

Etiology and Pathogenesis

The overtaking of the drug damages the anti-pathogenic *Qi* and causes attack of the heart by pathogenic toxin.

Essentials of Diagnosis

2-3 quaters after taking aconite, the patient will have numbness and hotness sensation of the mouth and tongue followed by vomiting of filth and salivation. In mild case, there may be dizziness, blurred vision, dysphoria, and in severe case, there may appear slurred speech, dysphagia, blepharoptosis, paralysis of limbs, shortness of breath, speeded heart beating, or even sudden convulsion, coma, cyanosis, platycoria, and respiratory failure, leading to death.

Treatment Based on Differentiation of Syndromes

1. Rhizoma Zingiberis Recens 30 g, Flos Lonicerae 30 g, Semen Phaseoli Radiati 90 g.

Decoct the above ingredients in water for oral dose.

2. Semen Phaseoli Radiati 90 g, Rhizoma Coptidis 9 g, Semen Sojae Nigrum 90 g, Herba Coriandrum 3 g, Radix Glycyrrhizae 30 g.

Decoct the ingredients in water for oral use 3 doses a day until all the symptoms subside and then 1 dose a day contineously for 1 week.

11.3 Dog button Poisoning

The main compositions of dog button are strychnine and brucine. The poisoning is usually caused by accidental use or over dosage in treatment.

Etiology and Pathogenesis

Ma Qian Zi(dog button), which bitter in flavour and cold in property, is mostly used externally or prepared into pills, and not decoction. Over dosage will disturb mental activities, leading to convulsion, or even dissociation of *Yin* and *Yang*, endangering one's life.

Essentials of Diagnosis

Few minutes after taking the drug, there first appear tremor, compression sensation in the chest and hypersensitivity, and then spasm of masseter and muscles of the neck, and sometimes nausea and vomiting. During the spasm, the patient may have consciousness with weeping and laughing without reason followed by staring of the eyes, gradual spasm of respiratory muscle, general cyanosis, and platycori. With the external stimulation by sound, light and wind, etc. the patient may have immediate tonic spasm that lasts for few minutes each time. The pulse is thready and the tongue dark purple.

Treatment Based on Differentiation of Syndromes

1. Immediate gastric lavage is necessary. Apply 30 ml of ginger juice mixed with 3 000 ml of clean water for repeated gastric lavage better with a manual gastric lavage machine. After gastric lavage is finished, perfuse the stomach with universal antidotic powder (activated carbon 10 g and magnesium sulfate 20 g) once.

2. High clean enema: Use about 500 ml of decoction of either *Da Huang* (Radix et Rhizoma Rhei) or *Fan Xie Ye* (Folium Cassiae) for high enema.

3. Orally take 30−100 ml of sesame oil to protect gastric mucosa and promote purgation.

4. Detoxification with drugs: Decoct *Huang Qin* (Radix Scutellariae) 30 g and *Gan Cao* (Radix Glycyrrhize) 15 g in water for oral use, and continue the management for several times.

5. In case of spasm or convulsion, *An Gong Niu Huang Wan* (20) could be used.

6. Other emergency management:

(1) Acupuncture: Needle point Hegu(LI4), Shuigou(DU26), etc.

(2) Muscular injection of diazepam for sedation.

(3) Give intravenous infusion or liver−protecting therapy according to the pathological conditions.

11.4 Heliosis

According to its clinical manifestations, heliosis is divided into *Shu Feng* (summer−heat spasm), *Shu Jue* (syncope due to summer−heat), *Shu Jing* (convulsion due to summer−heat) and *Yang*

Shu(summer-heat syndrome of *Yang* type), etc. In modern medicine, it is known as heat stroke, heat exhaustion, sunstroke and heat cramp.

Etiology and Pathogenesis

In hot summer days, the pathogenic summer-heat is liable to consume both *Qi* and *Yin*, giving rise to this disease. Especially those senile and asthenic patients whose *Yin* and body fluid are deficient due to declined spleen and stomach, and parturients whose delivery badly consumes *Yuan*(source) *Qi*, are more liable to be affected by summer-heat, resulting in the occurrence of the disease.

Treatment Based on Differentiation of Syndromes

1. Consumption of both *Qi* and *Yin* by summer-heat

Symptoms and signs: dizziness, mental restlessness, pale complexion, sweating, lassitude, shortness of breath, cold limbs, or even coma, red tongue with little coating, and thready, rapid and forceless pulse.

Principle of treatment: Tonifying *Qi* and nourishing *Yin*

Recipe: *Sheng Mai San*(26) with modification.

Ingredients:

Radix Pseudostellariae	60 g
Radix Panacis Quinquefolii	30 g
Radix Ophiopogonis	30 g
Radix Asparagi	30 g
Fructus Schisandrae	15 g
Radix Trichosanthis	12 g
Rhizoma Polygonati Odorati	18 g
Radix Paeoniae Alba	15 g
Radix Glycyrrhizae Praeparata	12 g

Decoct the above ingredients in water for oral dose.

2. Mist of the pericardium by summer—heat

Symptoms and signs: High fever, no sweating, red face and face and eyes, dry mouth, coarse breathing, thirst with much drinking, dysphoria, or even coma, red tongue with yellow coating, and thready and rapid or hidden pulse.

Principle of treatment: Clearing away summer—heat and cooling the *Ying* system for resuscitation.

Recipe: *Bai Hu Tang* (10) with modification.

Ingredients:

Rhizoma Anemarrhenae	12 g
Radix Rehmanniae	15 g
Gypsum Fibrosum	30 g
Semen Oxyzae Glutin Osae	30 g
Radix Scrophulariae	18 g
Rhizoma Coptidis	9 g
Herba Lophatheri	9 g
Herba Elsholtziae seu Moslae	12 g
Herba Agastachis	12 g
Exocarpium Citrulli	60 g

Decoct the above ingredients in water for oral dose.

12 Common Emergent Cases of Gynecology

Traditional Chinese medicine has its own characteristics. Only the most commonly seen emergency cases are to be introduced in this chapter. The rest could be managed with reference to Gynecology.

12.1 Dysmenorrhea

Dysmenorrhea is one of the most commonly seen diseases in gynecology. It corresponds to *Tong Jing*(painful menstruation) and *Jing Xing Fu Long*(abdominal pain during period) in traditional Chinese medicine.

Etiology and Pathogenesis

Stagnation of cold and damp, stagnation of *Qi* and blood, deficiency of *Qi* and blood and deficiency of the liver and kidney may all lead to disturbance in *Qi* and blood circulation, giving rise to dysmenorrhea.

Essentials of Diagnosis

There appear lower abdominal pain and lumbago before, during or after period. In severe case, the pain is intolerable.

Treatment Based on Differentiation of Syndromes

1. Stagnation of cold and stasis of blood

Symptoms and signs: Lower abdominal coldness and pain or colic pain before or during menstruation which is alleviated by warmth and aggravated by pressure, oligomenorrhea, dark red or purple menstrual flow with clots, accompanied by aversion to cold, loose stool, and cold limbs. The tongue is bluish purple with white and moist or sticky coating, and the pulse is deep and tense.

Principle of treatment: Warming the channel and activating blood circulation.

Recipe: *Dang Gui Si Ni Tang*(150) with modification.

Ingredients:

Radix Angelicae Sinensis	12 g
Ramulus Cinnamomi	9 g
Radix Paeoniae	12 g
Herba Asari	3 g
Caulis Aristolochiae Manshuriensis	6 g
Radix Glycyrrhizae Praeparata	9 g
Fructus Ziziphi Jujubae	10 pieces

Decoct the above ingredients in water and take the decoction orally when warm.

Modification: In case of prominence of cold, add Fructus evodiae 10 g, Rhizoma Zingiberis Recens 10 g, Folium Artemisiae Argyi 10 g and Fructus Foenicuii 10 g.

In case of cold limbs, add Radix Aconiti Praeparata 12 g and Cortex Cinnamomi 6 g.

In case of abdominal pain which is aggravated by pressure with blood clots in the flow, add Pollen Typhae 10 g and stir-heated Faeces Trogopterorum 10 g.

For general lassitude, pallor, poor appetite and loose stool, add Radix Codonopsis Pilosulae 12 g, stir-heated Rhizoma Atractylodis Macrocephalae 12 g, Poria 12 g and Rhizoma Pinelliae 12 g.

2. Stagnation of *Qi* and stasis of blood

Symptoms and signs: Lower abdominal distending pain before or during period which is aggravated by pressure, oligomenorrhea or uneven menstrual flow which is dark purple with clots, the pain is alleviated after the clots discharged. It is often accompanied with distention and pain in the breasts or intercostal and hypochondriac region. The tongue is dark pruple with stagnant spots on the border. The coating is thin. The pulse is deep and wiry or deep and hesitant.

Principle of treatment: Activating *Qi* and blood circulation, dispersing stasis of blood and arresting pain.

Recipe: *Xue Fu Zhu Yu Tang*(32) with modification.

Ingredients:

Radix Angelicae Sinensis	9 g
Radix Paeoniae Rubra	9 g
stir-heated Semen Persicae	9 g
Flos Carthami	9 g
Radix Cyathulae	9 g
Fructus Meliae Toosendan	9 g
Rhizoma Cyperi Praeparata	12 g
Herba Leonuri	15 g
Rhizoma Ligustici Chuanxiong	6 g
Radix Bupleuri	6 g
Fructus Aurantii	6 g
powder of Rhizoma Corydalis (divided into 2 portions to be taken following its infusion)	6 g
Radix Glycyrrhizae	3 g

Decoct the above ingredients in water for oral dose.

3. Deficiency of *Qi* and blood

Symptoms and signs: Dull pain in lower abdomen during or after period which is alleviated by warmth and pressure, oligomenorrhea, pale and dilute menstrual flow, pale complexion, spiritlessness, lassitude, feeble speech, dizziness, palpitation, pale tongue with thin and white coating, and deficient and thready pulse.

Principle of treatment: Tonifying *Qi* and Nourishing blood.

Recipe: *Shi Quan Da Bu Tang*(151) with modification.

Ingredients:

Radix Astragali seu Hedysari Praeparata	15 g
Radix Codonopsis Pilosulae	15 g
stir-heated Rhizoma Atractylodis Macrocephalae	9 g
Poria	9 g
Radix Angelicae Sinensis	9 g
stir-heated Radix Paeoniae Alba	9 g
Cortex Cinnamomi	5 g
Radix Glycyrrhizae Praeparata	5 g
Natrii Sulfas (divided into 2 portions to be taken following its infusion)	6 g
Rhizoma Zingiberis Recens	2 slices
Fructus Ziziphi Jujubae	4 pieces

Decoct the above ingredients in water for oral dose.

4. Deficiency of *Yin* of the liver and kidney

Symptoms and signs: Oligomenorrhea with pale menstrual flow, lower abdominal pain, accompanied by soreness and weakness of lumbus and knees, dizziness, tinnitus, pale tongue with thin and white coating, and deep and thready pulse.

Principle of treatment: Nourishing and tonifying the liver and kidney.

Recipe: *Tiao Gan Tang*(152) with modification.

Ingredients:

Radix Rehmanniae Praeparata	24 g
stir-heated Rhizoma Dioscoreae	30 g
Radix Angelicae Sinensis	12 g
Radix Paeoniae Alba	12 g
Fructus Corni	9 g
Radix Morindae Officinalis	9 g
Folium Artemisiae Argyi	9 g
Radix Glycyrrhizae	6 g
Colla Corii Asini (to be melted for use)	12 g

Decoct the above ingredients in water for oral dose.

Modification: In case of severe lumbago, add Cortex Eucommiae 15 g, Radix Dipsaci 15 g and Rhizoma Cibotii 15 g.

In case of hypochondriac pain, add Rhizoma Cyperi 12 g, Radix Curcumae 12 g and Fructus Meliae Toosendan 12 g.

In case of lower abdominal pain on the sides, add Fructus Foeniculi and Semen Citri Reticulatae.

Notes: Generally treatment should be given 3-5 days before period for ideal therapeutic effect and contineous treatments of 2-3 cycles are needed for consolidation of the result.

12.2 Acute Pelvic Inflammation

Acute pelvic inflammation belongs to *Re Ru Xue Shi*(invasion of the uterus by pathogenic heat), *Dai Xia Bing*(leukorrhea), *Fu Nu Zheng Jia*(womon's abdominal mass), *Jing Shui Bu*

Tiao(irregular menstruation), *Jing Xing Fu Tong*(dysmenorrhea), etc.

Etiology and Pathogenesis

This disease is resulted from accumulation of toxic heat in the uterus caused by invasion of noxious factors when the uterine collaterals are deficient due to constitutional asthenia, menstrual discharge or delivery.

Essentials of Diagnosis

1. Abrupt onset of fever and chillness complicated with lower abdominal pain and increase of vaginal discharge.

2. Gynecological examination reveals tension of abdominal wall, tenderness and rebounding pain on the lower abdomen, congestion of the uterine neck and vagina, elevating pain, slightly enlarged uterus with obvious tenderness and poor mobility, thickened appendages with tenderness and palpable mass. If there has formed abscess, there may be fullness and fluctation sensation in the posterior fornix with haphalgesia.

3. Increase of total account of white blood cells and neutrophil.

Treatment Based on Differentiation of Syndromes

This disease pertains to the syndrome of accumulation of toxic heat.

Symptoms and signs: Chillness and fever, thirst, lower abdominal pain which is aggravated by pressure, leukorrhea with yellow and purulent discharge, or bloody and stinking discharge, constipation, scanty and deep yellow urine, red tongue with yellow or yellow and sticky coating, and wiry and rapid or slippery and rapid pulse.

Principle of treatment: Clearing away toxic heat, activating blood circulation and dispersing blood stasis.

Recipe: *Yin Qiao Hong Teng Tang*(153).

Ingredients:

Flos Lonicerae	30 g
Fructus Forsythiae	30 g
Caulis Sargentodoxae	30 g
Herba Patriniae	30 g
Semen Coicis	12 g
Cortex Moutan Radicis	9 g
Fructus Gardeniae	12 g
Radix Paeoniae Rubra	12 g
Semen Persicae	12 g
Rhizoma Corydalis	9 g
Fructus Meliae Toosendan	9 g
Resina Olibani	4.5 g
Resina Commiphorae Myrrhae	4.5 g

Decoct the above ingredients in water for oral dose.

Modification: In case of severe abdominal pain aggravated by pressure, add Pollen Typhae, Faeces Trogopterorum, Rhizoma Sparganii, Rhizoma Zedoariae and Radix Salviae Miltiorrhizae 12 g each.

In case of constipation, add Radix et Rhizoma Rhei 10 g and Natrii Sulfas 6 g (to be taken following its infusion). In case of loose stool with hot and stinking discharge, add Rhizoma Coptidis 12 g, Radix Scutellariae 12 g and Radix Puerariae 12 g.

In case of stinking leukorrhea, add Herba Artemisiae Scopariae, Cortex Phellodendri and Cortex Ailanthi 12 g each.

When the pathogenic toxin has transmitted into the interior, giving rise to coma, delirium and convulsion, etc. *An Gong Niu*

Huang Wan or *Zi Xue Dan* (1 pill) could be used by oral feeding at the same time as an emergency management.

12.3　Functional Uterine Bleeding

Functional uterine bleeding corresponds to *Beng Lou*(uterine bleeding) in traditional Chinese medicine.

Etiology and Pathogenesis

Stagnation of *Qi*, stasis of blood, heat in the blood, deficiency of the spleen and deficiency of the kidney may all dysfunction of the *Chong* and *Ren* Channels in controlling blood, giving rise to uterine bleeding.

Essentials of Diagnosis

1. This disease mostly occurs during puberty and climacterium.

2. Its main clinical manifestation is different kinds of vaginal irregular bleeding.

3. Profuse bleeding or long-time dribbling may give rise to various symptoms of anema.

4. Gynecological examination: anovulation or dysfunction of the corpus luteum.

Treatment Based on Differentiation of Syndromes

1. Uterine bleeding due to stasis of blood

Symptoms and signs: Sudden onset of massive vaginal bleeding or contineous dribbling with dark purple flow and clots, lower abdominal pain which is aggravated by pressure and alleviated by discharge of blood clots, dark red tongue with stagnant spots on the tip and border of the tongue, thin and white tongue coating, and deep and tense or deep and hesitant pulse.

Principle of treatment: Activating blood circulation, dispersing blood stasis, regulating *Qi* and arresting pain.

Recipe: *Tao Hong Si Wu Tang* (154) plus *Shi Xiao San*(155) with modification.

Ingredients:

Radix Angelicae Sinensis	12 g
Radix Rehmanniae Praeparata	15 g
Rhizoma Ligustici Chuanxiong	6 g
Radix Paeoniae Alba	12 g
Semen Persicae	9 g
Flos Carthami	9 g
stir-heated Pollen Typhae	9 g
Faeces Trogopterorum	9 g
Rhizoma Cyperi Praeparata	12 g

Decoct the above ingredients in water for oral dose.

Modification: In case of massive bleeding, use carbonized Pollen Typhae instead of Pollen Typhae and add carbonized Radix Rubiae and carbonized Herba Cephalanoploris 12 g each.

2. Uterine bleeding due to heat in the blood

Symptoms and signs: Massive vaginal bleeding with deep red flow complicated with sticky dark purple clots, dry mouth, red face, thirst with desire of cold drink, dysphoria, red tongue with yellow coating, and full and rapid or slippery and rapid pulse.

Principle of treatment: Clearing away heat, nourishing *Yin*, cooling the blood and stopping bleeding.

Recipe: *Qing Re Gu Jing Tang*(156) with modification.

Ingredients:

Radix Rehmanniae	24 g
Cortex Lycii Radicis	9 g
Radix Scutellariae	9 g

stir-heated Fructus Gardeniae 9 g
Plastrum Testudinis Pareparata 15 g
clacined Concha Ostreae 24 g
Radix Sanguisorbae 12 g
Nodus Nelumbinis Rhizomatis 9 g
carbonized Cortex Trachycarpi 9 g
Colla Corii Asini 9 g
Radix Ophiopogonis 15 g
Radix Glehniae 15 g
Radix Glycyrrhizae 3 g

Decoct the above ingredients in water for oral dose.

3. Uterine bleeding due to stagnation of *Qi*

Symptoms and signs: Menorrhagia or oligomenorrhea, dribbling of vaginal bleeding, dark menstrual flow with blood clots, accompanied with fullness and distention in the chest and hypochondriac region, distending pain in the breasts and lower abdomen, poor appetite, vomiting of filthy food, sour regurgitation, headache, dark red tongue with thin and white or thin and yellow coating, and wiry pulse.

Principle of treatment: Regulating the *Qi* of the liver, resolving stasis of blood and stopping bleeding.

Recipe: *Ping Gan Kai Yu Zhi Xue Tang*(157) with modification.

Ingredients:
Radix Bupleuri 12 g
Radix Angelicae Sinensis 9 g
stir-heated Radix Paeoniae Alba 12 g
stir-heated Rhizoma Atractylodis
    Macrocephalae 12 g

Cortex Moutan Radicis	9 g
Radix Rehmanniae	12 g
vinegar–prepared Rhizoma Cyperi	15 g
Poria	10 g
carbonized Cortex Trachycarpi	9 g
carbonized Spica Schizonepetae	6 g
Herba Menthae	3 g
powder of Radix Notoginseng (to be taken following its infusion)	3 g
Radix Glycyrrhizae	6 g

Decoct the above ingredients in water for oral dose.

Modification: If there appear filthy vomiting and sour regurgitation, add Fructus Evodiae and Rhizoma Coptidis 10 g each.

4. Uterine bleeding due to damp–heat

(1) Damp is more prominent than heat.

Symptoms and signs: Massive uterine bleeding or persistent dribbling with sticky fluid, white and greasy leukorrhea, accompanied with facial edema or sallowish complexion, dizziness, distention and heaviness in the head, fullness and stuffiness in the chest and epigastrium, greasiness in mouth, poor appetite, general lassitude, heaviness or pain in the lumbus, loose stool, dysuria, white and sticky tongue coating, and soft and slippery pulse.

Principle of treatment: Eliminating damp, clearing heat, regulating menstruation and stopping bleeding.

Recipe: *Tiao Jing Sheng Yang Chu Shi Tang*(158) with modification.

Ingredients:

Rhizoma Atractylodis	12 g

Radix Astragali seu Hedysari Praeparata	12 g
Radix Bupleuri	9 g
Radix Angelicae Sinensis	6 g
Rhizoma Cimicifugae	6 g
Rhizoma seu Radix Notopterygii	3 g
Radix Angelicae Pubescentis	6 g
Radix ledebouriellae	6 g
Rhizoma Ligustici	6 g
Fructus Viticis	6 g
Radix Glycyrrhizae Praeparata	3 g

Decoct the above ingredients in water for oral dose.

(2) Heat is more prominent than damp.

Symptoms and signs: Massive uterine bleeding or persistent dribbling, stinking, sticky and purplish red menstrual flow, accompanied with fever, spontaneous sweating, greasy and red complexion, bitterness and greasiness in mouth, thirst with no desire of drinking, mental restlessness, insomnia, hotness and pain in the lower abdomen which is aggravated by pressure, constipation or hesitant loose stool, difficult urination with deep yellow urine, deep red tongue with yellow and sticky coating, and slippery and rapid pulse.

Principle of treatment: Clearing away heat, excreting damp, regulating menstruation and stopping bleeding.

Recipe: *Huang Lian Jie Du Tang* (114) with modification.

Ingredients:

Radix Rehmanniae	15 g
Rhizoma Coptidis	6 g
Radix Scutellariae	9 g

Fructus Gardeniae	9 g
Cortex Phellodendri	6 g
carbonized Pollen Typhae	6 g
carbonized Folium Artemisiae Argyi	9 g

Decoct the above ingredients in water for oral dose.

5. Uterine bleeding due to deficiency of the spleen

Symptoms and signs: Massive uterine bleeding or persistent dribbling with light red and dilute menstrual flow, accompanied with pale complexion or facial edema, deficiency of *Qi* with dislike of speaking and lassitude, poor appetite, abdominal distention, loose stool, cold limbs, light red swollen tongue with tooth marks on the border, thin and moist or sticky tongue coating, and thready and forceless pulse.

Principle of treatment: Strengthening the spleen *Qi*, nourishing blood and stopping bleeding.

Recipe: *Gu Ben Zhi Beng Tang* (159) with modification.

Ingredients:

Radix Codonopsis Pilosulae	30 g
Radix Astragali seu Hedysari	24 g
Rhizoma Atractylodis Macrocephalae	15 g
Radix Rehmanniae Praeparata	15 g
Radix Polygoni Multifori	15 g
carbonized Radix Rubiae	9 g
carbonized Cortex Trachycarpi	12 g
calcined Os Draconis	24 g
calcined Concha Ostreae	24 g
baked Rhizoma Zingiberis	1.5 g
Radix Glycyrrhizae Praeparata	6 g

Decoct the above ingredients in water for oral dose.

Modification: In case of sudden massive uterine bleeding accompanied by dizziness, palpitation, grayish pale face and lips, cold limbs, feeble breathing, or even coma, and fading or large superficial and rootless pulse, *Du Shen Tang* should be used immediately: Decoct *Hong Shen*(Radix Ginseng Rubra) 30 g for oral use 2—3 times a day.

6. Uterine bleeding due to deficiency of the kidney

(1) Deficiency of kidney *Yang*

Symptoms and signs: Profuse uterine bleeding or contineous dribbling with light red menstrual flow, lower abdominal coldness and pain which are alleviated by hot compress, accompanied with dark face, spiritlessness, aversion to cold, cold limbs, soreness and weakness of the lumbus and knees, clear and long urination, loose stool, pale tongue with thin and white coating, and deep and thready pulse.

Principle of treatment: Warming and promoting kidney *Yang*, regulating menstruation and stopping bleeding.

Recipe: *You Gui Wan*(134) with modification.

Ingredients:

Radix Rehmanniae Praeparata	24 g
Rhizoma Dioscoreae	24 g
Fructus Corni	12 g
Fructus Lycii	12 g
Cortex Eucommiae	12 g
Semen Cuscutae	18 g
Radix Dipsaci	12 g
Colla Corii Asini	
(to be melted for use)	10 g

carbonized folium
   Artemisiae Argyi   12 g
   Radix Aconiti Praeparata   6 g
   Rhizoma Zingiberis Praeparata   3 g
   Os Sepiella seu Sepiae   12 g

Decoct the above ingredients in water for oral dose.

(2) Deficiency of kidney *Yin*

Symptoms and signs: Scanty or dribbling menstrual flow which is bright red in colour, accompanied by dizziness, tinnitus, mental restlessness with hotness sensation in the chest, palms and soles, insomnia, night sweating, soreness and weakness of the lumbus and knees, or heel pain, red tongue with little or no coating, and thready, rapid and forceless pulse.

Principle of treatment: Nourishing the kidney *Yin* and tonifying the kidney essence to astringe bleeding.

Recipe: *Liu Wei Di Huang Tang*(118) with modification.

Ingredients:

Radix Rehmanniae	30 g
Rhizoma Dioscoreae	30 g
Fructus Corni	12 g
Cortex Moutan Radicis	9 g
Poria	12 g
Rhizoma Alismatis	9 g
Fructus Ligustri Lucidi	12 g
Herba Ecliptae	12 g
Fructus Lycii	12 g
Colla Corii Asini	
(to be melted for use)	10 g
Plastrum Testudinis Praeparata	15 g

Herba Agrimoniae 30 g
Decoct the above ingredients in water for oral dose.

13 Common Emergent Cases of Paediatrics

Paediatrics has developed into a specialized branch in traditional Chinses medicine long time ago. Each stage of growth and development of children has its own characteristic in physiology, pathology and so does the treatment based on differentiation of syndrome. The corresponding treatment should be given according to differentiation. Only the most commonly observed emergency cases in the clinic are introduced in the chapter. The rest could be managed with reference to the book Paediatrics.

13.1 Neonatal Pneumonia

Neonatal pneumonia corresponds to *Ru Zi Ke Chuan*(nursling asthmatic cough) in traditional Chinese medicine.

Etiology and Pathogenesis

This disease is caused by fetal toxicosis interiorly of invasion of wind—cold or wind—heat exteriorly while the antipathogenic *Qi* is too weak to dispel the pathogens.

Essentials of Diagnosis

1. It may occur in all the four seasons, but more in winter and spring.

2. It is characterized by such manifestations as fever, cough, shortness of breath, nares flaring and very small rales or crepitant rales in the lung.

3. X—ray examination shows mist—dots or mist—flake shadow on the lobes of lung.

4. The total number of lercocytes is increased in bacterial pneumonia, and that of leucocytes and neutrophils in viral pneumonia is close to the normal or decreased.

Treatment Based on Differentiation of Syndromes

1. Invasion of the lung by wind—cold

Symptoms and signs: Chillness, fever, cough, no sweating, nares flaring, no thirst, white and dilute expectoration, normal tongue with thin and white or white and sticky coating, bluish red finger prints, especially at the wind pass, and superficial, tense and rapid pulse.

Principle of treatment: Relieving the exterior by dispersing the lung.

Recipe: *San Ao Tang*(160) with modification.

Ingredients:

Herba Ephedrae Praeparata	3 g
Semen Armeniacae Amarum	6 g
Radix Glycyrrhizae	3 g
Radix Peucedani	6 g
Radix Ledebouriellae	3 g
Radix Platycodi	6 g
Fructus Perillae	3 g
Semen Raphani	3 g
Semen Sinapis Albae	2 g

Decoct the above ingredients in water and divide the decoction into 2 portions for oral use.

Modification: In case of cold in both the interior and exterior, add Herba Asari 1 g.

2. Invasion of the lung by wind—heat

Symptions and signs: Fever, no chillness, cough, shortness of breath, redness in the throat, red face, thirst, red tongue with thin and yellow coating, and superficial and rapid pulse.

Principle of treatment: Relieving the exterior with the drugs pungent in flavour and cool in property, dispersing the lung and resolving the phlegm.

Recipe: *Ma Xing Shi Gan Tang*(8) with modification.

Ingredients:

Herba Ephedrae Praeparata	3 g
Semen Armeniaecae Amarum	6 g
Gypsum Fibrosum (bo be decocted first)	15 g
Radix Glycyrrhizae	3 g
Radix Scutellariae	6 g
Herba Houttuyniae	9 g
Fructus Forsythiae	6 g
Fructus Trichosanthis	9 g
Radix Isatidis	9 g
Radix Platycodi	6 g
Flos Lonicerae	9 g

Decoct the above ingredients in water and divide the decoction into 2 portions for oral use.

Modification: In case of high fever with dysphoria, add Rhizoma Coptidis 1 g, and for upward attack of phlegm—heat, add powder of Cornu Antelopis 0.5 g and Ramulus Uncariae Cum Uncis 3 g.

3. Deficiency of heart *Yang*

Symptoms and signs: Pale complexion, cyanosis of the lips, rapid asthmatic breathing, cold limbs, mental restlessness, deficient and rapid or fading pulse.

Principle of treatment: Warming and tonifying the heart *Yang*, dispersing the *Qi* of the lung.

Recipe: *Shen Fu Tang*(29) with modification.

Ingredients:

Radix Ginseng	2 g
Radix Aconiti Praeparata	2 g
Rhizoma Zingiberis	2 g
Semen Armeniacae Amarum	9 g
Radix Glycyrrhizae	3 g

Modification: In case of cyanotic face, lips and tongue due to deficiency of heart *Yang*, add Radix Salviae Miltiorrhizae 6 g, Radix Angelicae Sinensis 3 g and Flos Carthami 2 g.

In case of profuse sweating with drop of the blood pressure, add calcined Os Draconis 30 g and calcined Concha Ostreae 30 g.

4. Deficiency of anti-pathogenic *Qi* with remained pathogenic factors.

(1) Deficiency of *Yin* with lingering of pathogenic factors

Symptoms and signs: Lower fever, dry and red lips, hidrosis, thirst, dry cough with little sputum, red and glossy tongue with little coating, and thready and rapid pulse.

Principle of treatment: Nourishing *Yin* and Clearing away heat from the lung.

Recipe: *Sha Shen Mai Dong Tang*(5) with modification.

Ingredients:

Radix Glehniae	9 g
Radix Ophiopogonis	6 g
Rhizoma Polygonati Odorati	6 g
Folium Mori	6 g
Radix Trichosanthis	9 g

Cortex Lycii Radicis	6 g
Cortex Mori Radicis	6 g
Radix Stemonae Praeparata	6 g
Fructus Schisandrae	3 g

Decoct the above ingredients in water and divide the decoction into 2 portions for oral use.

Modification: If dry cough is more prominent, add Fructus Chebulae 3 g.

In case of severe choking cough, add Cortex Lycii Radicis 12 g and Cortex Mori Radicis 15 g.

In case of obvious low fever, the above ingredients could be used together with *Qing Hao Bie Jia Tang*(143).

(2) Deficiency of *Qi* of the lung and spleen

Symptoms and signs: Low fever up and down, spiritlessness and lassitude, pallor, hidrosis, feeble cough, poor appetite, loose stool, pale tongue with white and slippery coating, and thready pulse.

Principle of treatment: Tonifying the *Qi* of the spleen.

Recipe: *Liu Jun Zi Tang*(161) with modification.

Ingredients:

Radix Codonopsis Pilosulae	9 g
Rhizoma Atractylodis Macrocephalae	6 g
Poria	6 g
Radix Glycyrrhizae	3 g
Pericarpium Citri Reticulatae	3 g
Rhizoma Pinelliae	3 g
Radix Stemonae Praeparata	6 g
Fructus Schisandrae	3 g

Radix Ophiopogonis　　　　　　　　　　6 g
Radix Platycodi　　　　　　　　　　　　6 g

Decoct the above ingredients in water and dividedthe decoction into 2 portions for oral use.

Modification: In case of hidrsis due to deficiency, add Radix Astragali seu Hedysari 12 g.

In case of poor appetite, add Massa Fermentata Medicinalis 6 g, Fructus Crataegi 6 g or Fructus Hordei Germinatus 6 g.

In case of severe cough, add Radix Asteris 6 g and Flo Farfarae 6 g.

13.2　Neonatal Septicemia

Neonatal septicemia corresponds to *Xue Du Nei Xian*(sinking of pathogenic toxin into the interior) or *Bao Du Zou Huang*(herpes toxin affecting the blood) in traditional Chinese medicine.

Etiology and Pathogenesis

Owing to the delicacy of the *Zang Fu* organs, immaturity of the *Qi* and body, and deficiency of the defensive system, the newborn is liable to be affected by toxic factors which tend to enter the nutrient blood, turns into heat and fire, that invade the pericardium, giving rise to collapse of antipathogenic *Qi*.

Essentials of Diagnosis

1. Spiritlessness, bluish gray complexion, normal body temperature, or high fever, dysphoria, coma and convulsion.

2. Jaundice progressively getting worse, enlargement of the liver and spleen, no eating, no crying, or scattered bleeding spots on the skin.

3. Damage of the umbilicus, or sken membrane, and puru-

lent focus on the oral mucous.

4. The culture of blood bacterium in positive reaction, and increase or decrease of white blood cells.

Treatment Based on Differentiation of Syndromes

1. Hyperactivity of pathogenic toxin

Symptoms and signs: Abrupt onset with high fever, dysphoria, jaundice, enlargement of the liver and spleen, or even coma, convulsion, sweating, constipation, yellow urine, red tongue with yellow coating, and thready and rapid pulse.

Principle of treatment: Clearing away heat, eliminating toxin and cooling the blood.

Recipe: *Huang Lian Jie Du Tang*(114) with modification.

Ingredients:

Rhizoma Coptidis	3 g
Radix Scutellariae	6 g
Cortex Phellodendri	3 g
Flos Lonicerae	9 g
Fructus Forsythiae	9 g
Radix Rehmanniae	6 g
Radix Paeoniae Rubra	6 g
Cortex Moutan Radicis	6 g
Radix Glycyrrhizae	1.5 g

Decoct the above ingredients in water to be taken orally in two times.

Modification: In case of coma, take *Zhi Bao Dan* with hot water.

In case of convulsion, take *Zi Xue Dan* with hot water.

2. Deficiency of *Qi* and blood

Symptoms and signs: Gradual onset of the disease, sallow or

bluish grayish face, spiritlessness, no eating, no crying, feeble breathing, body temperature failing to rise, cold limbs, cold sweating on the forehead, scattered bleeding spots on the body, pale tongue with thin and white coating, and thready and forceless pulse.

Principle of treatment: Tonifying *Qi*, warming *Yang*, strengthening the body resistance and eliminating pathogenic factors.

Recipe: *Shen Fu Tang*(29) plus *Si Ni Tang*(27) with modification.

Ingredients:
Radix Ginseng	2 g
Radix Astragali seu Hedysari	6 g
Radix Angelicae Sinensis	6 g
Radix Aconiti Praeparata	1 g
Rhizoma Zingiberis	1 g
Radix Glycyrrhizae	1.5 g

Decoct the above ingredients in water to be taken orally in two times.

Strengthen the anti-pathogenic *Qi* first when necessary and use the drugs to eliminate toxic heat when the *Yang Qi* is recovered.

13.3 Toxic Dysentery

Toxic dysentery corresponds to *Yi Li*(fulminant dysentery) in traditional Chinese medicine.

Etiology and Pathogenesis

This disease is mainly caused by insanitary food-intake which results in invasion of damp-heat and pestilential toxin via mouth, fighting with anti-pathogenic *Qi* in the stomach and intestine.

Essentials of Diagnosis

1. It is more common in spring and autumn.
2. It is more liable to affect the children aged from 2 to 7.
3. Abrupt onset with rapid change and sudden rise of body temperature up to 40 °C.
4. In severe case, there may appear coma, convulsion, piebald skin and signs of cerebral edema.
5. Laboratory examination: The result of stool culture is positive, Enteroscope could find a great number of pus cells, red and white blood cells, phagocytes and increased number of leukocytes.

Treatment Based on Differentiation of Syndromes

1. Invasion of the interior by excessive heat

Symptoms and signs: Sudden chillness and fever, dysphoria, delirium, thirst, infantile convulsion, stinking stool complicated with pus and blood, short and deep yellow urination, red tongue with yellow and sticky coating, and slippery and rapid pulse.

Principle of treatment: Clearing away toxic heat and dispelling wind for resuscitataion.

Recipe: *Ge Gen Qin Lian Tang*(162) with modification.

Ingredients:

Radix et Rhizoma Rhei	2 g
Rhizoma Coptidis	5 g
Radix Scutellariae	6 g
Cortex Phellodendri	6 g
Fructus Gardeniae	5 g
Radix Puerariae	9 g

Flos Lonicerae	12 g
Cortex Moutan Radicis	6 g
Radix Paeoniae Rubra	9 g

Decoct the above ingredients in water for oral dose.

Modification: If there appear pus and blood in the stool, add Radix Pulsatillae 9 g and Herba Euphorbiae Humifusae 15 g.

In case of coma, add *Zi Xue Dan* to be taken with the decoction.

If the toxic heat is too severe to be excreted as dysentery, add *Yuan Ming Fen*(Natrii Sulfas) 6 g.

2. The syndrome with tenseness in the interior and flaccidness in the exterior (unconsciousness with external prostration)

Symptoms and signs: In addition to sudden unconsciousness, there also appear pallor, cold limbs, cold sweating, and thready fading pulse.

Principle of treatment: Recuperating depleted *Yang* and rescuing the patient from collapse.

Recipe: *Si Ni Tang* plus *An Gong Niu Huang Wan*(20)
Ingredients:

prepared Aconiti Praeparata	3 g
Rhizoma Zingiberis	5 g
Radix Glycyrrhizae	2 g

Modification: If the external collapse of *Yang Qi* is the Principal, treatment should be given to restore *Yang* and save the collapse first without application of *An Gong Niu Huang Wan*. After the *Yang* is restored, use the drugs cool in property to clear away heat and promote resuscitation. Besides, acupuncture could be used in combination.

14 Acute Diseases of Ear, Nose and Throat

Treatment of diseases of ear, nose and throat by Chinese medicine has be specialized since long time ago with it own characteristics and remarkable therapeutic results. In this chapter, only the most commonly seen acute diseases are introduced, and the rest could be managed with reference to the book Otorhinolaryngology.

14.1 Acute Nasosinusitis

Acute nasosinusitis corresponds to *Bi Yuan* (rhinorrhea with turbid discharge) in traditional Chinese medicine.

Etiology and Pathogenesis

Hyperactivity of heat and fire leads to excessive heat in the lung, gallbladder and spleen channels, giving rise to the disease.

Essentials of Diagnosis

1. This disease is characterized by nasal obstruction, profuse nasal discharge, accompanied by headache, fever and hyposmia.

2. Local examination: Redness and swelling of nasal conchae, and purulent nasal discharge in the nose.

Treatment Based on Differentiation of Syndromes

1. Internal treatment:

(1) Invasion of the lung channel by wind—heat

Symptoms and signs: Nasal obstruction, hyposmia, profuse yellowish white and sticky nasal discharge, pain in the forehead and zygoma, accompanied by fever, chillness, cough and dry mouth, red tongue with thin and white coating, and superficial and rapid or superficial, slippery and rapid pulse.

Principle of treatment: Clearing heat, dispelling wind and removing obstruction from the nose with aromatics.

Recipe: *Can Er Zi San*(163) with modification.

Ingredients:

Radix Angelicae Dahuricae	6 g
Herba Menthae	9 g
Flos Magnoliae	6 g
Fructus Xanthii	9 g
Radix Scutellariae	9 g
Flos Chrysanthemi	12 g
Fructus Forsythiae	30 g
Radix Puerariae	12 g

Decoct the above ingredients in water for oral dose.

Modification: In case of severe headache, add Rhizoma Ligustici 12 g for vertex headache, Fructus Viticis 15 g for superaorbital headache, and Radix Bupleuri 12 g for temporal headache.

In case of profuse yellow and turbid nasal discharge, add Semen Benincasae 30 g, Semen Trichosanthis 30 g and Herba Hoyttuyniae 30 g.

For profuse sputum, add stir-heated Semen Armeniacae Amarum 12 g, Radix Platycodi 12 g, Semen Trichosanthis 30 g and Semen Benincasae 30 g.

(2) Accumulation of heat in the gallbladder channel

Symptoms and signs: Nasal obstruction, Hyposmia, profuse yellow, purulent and stinking nasal discharge, severe pain in the forehead, ortemporal region or zygomatic region, accompanied by fever, bitter taste in mouth, dry throat, dizziness, tinnitus, deafness, red tongue with yellow coating, and wiry and rapid pulse.

Principle of treatment: Clearing heat from the gallbladder, excreting dampness and removing obstruction from the nose.

Recipe: *Long Dan Xie Gan Tang*(97) with modification.

Ingredients:

Radix Gentianae	9 g
Fructus Gardeniae	9 g
Radix Scutellariae	9 g
Radix Bupleuri	12 g
Rhizoma Alismatis	9 g
Caulis Aristolochiae Manshuriensis	6 g
Semen Plantaginis	
(to be wrapped for decocting)	30 g
Radix Angelicae Sinensis	12 g
Radix Rehmanniae	15 g
Folium Mori	12 g
Flos Chrysanthemi	12 g
Radix Glycyrrhizae	3 g

Decoct the above ingredients in water for oral dose.

Modification: In case of severe nasal obstruction, add Radix Angelicae Dahuricae10 g, Flos Magnoliae 10 g, Herba Menthae 10 g and Fructus Xanthii 10 g.

(3) Accumulation of damp—heat in the spleen channel

Symptoms and signs: Severe nasal obstruction, anosmia, profuse and persistent yellow, sticky and turbid nasal discharge, accompanied by severe heaviness, distention and pain of the head, epigastric and abdominal fullness and distention, heaviness of the limbs, poor appetite, yellow urine, red tongue with yellow and sticky rongue coating, and slippery and rapid or soft pulse.

Principle of treatment: Clearing away heat from the spleen and eliminating turbid damp.

Recipe: *Huang Qin Hua Shi Tang*(164) with modification.

Ingredients:

Radix Scutellariae	9 g
Talcum	24 g
Caulis Aristolochiae Manshuriensis	6 g
Fructus Forsythiae	30 g
Poria	12 g
Polyporus Umbellatus	9 g
Pericarpium Arecae	9 g
Semen Amomi Cardamomi (smashed)	6 g
Rhizoma Acori Graminei	12 g
Herba Agastachis	9 g

Decoct the above ingredients in water for oral dose.

2. External treatment:

(1) Nasal drip: Use *Di Bi Ling* made of *E Bu Shi Cao* (Herba Centipedae) and *Xin Yi*(Flos Magnoliae), or *Xin Yi Di Bi Ji*(nasal drops made of Flos Magnoliae) or 25% *Mu Dan Pi Ye*(medicinal liquid made of Cortex Moutan Radicis) 1–2 drops each time and 3–4 times a day.

(2) Nasal insufflation: Insufflate small amount of *Bing Lian San*(mixed powder of Huang Lian–Rhizoma Coptidis, *Xin Yi Hua*–Flos Magnoliae and *Bing Pian*–Borneolum Syntheticum) into the nasal cavity, 3–4 times a day.

3. Acupuncture treatment:

(1) Invasion of the lung channel by wind-heat

Puncture point Yintang(EX-HN3), Yingxiang(LI20), Hegu(LI4), Fenglong(ST40), Tongtian(BL7) and Lieque(LU7) with selection of 2-3 points each time. Strong stimulation is advisable and needles should retained in the points for 10-15 minutes. Besides, injection of 0.5 ml of *Yu Xing Cao Zhu She Ye*(Injection of Herba Houttuyniae) into point Feishu(UL13) is also effective.

(2) Accumulation of heat in the gallbladder channel

Puncture point Fengchi(GB20), Yanglingquan(GB34), Xuanzhong(GB39), Naokong(GB19) and Xingjian(LR2) with strong stimulation, and the needles are retained in the points for 10-15 minutes.

(3) Accumulation of damp-heat in the spleen channel

Puncture point Yingxiang(LI20), Yintang(EX-HN3), Shangxing(DU23), Hegu(LI4), Zanzhu(BL2), Tongtian(BL7), Fengchi(GB20) and Zusanli(ST36) with strong stimulation, and the needles are retained in the points for 10-15 minutes.

14.2 Acute Tonsillitis

Acute tonsillitis corresponds to *Feng Re Ru E*(tonsillitis caused by pathogenic wind-heat) in traditional Chinese medicine.

Etiology and Pathogenesis

Invasion of pathogenic wind-heat stirs up the fire-heat in the lung and stomach, which goes up to affect the throat, giving rise to this disease.

Essentials of Diagnosis

1. This disease takes sorethroat and dysphagia as its main manifestations.

2. Local examination: Congestion and swelling of the throat, red and swollen tonsil with possible yellowish white pus spots or curdy flakes which are easy to scrape off.

Treatment Based on Differentiation of Syndromes

1. Internal treatment:

(1) Invasion of the lung by wind—heat

Symptoms and signs: Dryness and burning pain in the throat, redness and swelling of the tonsil and its surrounding tissues, accompanied by fever, chillness, headache, cough, red tip and border of the tongue, thin and white or slightly yellow coating, and superficial and rapid pulse.

Principle of treatment: Dispelling wind, clearing heat, eliminating toxin and removing swelling.

Recipe: *Shu Feng Qing Re Tang*(165) with modification.

Ingredients:

Herba Schizonepetae	9 g
Radix Ledebouriellae	12 g
Flos Lonicerae	30 g
Fructus Forsythiae	12 g
Radix Scutellariae	12 g
Radix Paeoniae Rubra	12 g
Radix Scrophulariae	15 g
Bulbus Fritillariae Thunbergii	9 g
stir—heated Fructus Arctii	12 g
Radix Trichosanthis	12 g
Radix Platycodi	12 g
Cortex Mori Radicis	12 g
Radix Glycyrrhizae	3 g

Decoct the above ingredients in water for oral dose.

(2) Excessive heat in the lung and stomach

Symptoms and signs: Severe sorethroat radiating to the ear root and mandibular region, dysphagia, accompanied by high fever, thirst, foul breathing, sticky and yellow expectoration, abdominal distention, constipation, red tongue with yellow and sticky tongue coating, and full and rapid pulse.

Principle of treatment: Clearing away heat, comforting the throat, eliminating Toxin and removing swelling.

Recipe: *Qing Yan Li Ge Tang*(166) with modification.

Ingredients:

Herba Schizonepetae	6 g
Radix Ledebouriellae	9 g
Herba Menthae	9 g
Fructus Gardeniae	9 g
Radix Scutellariae	12 g
Flos Lonicerae	30 g
Fructus Forsythiae	12 g
Rhizoma Coptidis	6 g
stir−heated Fructus Arctii	12 g
Radix Platycodi	12 g
Radix Scrophulariae	18 g
Radix Rehmanniae (to be decocted later)	6−9 g
Natrii Sulfas (to be melted)	9 g

Decoct the above ingredients in water for oral dose.

Modification: In case of pain of mandibular lymph node, add Rhizoma Belamcandae 12 g, Fructus Trichosanthis 12 g and Bulbus Fritillariae Cirrhosae 12 g.

In case of high fever, add Gypsum Fibrosum 12 g and

Concretio Silicea Bambusae 12 g.

In case of redness and swelling of the tonsil, suck *Liu Shen Wan* 5—10 pills each time and 3 times a day.

2. External treatment:

(1) Insufflate medicinal powder onto the tonsil 3—5 times a day. In case of mild redness, swelling and pain of the throat, use *Bing Peng San*, and in severe case with purulent spots, use *Zhu Huang San*. In case of severe sorethroat with remarkable gerenal symptoms, or profuse purulent secretion, or repeated attacks, use *Xi Lei San*.

(2) Gargling: Decoct Herba Schizonepetae, Flos Chrysanthemi, Flos Lonicerae and Radix Glycyrrhizae in water for gargling 3—5 times daily.

(3) Sucking: Suck *Tie Lei Wan* or *Run Hou Wan* in mouth 1 pill each time and several times a day. Let the pill melt gradually and swallow it slowly.

3. Acupuncture treatment:

Points: Hegu(LI4), Neiting(ST44), Quchi(LI11), Tiantu(RN22), Shaoze(SI1) and Yuji(LU10).

Puncture 3—4 points of the above with strong stimulation. Needles are retained in the points for 10—15 minutes and treatment should be given 1—2 times a day. Besides, it is also advisable to select point Pishu(BL20), Quchi(LI11) and the point 5 *Fen* medial to point Jianjing(GB21) for injection with 2ml of either *Yu Xing Cao Zhu She Ye*(Injection of Herba Houttuyniae) or *Chai Hu Zhu She Ye*(Injection of Bupleuri), once a day.

14.3 Acute Purulent Otitis Media

Acute purulent otitis media is one of the most commonly encountered acute diseases of the ear. It corresponds to *Ji Xing Nong Er* (otitis media suppurativa) in traditional Chinese medicine.

Etiology and Pathogenesis

This disease is resulted from hyperactivity of the fire in the liver and gallbladder and invasion by exogenous pathogenic wind, heat and damp.

Essentials of Diagnosis

1. This disease is characterized by acute perforation of the dar drum, aural discharge fo pus, and hypoacusis. It may also accompanied by general symptoms such as chillness and fever.

2. Local examination: Hyperemia of the ear drum, perforation of the tense part of tympanic membrane, secretion of yellow. sticky and shining pus, and conduction deafness.

Treatment Based on Differentiation of Syndrome

1. Internal treatment:

(1) Preponderance of wind—heat in the interior

Symptoms and signs: Stuffiness, distention and obstruction in the ear, accompanied by fever, chillness, headache, nasal obstruction and discharge. In infants, there may be dysphoria, cry, or even nausea and vomiting, high fever, convulsion, coma and neck rigidity. The tongue is red with thin and white or slightly yellow coating, and the pulse superficial and rapid.

Principle: Dispelling wind and heat, eliminating toxin, and removing swelling.

Recipe: *Yin Qiao San*(2) with modification.

Ingredients:

Flos Lonicerae	30 g
Fructus Forsythiae	15 g

Herba Schizonepetae	9 g
Herba Menthae	9 g
Flos Chrysanthemi	12 g
Herba Taraxaci	15 g
Herba Violae	15 g
Radix Platycodi	12 g
Radix Glycyrrhizae	6 g
stir-heated Fructus Arctii	12 g
Squama Manitis	10 g
Spina Gleditsiae	9 g

Decoct the above ingredients in water for oral dose.

Modification: In case of nausea and vomiting, high fever and convulsion in infant, add Concretio Silicea Bambusae), Caulis Bambusae in Taen, Ramulus Uncariae cum Uncis, Periostracum Cicadae and Radix Paeoniae Alba 10 g each.

In case of coma and neck rigidity, treatment should be given to clear away heat from the *Ying* system, cool the blood, and eliminate toxin for resuscitation with reference to Chapter 1 First-aid for Imminent Symptoms and Signs.

(2) Accumulation of damp-heat in the liver and gallbladder

Symptoms and signs: This syndrome mostly occurs to adults with such manifestations as severe pain, hypoacusis, headache, restlessness, red face and eyes, bitter mouth, dry throat, deep yellow urine, constipation, red tongue border, yellow and sticky tongue coating, and wiry and rapid pulse.

Principle of treatment: Clearing away damp-heat from the liver and gallbladder, eliminating toxin and dispelling pus.

Recipe: *Long Dan Xie Gan Tang*(97) with modification.

Ingredients:

Radix Gentianae	9 g
Charred Fructus Gardeniae	9 g
Radix Scutellariae	12 g
Radix Bupleuri	12 g
Semen Plantaginis (to be wrapped for decocting)	12 g
Caulis Aristolochiae Manshuriensis	6 g
Rhizoma Alismatis	12 g
Radix Rehmanniae	15 g
Radix Angelicae Sinensis	12 g
Rhizoma Anemarrhenae	12 g
Cortex Phellodendri	9 g
Fructus Xanthii	6 g
Flos Magnoliae	9 g
Radix Glycyrrhizae	6 g

Decoct the above ingredients in water for oral dose.

Modification: At the initial stage, add Squama Manitis 10 g and Spina Gleditsiae 10 g.

Where there appears secretion of pus, add Radix Angelicae Dahuricae, Radix Platycodi, Radix Astragali seu Hedysari and Rhizoma Ligustici Chuanxiong 12 g each.

For constipation, add Radix et Rhizoma Rhei 10 g and Natrii Sulfas 6 g.

2. External treatment:

(1) Clear away the pus from the auditory canal and clean it with hydrogen dioxide solution.

(2) Ear drop: Use *Huang Lian Di Er Ye*(Ear Drop of Rhizoma Coptidis), or *Yu Xing Cao Ye*(Solution of Herba Houttuyniae) or *Yin Hua Zhu She Ye*(Injection of Flos

Loniceras) to drop the ear 2—3 drops each time and 3 times a day.

(3) Insufflation: Clear away the pus secretion from the auditory canal and insufflate little powder of *Lan Er San* made of *Chuan Xin Lian*(Herba Andrographitis) 12 g, *Zhu Dan Zhi*(Pig's Bile) 10 g and *Ku Fan* (Alumen Exsiccatum) 10 g or *Hong Mian San* made of *Ku Fan*(Alumen Exsiccatum)12 g, *Gan Yan Zhi*(Dry Rouge) 6 g and *She Xiang*(Moschus) 0.05 g into the canal.

14.4　Meniere's Disease

Meniere's Disease corresponds to *Xuan Yun*(vertigo) in traditional Chinese medicine. Since *Xuan Yun* is a special symptom, it is also known as *Er Xuan Yun*(auditory vertigo).

Etiology and Pathogenesis

Deficiency of *Qi* and blood, deficiency of the brain that deprives the ears of nourishment, deficiency of *Yang* giving rise to edema, hyperactivity of liver *Yang* and stagnation of turbid phlegm in the head may all lead to vertigo.

Essentials of Diagnosis

1. This disease is characterized by sudden attack of vertigo with spinning sensation accompanied with nausea and vomiting.

2. Examination: There may be spontaneous nystagmus.

Treatment Based on Differentiation of Syndrome

1. Deficiency of the sea of marrow

Symptoms and signs: Frequent attacks of vertigo, tinnitus during attacks, hypoacusis, accompanied by soreness and weakness of the lumbus and knees, mental restlessness, insomnia, dreamdisturbed sleep, seminal emission, poor appetite, hotness sensation in the palms and soles, red tongue with little coating, and wiry, thready and rapid pulse.

Principle of treatment: Tonifying the kidney *Yin*, replenishing essence and promoting marrow.

Recipe: *Qi Ju Di Huang Wan*(167) with modification.

Ingredients:

Radix Rehmanniae Praeparata	24 g
Fructus Corni	12 g
Rhizoma Dioscoreae	12 g
Cortex Moutan Radicis	9 g
Poria	9 g
Rhizoma Alismatis	9 g
Fructus Lycii	15 g
Flos Chrysanthemi	12 g
Radix Polygoni Multiflori Praeparata	18 g
Radix Paeoniae Alba	12 g
Concha Haliotidis	24 g
Concha Ostreae	24 g
Colla Cornu Cervi	9 g
Colla Plastrum Testudinis	10 g

Decoct the above ingredients in water for oral dose.

2. Deficiency of *Qi* and blood

Symptoms and signs: Vertigo accompanied with pale complexion, lassitude, sleepiness, indifference of expression, deficiency of *Qi*, dislike of speaking, shortness of breath after exertion, palpitation, pale lips and nails, poor appetite, loose stool, pale tongue with thin and white coating, and thready and forceless pulse.

Principle of treatment: Tonifying *Qi*, nourishing blood, strengthening the spleen and calming the mind.

Recipe: *Gui Pi Tang*(50) with modification.

Ingredients:

Radix Astragali seu Hedysari	24 g
Radix Codonopsis Pilosulae	18 g
Radix Angelicae Sinensis	12 g
Arillus Longan	12 g
Semen Ziziphi Spinosae	24 g
Rhizoma Atractylodis Macrocephalae	15 g
Poria	12 g
Radix Aucklandiae	9 g
Radix Polygalae	9 g
Radix Glycyrrhizae Praeparata	9 g
Radix Polygoni Multiflori Praeparata	18 g
Radix Rehmanniae Praeparata	15 g
Radix Paeoniae Alba	12 g
Fructus Tribuli	9 g

Decoct the above ingredients in water for oral dose.

3. Edema due to deficiency of *Yang*

Symptoms and signs: Vertigo accompanied by palpitation, aversion to cold, cold limbs, spiritlessness, lumbago, cold back, frequency of long urination at night, pale tongue with white and moist coating, and deep, thready and forceless pulse.

Principle of treatment: Warming and tonifying the kidney *Yang*, dispelling cold and promoting diuresis.

Recipe: *Zhen Wu Tang*(45) with modification.

Ingredients:

Radix Aconiti Praeparata	6 g
Poria	12 g
Rhizoma Atractylodis Macrocephalae	15 g

Radix Paeoniae Alba	12 g
Rhizoma Zingiberis Recens	3 pieces

Decoct the above ingredients in water for oral dose.

Modification: In case of excessive cold, add Pericarpium Zanthoxyli 12 g, Herba Asari 3 g, Ramulus Cinnamomi 12 g and Radix Morindae Officinalis 12 g.

4. Hyperactivity of liver *Yang*

Symptoms and signs: Attack of vertigo due to emotional change, irritability, mental restlessness, red face, accompanied with headache, bitter mouth, dry throat, fullness and discomfort in the chest and hypochondriac region, dream-disturbed sleep, insomnia, red tongue with yellow coating, and wiry and rapid pulse.

Principle of treatment: Nourishing *Yin*, soothing *Yang*, calming the liver and dispelling wind.

Recipe: *Tian Ma Gou Teng Yin*(168) with modification.

Ingredients:

Rhizoma Gastrodiae	9 g
Ramulus Uncariae cum Uncis	15 g
Concha Haliotidis	24 g
Fructus Gardeniae	9 g
Radix Scutellariae	9 g
Radix Cyathulae	12 g
Cortex Eucommiae	12 g
Ramulus Loranthi	18 g
Caulis Polygoni Multiflori	18 g
Poria cum Ligno Hospite	12 g

Decoct the above ingredients in water for oral dose.

Modification: In case of severe vertigo, add Os Draconis 30 g and Concha Ostreae 30 g.

In case of bitter mouth and dry throat, add Radix Gentianae 12 g and Cortex Moutan Radicis 12 g.

5. Stagnation of phlegm—damp in the middle *Jiao*

Symptoms and signs: Vertigo with heaviness and distention in the forehead, stuffiness and distention in the ear, tinnitus, hypoacusis, Stuffy chest, severe nausea and vomiting of phlegmy fluid, accompanied with palpitation, poor appetite, lassitude, white and sticky tongue coating, and soft and slippery or wiry and slippery pulse.

Principle of treatment: Strengthening the spleen, regulating the middle *Jiao*, promoting diuresis and opening the mind.

Recipe: *Ban Xia Bai Zhu Tian Ma Tang*(169) with modification.

Ingredients:

ginger juice—prepared Rhizoma Pinelliae	12 g
Pericarpium Citri Reticulatae	12 g
Rhizoma Atractylodis Macrocephalae	15 g
Poria	12 g
Rhizoma Dioscoreae	20 g
Rhizoma Alismatis	9 g
Rhizoma Gastrodiae	9 g
Ramulus Uncariae cum Uncis	12 g
Fructus Ziziphi Jujubae	3 pieces
Radix Glycyrrhizae Praeparata	6 g

Decoct the above ingredients in water for oral dose.

Modification: In case of frequent nausea and vomiting, add Ochra Haematitum 30 g and Flos Inulae 10 g. Drink the decoction frequently with small amount each time.

In case of severe tinnitus and deafness, add Magnetitum 30 g and Rhizoma Acori Graminei 12 g.

For sallowish complexion, shortness and breath and lassitude, add Radix Codonopsis Pilosulae 30 g, Radix Astragali seu Hedysari 30 g and Semen Coicis 30 g.

15 Common Emergent Cases of Dermatology

Traditional Chinese medicine has its own systemic treatments for skin diseases with therapeutic characteristics and results. Only some most commonly seen emergency cases are introduced in this chapter and the rest could be managed with reference to the book Dermatology.

15.1 Herpes Zoster

Herpes Zoster is known as *Chan Yao Huo Dan*(erysipelas around the waist), *She Chuan Chuang*(herpes zoster complicated with infection), *Zhi Zhu Pao*(spider-like herpes) and *Chuan Yao Long*(serpiginous-belt eruption), etc. in traditional Chinese medicine.

Etiology and Pathogenesis

This disease is usually resulted from stagnation of the liver fire, invasion of the spleen by dampness, stagnation of *Qi* and blood as well as affection by exogenous pathogenic noxious factor.

Essentials of Diagnosis

1. This disease mostly occurs in spring and autumn.

2. It may appear in any part of the body, but more in the lumbar region.

3. It is often accompanied by general prodromal symptoms such as fever and headache.

4. Skin eruption: Herpes zosters densely distributed in bands along the distribution of the nerve affected with severe pain. It

usually affects the patients unilaterally.

Treatment Based on Differentiation of Syndromes

1. Internal treatment:

(1) Excessive heat in the liver and gallbladder

Symptoms and signs: Red skin eruption, millet-like herpes zosters densely distributed in patches with burning pain, accompanied by bitter mouth, dry throat, thirst, irritability, yellow urine, dry stool, red tongue with yellow coating, and wiry, slippery and slightly rapid pulse.

Principle of treatment: Clearing away heat, promoting diuresis, eliminating toxin and arresting pain.

Recipe: *Long Dan Xie Gan Tang*(97) with modification.

Ingredients:

Radix Gentianae	9 g
Fructus Gardeniae	9 g
Radix Scutellariae	9 g
Radix Rehmanniae	15 g
Folium Isatidis	15 g
Fructus Forsythiae	12 g
Rhizoma Alismatis	9 g
Semen Plantaginis (to be wrapped for decocting)	12 g
Rhizoma Corydalis	9 g
Radix Glycyrrhizae	6 g

Decoct the above ingredients in water for oral dose.

Modification: If the herpes zosters appear on the head and face, add Flos Chrysanthemi 12 g and Ramulus Uncariae cum Uncis 30 g. If it appears on the chest and lumbar region, add Radix Bupleuri 12 g. If it appears on the upper limb, add

Rhizoma Curcumae Longae 12 g. If it appears on the lower limb, add Radix Achyranthis Bidentatae 12 g.

In case of remarkable heat in the blood which leads to necrosis of bloody vesicles, add Rhizoma Imperatae 30 g, Radix Paeoniae Rubra 12 g and Cortex Moutan Radicis 10 g. In case of severe infection, add Flos Lonicerae 30 g, Herba Taraxaci 30 g and Radix Isatidis 30 g. And for constipation, add Radix et Rhizoma Rhei 10 g.

In case of asthenia due to senility, add Radix Astragali seu Hedysari and Radix Codonopsis Pilosulae.

(2) Preponderance of dampness due to deficiency of spleen

Symptoms and signs: Pale affected area with light yellowish blisters which are liable to break with ulceration, secretion and severe pain, accompanied with abdominal distention, poor appetite, loose stool, pale and swoolen tongue with white and sticky or white and thick coating, and deep and slow or slippery pulse.

Principle of treatment: Strengthening the spleen, promoting diuresis, eliminating toxin and arresting pain.

Recipe: *Chu Shi Wei Ling Tang*(170) with modification.

Ingredients:

Rhizoma Atractylodis	6 g
Cortex Magnoliae Officinalis	6 g
Pericarpium Citri Reticulatae	9 g
Polyporus Umbellatus	12 g
Talcum	12 g
stir—heated Atractylodis Macrocephalae	12 g
stir—heated Cortex Phellodendri	12 g
Radix Isatidis	15 g
Rhizoma Corydalis	9 g

Radix Paeoniae Rubra 12 g
Semen Plantaginis
(to be wrapped for decocting) 12 g
Rhizoma Alismatis 9 g
Radix Glycyrrhizae 6 g

Decoct the above ingredients in water for oral dose.

(3) Stagnation of *Qi* and blood

Symptoms and signs: Herpes with dark red base and bloody fluid, intolerable pain, dark tongue with possible stagnant spots, white tongue coating, and wiry and thready pulse.

Principle of treatment: Activating *Qi* and blood circulation, dispersing blood stasis and arresting pain.

Recipe: *Huo Xue San Yu Tang*(171) with modification.

Ingredients:

Caulis Spatholobi 15 g
Ramulus Euonymi Alatae 15 g
Flos Carthami 12 g
Semen Persicae 9 g
Rhizoma Corydalis 9 g
Fructus Meliae Toosendan 12 g
Radix Aucklandiae 12 g
Pericarpium Citri Reticulatae 9 g
Luffa Cylindrica 12 g
Caulis Lonicerae 15 g

Decoct the above ingredients in water for oral dose.

2. External treatment:

(1) If there are blisters, mix *Xiong Huang Jie Du San*(powder of Realgar for Detoxification) 30 g and *Hua Du San*(Powder for Detoxification 3 g with water evenly for external use, or pound

fresh *Ma Chi Xian*(Herba Portulacae) (or outer leaf of cabbage) for external use.

(2) In case of mild ulceration, mix *Qu Shi San*(Powder for Removing Dampness) with vegetable oil evenly for external use.

(3) In case of after neuralgia, apply *Hei Se Ba Gao Gun* or *Tuo Se Ba Gao Gun* for hot compress.

15.2 Acute Erysipelas

Acute erysipelas has different names in Chinese medical classics due to different locations. When it appears on the head and face, it is known as *Bao Tou Huo Dan*(Erysipelas on the head); When it appears on the hypochondrium, lumbus and hips, it is known as *Nei Fa Dan Du*(erysipelas in the lower part of the trunk); When it appears on the legs, it is known as *Tui You Feng*(erysipelas on legs); When it appears on the tibia and ankle, it is known as *Liu Huo*(erysipelas on the shank); Besides, infantile erysipelas is known as *Chi You Dan*(wandering erysipelas).

Essentials of Diagnosis

1. Abrupt onset of the disease.

2. It may occur to any part of the body, but more to the face and legs.

3. It may be accompanied by general symptoms such as chillness, fever, headache, nausea, vomiting, and joint pain. Besides, there may be swelling and pain of the local lymph nodes.

4. Skin eruption: Irregular bright red eruption swelling up with evident border and its colour fading when pressed, burning pain, tenderness, quick spreading of the eruption, and blisters, bullae or necrosis in severe case.

Treatment Based on Differentiation of Syndromes

1. Internal treatment:

The syndromes of acute erysipelas are accumulation of damp-heat in the interior and hyperactivity of toxic heat.

Accumulation of toxic heat in the interior

Symptoms and signs: Evident redness, swelling and pain of the sken, accompanied by chillness, fever, headache, excessive thirst, constipation, scanty and deep yellow urine, red tongue with yellow coating, and superficial and rapid or full and rapid pulse.

Principle of treatment: Clearing away toxic heat, cooling the blood and removing heat from the *Ying* system.

Recipe: *Wu Wei Xiao Du Yin*(172) with modification.

Ingredients:

Flos Lonicerae	30 g
Flos Chrysanthemi	15 g
Herba Taraxaci	15 g
Herba Violae	15 g
Folium Isatidis	24 g
Rhioma Paridis	15 g
Cortex Moutan Radicis	10 g
Radix Paeoniae Rubra	10 g
Radix Isatidis	20 g

Decoct the above ingredients in water for oral dose.

Modification: In case of erysipelas on the face, add Fructus Arctii 10 g, Herba Menthae 10 g and Chrysanthemi 10 g. In case of erysipelas on the hypochondriac region, lumbus and hips, add Radix Bupleuri, Radix Scutellariae and Fructus Gardeniae. In case of erysipelas on the lower limbs, add Cortex Phellodendri, Polyporus Umbellatus, Rhizoma Dioscoreae Hypoglaucae and

Radix Achyranthis Bidentatae.

If there appears high fever, add Gypsum Fibrosum 30 g, Rhizoma Anemarrhenae 12 g and Radix Trichosanthis 12 g.

In case of high fever and coma due to invasion of the blood and *Ying* systems by heat, use *Xi Jiao Di Huang Tang*(25) or *Qing Ying Tang*(19) with modification instead with combination of *Zi Xue Dan* and *An Gong Niu Huang Wan* 1 pill each time and 2—3 time a day.

2. External treatment:

(1) Mix *Ru Yi Jin Huang San* 30 g and *Hua Du San* 1.5 g with cool tea evenly for external use.

(2) Mix *Hua Du San* 1.5 g and *Ru Yi Jin Huang San* 20 g with pounded fresh outer leaf of cabage, or *Ma Chi Xian*(Herba Portulacae), or *Lu Dou Ya*(mung bean sprout) evenly for external use.

16 The Main Approaches of Emergency Treatment in TCM

Comprehensive approaches are usually adopted for emergency cases in traditional Chinese medicine such as emetic method, purgative method, acupuncture, Chinese massage therapy, venous transfusion, aerosol inhalation, administration per anum, etc. The single or combinationd of the above could be used according to pathological conditions.

16.1 Emetic and Purgative Method

Emetic method

Emetic method is to cause vomiting by using emetics or certain mechanical stimulation so as to help excrete toxic substance in the throat or stomach.

1. Indication: Obstruction of the respiratory tract due to stagnation of phlegm in the throat, fullness, distention and pain in epigastrium due to food retention, accidental intake of harmful substance which is still remained in the stomach for short time.

2. Common prescription and method

(1) *Gua Di San*(175)

Ingredients:

Gua Di(Musk−mellon Pedicel)
   (to be baked until it gets light brown)   1 g
Chi Xiao Dou(Semen Phaseoli)   1 g
Administration: Grind the above ingredients into fine pow-

der which is mixed evenly and taken with the decoction of *Dou Chi*(Semen Sojae Praeparatum)9 g, 1-3 g each time.

Function: Inducing vomiting of phlegmy fluid and food retention.

Indication: Acute poisoning, retention of phlegm and food, fullness and stuffiness in the chest, dysphoria, Nausea due to upward perversion of *Qi*, and superficial, tense and forceful pulse at the *Cun* region.

(2) *Yan Tang Tan Tu Fang*

Ingredients: *Shi Yan*(Common Salt)

Administration: Orally take 2 000 ml of saturated salt solution, then wash and sterilize the operator's finger and stick a piace of coffon stick into the patient's throat with a piece of pad between the upper and lower teeth to avoid bite wound, and irritate the throat with the cotton stick to induce vomiting.

Function: Inducing vomiting of phlegmy fluid and food retention.

Indication: Retention of food, nausea with no vomiting, feeling of diarrhea with no bowel, fullness in the chest with dysphoria.

Purgative method

Purgative method is to use purgatives to induce bowel movement so as to eliminate stasis, clear away excessive heat, remove water retention and constipation due to deficiency of yang.

1. Indication: Epigastric and abdominal fullness and distention, and constipation due to stagnation of heat in the epigastrium and abdomen.

2. Common prescription and method

(1) *Fan Xie Ye Dao Xie Fang*

Ingredients: *Fan Xie Ye*(Folium Cassiae) 10 g

Administration: Take it orally instead of tea.

Function: Removing constipation with purgative.

Indication: Constipation with abdominal fullness, distention and pain.

(2) *Da Cheng Qi Tang*

Ingredients:

Radix et Rhizoma Rhei	12 g
Cortex Magnoliae Officinalis (prepared)	15 g
Fructus Citri Aurantii Immaturus	12 g
Natrii Sulfas	9 g

Administration: Decoct the above ingredients except with put in later, and melt in the decoction for oral use.

Function: Removing constipation due to heat with drastic purgation.

Indication: Stagnation of heat in the epigastrium and abdomen, constipation, abdominal fullness and pain aggravated by pressure, palpable mass, yellow, dry and thorny coating or charring black and thorny coating with cracks, and deep and excess pulse.

16.2 Acupuncture Massage Therapy

Acupuncture

Acupuncture is to insert acupuncture needles into certain acupoints to treat diseases.

1. Common indication

(1) Drowning

Principle of treatment: Dispersing the lung and opening the mind for resuscitation.

Commonly used points: Huiyang(BL35), Suliao(DU25), Neiguan(PC6) and Yongquan(KI1).

Manipulation: Strong stimulation is usually used.

(2) Heatstroke

Principle of treatment: Clearing away summer-heat, regulating the stomach and saving collapse for resuscitation.

Points for mild case: Dazhui(DU14), Quchi(LI11), Hegu(LI4) and Neiguan(PC6)

Manipulation: Reduce point Dazhui, Quchi and Hegu, and puncture point Neiguan evenly.

Points for severe case: Baihui(DU20), Shuigou(DU26), Shixuan(EX-UE11), Quze(PC3), Weizhong(BL40), Yanglingquan(GB34), Chengshan(BL57), Shenque(RN8) and Guanyuan(RN4).

Manipulation: Reinforcing Shenque and Guanyuan and reducing the rest points.

(3) Convulsion

Convulsion with fever

Principle of treatment: Dredging and regulating the *Du* channel, and clearing away pathogenic heat.

Commonly used points: Yintang(EX-HN3), Taiyang(EX-HN5), Sifeng(EX-UE10), Shixuan(EX-UE11), Dazhui(DU14), Hegu(LI4), Shenzhu(DU12) and Quchi(LI11).

Manipulation: Strong stimulation is generally used.

Convulsion with no fever

Principle of treatment: Dredging and regulating the *Du* channel, and calming convulsion.

Commonly used points: Dazhui(DU14), Jingsuo(DU8), Houxi(SI3) and Yanglingquan(GB34).

Manipulation: Strong stimulation is usually used.

(4) Shock

Principle of treatment: Restoring *Yang* for resuscitation.

Commonly used points: Suliao(DU25), Neiguan(PC6).

Manipulation: Manipulate the neddles contineously with moderate stimulation first. When the blood pressure becomes stable, manipulate the needles intermittently or just leave the needles in the points.

(5) Syncope

Principle of treatment: Relieving syncope and regulating the spleen and stomach.

Commonly used points: Shuigou(DU26), Zhongchong(PC9), Zusanli(ST36).

Manipulation: Puncture point Shuigou with strong stimulation for a short time first and then needle point Zhongchong and Zusanli.

Chiness Massage Therapy

Chinese massage is such a kind of therapeutic method that resorts to massager's hands and limbs to help the patient with passive exercise. It has the function to regulate *Qi* and blood, remove obstruction from the channels, promote metabolism, highten anti-pathogenic ability and improve local blood circulation and nourishment.

1. Massage manipulation:

(1) Pushing manipulation: A massage technique performed by pushing and squeezing the patient's muscle streight and forcefully with one finger or palm.

(2) Grasping manipulation: A massage maneuver by grasping and squeezing or grasping, lifting and rapidly releasing the af-

fected muscle with one or both hands of the operator.

(3) Pressing manipulation: A massage technique performed by pressing certain acupoints or certain parts of the body surface downward or inward with finger or palm of the operator.

(4) Rubbing manipulation: Rubbing the affected part or acupoints with a thumb or palm repeatedly.

(5) Kneading manipulation: A massage manipulation performed by pressing certain part of the body by the thumb or the root of the palm and kneading in a circular motion with the help of wrist point or metacarpophalangeal joint.

(6) Finger-nail pressing: A massage manipulation performed by pressing right on certain acupoint with a finger nail to produce a stimulation in a certain degree.

(7) Foulage manipulation: Holding the limbs, lumbus or back tightly, kneading and pressing the muscles there in a circular motion repeatedly with both palms.

(8) Rotating manipulation: A massage manipulation performed by holding joint of certain part of the body (such as shoulder, neck, elbow, wrist, hips, knee, ankle, etc.) and rotating the joint in a circular motion s as to promote its ability in movement.

(9) Rolling manipulation: A massage manipulation performed by pressing certain part of the body surface with the dorsum of the ulnar side of the practioner's hand and rolling the hand back and forth, and left and right with the wrist force.

(10) Shaking manipulation: A massage manipulation performed by holding and pulling the distal end of the diseased joint, and shaking it abruptly up and down within physiological limit.

2. Precaution

(1) Under the circumstances of local skin lesion of dermatosis, Chinese massage is forbidden.

(2) It is not advisable to massage the abdomen right after food intake and evacuation of urine is necessary before the lumbus or abdomen is manipulated.

(3) It is not advosable to massage the lumbus or abdomen during menstrual period or pregnancy.

16.3 Venous Transfusion

With the improvement of the forms of Chinese herbal medicine, injection and venous transfusion have been extensively used in the emergency management and treatment in Chinese medicine. At present, Chinese herbal medicines have been prepared for the use of venous transfusion, intravenous injection and muscular injection. In comparison with oral administration, injection and transfusion are characterized by convenience in application, accuracy of dosage, fast function, high therapeutic effect and no affection to the function of the stomach and spleen in digestion and transportation. In comparison with the western drugs, they are characterized by extensive indications, long active period, less toxic and side effect and no production of drug-resistance.

1. *Dan Shen Zhu She Ye*(Injection of Radix Salviae Miltiorrhizae)

Ingredients: Radix Salviae Miltiorrhizae 1 500 g add water for injection up to 1 000 ml

Function and indication: It is taken as a vasodilator with the function to dilate coronary artery and increase the coronary blood flow. It is indicated in angina pectoris, myocardiac infarction, nephrotic syndrome, ascites and prevention of cerebral thrombosis.

Administration and dosage: Muscular injection 2—8 ml daily and 2—4 ml each time; Intravenous injection 4 ml each time by using 20 ml of 50% glucose solution for dilution, 1—2 times a day; Add 10 ml of the injection to 100—500 ml of 5% glucose solution for intravenous drip once a day.

2. *Sheng Mai Zhu She Ye*(Pulse—activating Injection)

Ingredients:

Radix Ginseng	1 000 g
Radix Ophiopogonis	3 120 g
Fructus Schisandrae	1 560 g
add water for injection up to	10 000 ml

Function and indication: Activating pulse and restoring consciousness. It is indicated in exhaustion of the kidney *Yang*, failure of *Yin* in astringing *Yang* or floating of *Yang* due to deficiency, which gives rise to coma, arrhythmia and asthenia due to prolonged illness.

Administration and dosage: Use 2—4 ml of the injection for muscular injection once a day; Add 10 or 20 ml of the injection to 250—500 ml of 10% glucose solution for intravenous drip.

3. *Qing Kai Ling Zhu She Ye*(Injection for Heat—clearing and Resuscitation)

Ingredients: Calculus Bovis, Cornu Bubali, Radix Scutellariae, Flos Lonicerae, Fructus Gardeniae.

Function and indication: Clearing away toxic heat and calming the mind. It is indicated in invasion of the interior by pathogenic heat giving rise to coma, delirium, fever, dysphoria, and convulsion. It may also obtain good therapeutic results in the treatment of virus hepatitis, subacute fatal hepatitis and chronic fatal hepatitis. Besides, it is also indicated in infection of upper repiratory tract, pneumonia, and high fever.

Administration and dosage: Add 20-40 ml of the injection to 100 or 200 ml of 10% glucose solution for intravenous drip; Use 2-4 ml of the injection for muscular injection once a day.

Precaution: The injection can not be used any more if there appears sediment or turbidity. If there appears turbidity when it is mixed with saline or glucose solution, it can not be used either.

16.4 Aerosol Inhalation

To cooperate the treatment of acute asthmatic cough and prevention of infectious diseases, the suitable medicine can be given to the patient through respiratory tract by means of aerosol inhalation.

1. Rice-vinegar spray

Put 500 g of boiling old rice-vinegar into a sprayer, let the patient sit in a room with its door and window closed and spray the vinegar in to the room. This method can help prevent influenza.

2. Compound aerosol of Herba Asari

Ingredients:

Volatile oil of *Xi Xin*
   (Herbal Asari)   50 ml
Bing Pian(Borneolum)   16 g
95% Alcohol   600 ml

Trichlorodifluoromethane quantum satis

Function and indication: Relaxing the chest and arresting pain. It is indicated in coronary disease and acute attack of angina pectoris.

Administration and dosage: Hold the sprayer upside down with its mouth directing the mouth cavity, press the top of the valve and spray the medicinal aerosol into the mouth and then ask the patient to keep the mouth closed for few minutes. This treatment should be given to the patient twice a day or during attacks.

16.5 Administration Per Anus

This method is to administrate the medicine through the anus so as to obtain therapeutic effect. It is not limited by dysfunction of the upper digestive tract or dysphagia. The drug is absorbed fast and functions quickly. For example, application of enema with decoction of *Sheng Da Huang*(Radix et Rhizoma Rhei) 12 g, *Shu Fu Zi*(Radix Aconiti Praeparata 12 g and *Mu Li*(Concha Ostreae) 30 g for edema, urine retention and obstruction and rejection due to renal failure may lower the urea nitrogen in the blood.

Index of Recipes

1. *Jing Fang Jie Biao Tang* 荆防解表汤

Herba Schizonepetae	荆芥	(*Jing Jie*)
Radix Ledebouriellae	防风	(*Fang Feng*)
Rhizoma seu Radix Notopterygii	羌活	(*Qiang Huo*)
Radix Angelicae Dahuricae	白芷	(*Bai Zhi*)
Bulbus Allii Fistulosi	葱白	(*Cong Bai*)
Rhizoma Zingiberis Recens	生姜	(*Sheng Jiang*)
Radix Glycyrrhizae	甘草	(*Gan Cao*)

2. *Yin Qiao San* 银翘散

Flos Lonicerae	银花	(*Yin Hua*)
Fructus Forsythiae	连翘	(*Lian Qiao*)
Radix Platycodi	苦桔梗	(*Ku Jie Geng*)
Fructus Arctii	牛蒡子	(*Niu Bang Zi*)
Spica Schizonepetae	荆芥穗	(*Jing Jie Sui*)
Herba Menthae	薄荷	(*Bo He*)
Semen Sojae Praeparatum	豆豉	(*Dou Chi*)
Herba Lophatheri	竹叶	(*Zhu Ye*)
Radix Glycyrrhizae	甘草	(*Gan Cao*)
Rhizoma Phragnitis	苇根	(*Wei Gen*)

3. *Huo Xiang Zheng Qi San* 藿香正气散

Herba Agastaches	藿香	(*Huo Xiang*)

Poria	茯苓	(Fu Ling)
Folium Perillae	柴苏	(Zi Su)
Pericarpium Arecae	大腹皮	(Ba Fu Pi)
stir—heated Rhizoma Atractylodis Macrocephalae	炒白术	(Chao Bai Zhu)
Massa Pinelliae Fermentatae	半夏曲	(Ban Xia Qu)
Radix Angelicae Daluricae	白芷	(Bai Zhi)
Pericarpium Citri Reticulatae	陈皮	(Chen Pi)
Cortex Magnoliae Officinalis	厚朴	(Hou Po)
Radix Platycodi	桔梗	(Jie Gen)
Radix Glycyrrhizae Praeparata	炙甘草	(Zhi Gan Cao)

4. *Sang Xing Tang* 桑杏汤

Folium Mori	桑叶	(Sang Ye)
Semen Armeniacae Amarum	杏仁	(Xing Ren)
Radix Adenophorae Strictae	沙参	(Sha Shen)
Bulbus Fritillariae Thunbergii	象贝	(Xiang Bei)
Semen Sojae Praeparatum	香豉	(Xiang Chi)
Fructus Gardeniae	栀子	(Zhi Zi)
Exocarpium Pyrus	梨皮	(Li Pi)

5. *Sha Shen Mai Men Dong Tang* 沙参麦门冬汤

Radix Adenophorae Strictae	沙参	(Sha Shen)
Rhizoma Polygonati Odorati	玉竹	(Yu Zhu)
Radix Glycyrrhizae	生甘草	(Sheng Gan Cao)
Folium Mori	冬桑叶	(Dong Sang Ye)
Radix Ophiopogonis	麦冬	(Mai Dong)
Semen Dolichoris	生扁豆	(Sheng Bian Dou)

Radix Trichosanthis　　　花粉　(Hua Fen)

6. Xin Jia Xiang Ru Yin 新加香薷饮
Herba Elsholtziae seu Moslae　　香薷　(Xiang Ru)
Cortex Magnoliae Officinalis　　厚朴　(Hou Po)
Flos Dolichoris (fresh)　　鲜扁豆花 (Xian Bian Dou Hua)
Flos Lonicerae　　银花　(Yin Hua)
Fructus Forsythiae　　连翘　(Lian Qiao)

7. Xiao Chai Hu Tang 小柴胡汤
Radix Bupleuri　　柴胡　(Chai Hu)
Radix Scutellariae　　黄芩　(Huang Qin)
Rhizoma Pinelliae　　半夏　(Ban Xia)
Radix Codonopsis Pilosulae　　党参　(Dang Shen)
Radix Glycyrrhizae Praeparata　　炙甘草 (Zhi Gan Cao)
Rhizoma Zingiberis Recens　　生姜　(Sheng Jiang)
Fructus Ziziphi Jujubae　　大枣　(Da Zao)

8. Ma Xing Shi Gan Tang 麻杏石甘汤
Herba Ephedrae　　麻黄　(Ma Huang)
Semen Ameniacae Amarum　　杏仁　(Xing Ren)
Gypsum Fibrosum　　石膏　(Shi Gao)
Radix Glycyrrhizae　　甘草　(Gan Cao)

9. Liang Ge San 凉膈散
Radix et Rhizoma Rhei　　大黄　(Da Huang)
Natrii Sulfas　　芒硝　(Mang Xiao)
Radix Glycyrrhizae　　甘草　(Gan Cao)
Fructus Gardeniae　　栀子　(Zhi Zi)

Radix Scutellariae 黄芩 (*Huang Qin*)
Herba Menthae 薄荷 (*Bo He*)
Fructus Forsythiae 连翘 (*Lian Qiao*)

10. *Bai Hu Tang* 白虎汤
Gypsum Fibrosum 石膏 (*Shi Gao*)
Rhizoma Anemarrhenae 知母 (*Zhi Mu*)
Radix Glycyrrhizae 甘草 (*Gan Cao*)
Semen Oryzae Sative 粳米 (*Gen Mi*)

11. *Da Cheng Qi Tang* 大承气汤
Radix et Rhizoma Rhei 生大黄 (*Sheng Da Huang*)
Cortex Magnoliae Officinalis 厚朴 (*Hou Po*)
Fructus Aurantii Immaturus 枳实 (*Zhi Shi*)
Natrii Sulfas 芒硝 (*Mang Xiao*)

12. *Qian Jin Wei Jing Tang* 千金苇茎汤
Rhizoma Phragnitis 苇茎 (*Wei Jing*)
Semen Benincasae 冬瓜仁 (*Dong Gua Ren*)
Semen Coicis 薏苡仁 (*Yi Yi Ren*)
Semen Persicae 桃仁 (*Tao Ren*)

13. *San Ren Tang* 三仁汤
Semen Coicis 生薏仁 (*Sheng Yi Ren*)
Semen Armeniacae Amarum 苦杏仁 (*Ku Xing Ren*)
Semen Amomi Cardamomi 白蔻仁 (*Bai Kou Ren*)
Cortex Magnoliae Officinalis 厚朴 (*Hou Po*)
Medulla Tetrapanacis 白通草 (*Bai Tong Cao*)
Talcum 滑石 (*Hua Shi*)

ginger juice—prepared
    Rhizoma Pinelliae   姜半夏(*Jiang Ban Xia*)
Caul Bambusae in Taeniam   淡竹叶(*Dan Zhu Ye*)

14. *Hao Qin Qing Dan Tang* 蒿芩清胆汤

Herba Artemisiae Chinghao   青蒿   (*Qing Hao*)
Radix Scutellariae   黄芩   (*Huang Qin*)
Fructus Aurantii Immaturus   枳实   (*Zhi Shi*)
Herba Lophatheri   淡竹茹(*Dan Zhu Ru*)
made of Indigo Naturalis,
    Talcum and Radix
    Glycyrrhizae (to be wrapped) 碧玉散(*Bi Yu San*)
Pericarpium Citri Reticulatae   陈皮   (*Chen Pi*)
ginger Juice—prepared
    Rhizoma Pinelliae   姜半夏(*Jiang Ban Xia*)
Poria Rubra   赤茯苓(*Chi Fu Ling*)

15. *Ba Zheng San* 八正散

Herba Polygoni Avicularis   萹蓄   (*Bian Xu*)
Herba Dianthi   瞿麦   (*Qu Mai*)
Semen Plantaginis   车前子(*Che Qian Zi*)
Talcum   滑石   (*Hua Shi*)
Fructus Gardeniae   生山栀(*Sheng Shan Zhi*)
Radix et Rhizoma Rhei   熟大黄(*Shu Da Huang*)
Caulis Akebiae   木通   (*Mu Tong*)
tip of Radix Glycyrrhizae   甘草梢(*Gan Cao Shao*)
Medulla Junci   灯心   (*Deng Xin*)

16. *Bai Tou Weng Tang* 白头翁汤

Radix Pulsatillae 白头翁 (*Bai Tou Weng*)
Cortex Phellodendri 黄柏 (*Huang Bai*)
Rhizoma Coptidis 黄连 (*Huang Lian*)
Cortex Fraxini 秦皮 (*Qin Pi*)

17. *Yu Nü Jian* 玉女煎

Gypsum Fibrosum 生石膏 (*Sheng Shi Gao*)
Radix Rehmanniae
  Praeparata 熟地 (*Shu Di*)
Radix Ophiopogonis 麦冬 (*Mai Dong*)
Rhizoma Anemarrhenae 知母 (*Zhi Mu*)
Radix Achyranthis Bidentatae 牛膝 (*Niu Xi*)

18. *Qing Ying Tang* 清营汤

Cornu Rhinoceri 犀角 (*Xi Jiao*)
Radix Rehmanniae 生地 (*Sheng Di*)
Radix Scrophulariae 玄参 (*Xuan Shen*)
Radix Ophiopogonis 麦冬 (*Mai Dong*)
Flos Lonicerae 银花 (*Yin Hua*)
Fructus Forsythiae 连翘 (*Lian Qiao*)
Radix Salviae Miltiorrhizae 丹参 (*Dan Shen*)
Rhizoma Coptidis 黄连 (*Huang Lian*)
Leave Bud of Herba
  Lophatheri 竹叶心 (*Zhu Ye Xin*)

19. *Qing Gong Tang* 清宫汤

Cornu Rhinoceri 犀角 (*Xi Jiao*)
Radix Scrophulariae 玄参 (*Xuan Shen*)
Radix Ophiopogonis 麦冬 (*Mai Dong*)

Plumula Forsythiae	连翘心	(Lian Qiao Xin)
Leave Bud of Herba Lophatheri	竹叶心	(Zhu Ye Xin)
Plumula Nelumbinis	莲子心	(Lian Zi Xin)

20. *An Gong Niu Huang Wan* 安宫牛黄丸

Calculus Bovis	牛黄	(Niu Huang)
Radix Curcumae	郁金	(Yu Jin)
Cornu Rhinoceri	犀角	(Xi Jiao)
Radix Scutellariae	黄芩	(Huang Qin)
Rhizoma Coptidis	黄连	(Huang Lian)
Fructus Gardeniae	山栀	(Shan Zhi)
Realgar	雄黄	(Xiong Huang)
Cinnabaris	朱砂	(Zhu Sha)
Fructus Pruni	梅片	(Mei Pian)
Moschus	麝香	(She Xiang)
pearl	珍珠	(Zhen Zhu)

21. *Niu Huang Qing Xin Wan* 牛黄清心丸

Calculus Bovis	牛黄	(Niu Huang)
Rhizoma Coptidis	黄连	(Huang Lian)
Radix Scutellariae	黄芩	(Huang Qin)
Fructus Gardeniae	生山栀	(Sheng Shan Zhi)
Curcumae	郁金	(Yu Jin)
Cinnabaris	朱砂	(Zhu Sha)

22. *Ling Yang Gou Teng Tang* 羚羊钩藤汤

Cornu Antelopis	羚羊角	(Ling Yang Jiao)
Folium Mori	桑叶	(Sang Ye)

Ramulus Uncariae cum Uncis　钩藤　(*Gou Teng*)
Bulbus Fritillariae Cirrhosae　川贝　(*Chuan Bei*)
Radix Rehmanniae(fresh)　鲜生地(*Xian Sheng Di*)
Caulis Bambusae in Taeniam　淡竹茹　(*Dan Zhu Ru*)
Flos Chrysanthemi　菊花　(*Ju Hua*)
Radix Paeoniae Alba　生白芍(*Sheng Bai Shao*)
Poria　茯苓　(*Fu Ling*)
Radix Glycyrrhizae　生甘草(*Sheng Gan Cao*)

23. *Da Ding Feng Zhu* 大定风珠

Radix Rehmanniae　干生地　(*Gan Sheng Di*)
Radix Paeoniae Alba　生白芍(*Sheng Bai Shao*)
Radix Ophiopogonis　麦冬　(*Mai Dong*)
Concha Ostreae　生牡蛎　(*Sheng Mu Li*)
Carapax Trionycis　生鳖甲　(*Sheng Bie Jia*)
Plastrum Testudinis　生龟板(*Sheng Gui Ban*)
Radix Glycyrrhizae
　Praeparata　炙甘草　(*Zhi Gan Cao*)
Colla Corii Asini　阿胶　(*E Jiao*)
Fructus Cannabis　火麻仁(*Huo Ma Ren*)
Fructus Schisandrae　五味子(*Wu Wei Zi*)
Fresh Egg Yolks　鸡子黄(*Ji Zi Huang*)

24. *Hua Ban Tang* 化斑汤

Gypsum Fibrosum　生石膏(*Sheng Shi Gao*)
Rhizoma Anemarrhenae　知母　(*Zhi Mu*)
Radix Glycyrrhizae　甘草　(*Gan Cao*)
Semen Oryzae Sativae　粳米　(*Geng Mi*)
Cornu Rhinoceri　犀角　(*Xi Jiao*)

Radix Scrophulariae 玄参 (Xuan Shen)

25. Xi Jiao Di Huang Tang 犀角地黄汤
Cornu Rhinoceri 犀角 (Xi Jiao)
Radix Rehmanniae 生地 (Sheng Di)
Radix Paeoniae Alba 芍药 (Shao Yao)
Cortex Moutan Radicis 丹皮 (Dan Pi)

26. Sheng Mai San 生脉散
Radix Ginseng 人参 (Ren Shen)
Radix Ophiopogonis 麦冬 (Mai Dong)
Fructus Schisandrae 五味子 (Wu Wei Zi)

27. Si Ni Tang 四逆汤
Radix Aconiti Praeparata 附子 (Fu Zi)
Rhizoma Zingiberis 干姜 (Gan Jiang)
Radix Glycyrrhizae Praeparata 炙甘草 (Zhi Gan Cao)

28. Ren Shen Bai Hu Tang 人参白虎汤
Gypsum Fibrosum 生石膏 (Sheng Shi Gao)
Rhizoma Anemarrhenae 知母 (Zhi Mu)
Radix Glycyrrhizae 甘草 (Gan Cao)
Semen Oryzae Sativae 粳米 (Geng Mi)
Radix Ginseng 人参 (Ren Shen)

29. Shen Fu Tang 参附汤
Radix Ginseng 人参 (Ren Shen)
Radix Aconiti Praeparata 附子 (Fu Zi)

30. *San Jia Fu Mai Tang* 三甲复脉汤
Radix Glycyrrhizae Praeparata 炙甘草 (*Zhi Gan Cao*)
Radix Rehmanniae 干地黄 (*Gan Di Huang*)
Radix ophiopogonis 麦冬 (*Mai Dong*)
Colla Corii Asini 阿胶 (*E Jiao*)
Radix Paeoniae Alba 芍药 (*Shao Yao*)
Concha Ostreae 牡蛎 (*Mu Li*)
Fructus Cannabis 麻仁 (*Ma Ren*)
Carapax Trionycis 鳖甲 (*Bie Jia*)
Plastrum Testudinis 龟板 (*Gui Ban*)

31. *Zhi Gan Cao Tang* 炙甘草汤
Radix Glycyrrhizae Praeparata 炙甘草 (*Zhi Gan Cao*)
Radix Codonopsis Pilosulae 党参 (*Dang Shen*)
Radix Rehmanniae 生地 (*Sheng Di*)
Radix Ophiopogonis 麦冬 (*Mai Dong*)
Colla Corii Asini 阿胶 (*E Jiao*)
Fructus Cannabis 麻仁 (*Ma Ren*)
Ramulus Cinnamomi 桂枝 (*Gui Zhi*)
Rhizoma Zingiberis Recens 生姜 (*Sheng Jiang*)
Fructus Ziziphi Jujubae 大枣 (*Da Zao*)

32. *Xue Fu Zhu Yu Tang* 血府逐瘀汤
Radix Angelicae Sinensis 当归 (*Dang Gui*)
Radix Rehmanniae 生地 (*Sheng Di*)
Radix Achyranthis Bidentatae 牛膝 (*Niu Xi*)
Flos Carthami 红花 (*Hong Hua*)
Semen Persicae 桃仁 (*Tao Ren*)
Radix Bupleuri 柴胡 (*Chai Hu*)

Fructus Aurantii	枳壳	(Zhi Ke)
Radix Paeoniae Rubra	赤芍	(Chi Shao)
Rhizoma Ligustici Chuanxiong	川芎	(Chuan Xiong)
Radix Platycodi	桔梗	(Jie Geng)
Radix Glycyrrhizae	甘草	(Gan Cao)

33. *Chang Pu Yu Jin Tang* 菖蒲郁金汤

Rhizoma Acori Graminei (fresh)	鲜石菖蒲	(Xian Shi Chang Pu)
Radix Curcumae	郁金	(Yu Jin)
stir-heated Fructus Gardeniae	炒山栀	(Chao Shan Zhi)
Fructus Forsythiae	连翘	(Lian Qiao)
Flos Chrysanthemi	菊花	(Ju Hua)
Herba Lophatheri	竹叶	(Zhu Ye)
Talcum	滑石	(Hua Shi)
Cortex Moutan Radicis	丹皮	(Dan Pi)
Fructus Arctii	牛蒡子	(Niu Bang Zi)
Succus Bambosue	竹沥	(Zhu Li)
Ginger Juice	姜汁	(Jiang Zhi)
powder	玉枢丹末	(Yu Shu Dan Mo)

34. *San Zi Yang Qin Tang* 三子养亲汤

Fructus Perillae	紫苏子	(Zi Su Zi)
Semen Sinapis Albae	白芥子	(Bai Jie Zi)
Semen Raphani	莱菔子	(Lai Fu Zi)

35. *Yin Chen Hao Tang* 茵陈蒿汤

Herba Artemisiae Capillaris	茵陈	(Yin Chen)
Fructus Gardeniae	栀子	(Zhi Zi)

Radix et Rhizoma Rhei　　大黄　(Da Huang)

36. *Xi Di Qing Luo Yin* 犀地清络饮
Succus Cornu Rhinocerotis　犀角汁 (*Xi Jiao Zhi*)
Cortex Moutan Radicis　牡丹皮 (*Mu Dan Pi*)
Fructus Forsythiae　连翘　(*Lian Qiao*)
Radix Paeoniae Rubra　赤芍药 (*Chi Shao Yao*)
Radix Rehmanniae(fresh)　鲜生地 (*Xian Sheng Di*)
Semen Persicae　桃仁　(*Tao Ren*)
Succus Bambosae　竹沥　(*Zhu Li*)
Ginger Juice　生姜汁 (*Sheng Jiang Zhi*)
Succus Rhizoma Acori
　Graminei　鲜菖蒲汁 (*Xian Chang Pu Zhi*)
Rhizoma Imperatae (fresh)　鲜茅根 (*Xian Mao Gen*)
Medulla Junci　灯心 (*Deng Xin*)

37. *Gua Lou Xie Bai Ban Xia Tang* 瓜蒌薤白半夏汤
Fructus Trichosanthis　瓜蒌　(*Gua Lou*)
Bulbus Allii Macrostemi　薤白　(*Xie Bai*)
Rhizoma Pinelliae　半夏　(*Ban Xia*)
Chinese spirit　白酒　(*Bai Jiu*)

38. *Gua Lou Xie Bai Bai Jiu Tang* 瓜蒌薤白白酒汤
Fructus Trichosanthis　瓜蒌　(*Gua Lou*)
Bulbus Allii Macrostemi　薤白　(*Xie Bai*)
Chinese spirit　白酒　(*Bai Jiu*)

39. *Zuo Gui Yin* 左归饮
Radix Rehmanniae Praeparata　熟地　(*Shu Di*)

Rhizoma Dioscoreae	山药	(Shan Yao)
Fructus Corni	山茱萸	(Shan Zhu Yu)
Semen Cuscutae	菟丝子	(Tu Si Zi)
Fructus Lycii	枸杞子	(Gou Qi Zi)
Colla Cornus Cervi	鹿角胶	(Lu Jiao Jiao)
Colla Plastri Testudinis	龟板胶	(Gui Ban Jiao)
Radix Achyranthis Bidentatae	淮牛膝	(Huai Niu Xi)

40. *Bu Yang Huan Wu Tang* 补阳还伍汤

Radix Astragali seu Hedysari	生黄芪	(Sheng Huang Qi)
tail of Radix Angelicae Sinensis	归尾	(Gui Wei)
Radix Paeoniae Rubra	赤芍药	(Chi Shao Yao)
Lumbricus	地龙	(Di Long)
Rhizoma Ligustici Chuanxiong	川芎	(Chuan Xiong)
Semen Persicae	桃仁	(Tao Ren)
Flos Carthami	红花	(Hong Hua)

41. *Ren Shen Gui Zhi Tang* 人参桂枝汤

Radix Ginseng	人参	(Ren Shen)
Rhizoma Zingiberis	干姜	(Gan Jiang)
Rhizoma Atractylodis Macrocephalae	白术	(Bai Zhu)
Ramulus Cinnamomi	桂枝	(Gui Zhi)
Radix Glycyrrhizae	甘草	(Gan Cao)

42. *Ban Xia Xie Xin Tang* 半夏泻心汤

Rhizoma Pinelliae	清半夏	(Qing Ban Xia)
Radix Scutellariae	黄芩	(Huang Qin)
Rhizoma Zingiberis	干姜	(Gan Jiang)

Rhizoma Coptidis 黄连 (*Huang Lian*)
Radix Ginseng 人参 (*Ren Shen*)
Radix Glycyrrhizae Praeparata 炙甘草 (*Zhi Gan Cao*)
Fructus Ziziphi Jujubae 大枣 (*Da Zao*)

43. *Ren Shen Si Ni Tang* 人参四逆汤
Radix Aconiti Praeparata 生附子 (*Sheng Fu Zi*)
Rhizoma Zingiberis 干姜 (*Gan Jiang*)
Radix Glycyrrhizae Praeparata 炙甘草 (*Zhi Gan Cao*)

44. *Ling Gui Zhu Gan Tang* 苓桂术甘汤
Poria 茯苓 (*Fu Ling*)
Ramulus Cinnamomi 桂枝 (*Gui Zhi*)
Rhizoma Atractylodis Macrocephalae 白术 (*Bai Zhu*)
Radix Glycyrrhizae 炙甘草 (*Zhi Gan Cao*)

45. *Zhen Wu Tang* 真武汤
Radix Aconiti Praeparata 熟附子 (*Shu Fu Zi*)
Rhizoma Zingiberis Recens 生姜 (*Sheng Jiang*)
Rhizoma Atractylodis Macrocephalae 白术 (*Bai Zhu*)
Poria 茯苓 (*Fu Ling*)
Radix Paeoniae Alba 白芍 (*Bai Shao*)

46. *Xie Bai San* 泻白散
Cortex Lycii Radicis 地骨皮 (*Di Gu Pi*)
Cortex Mori Radicis 桑白皮 (*Sang Bai Pi*)
Radix Glycyrrhizae 甘草 (*Gan Cao*)

Semen Oryzae Sativae 粳米 (Geng Mi)

47. *Ma Huang Fu Zi Xi Xin Tang* 麻黄附子细辛汤
Herba Ephedrae 麻黄 (Ma Huang)
Radix Aconiti Praeparata 附子 (Fu Zi)
Herba Asari 细辛 (Xi Xin)

48. *Bu Xin Dan* 补心丹
Radix Codonopsis Pilosulae 党参 (Dang Shen)
Radix Scrophulariae 元参 (Yuan Shen)
Radix Salviae Miltiorrhizae 丹参 (Dan Shen)
Poria 茯苓 (Fu Ling)
Fructus Schisandrae 五味子 (Wu Wei Zi)
Radix Polygalae 远志 (Yuan Zhi)
Radix Platycodi 桔梗 (Jie Geng)
Radix Angelicae Sinensis 当归 (Dang Gui)
Radix Asparagi 天冬 (Tian Dong)
Radix Ophiopogonis 麦冬 (Mai Dong)
stir-heated Semen Biotae 炒柏子仁 (Chao Bai Zi Ren)
stir-heated Semen Ziziphi
　Spinosae 炒酸枣仁 (Chao Suan Zao Ren)
Radix Rehmanniae 生地 (Sheng Di)

49. *An Shen Ding Zhi Wan* 安神定志丸
Rhizoma Acori Graminei 菖蒲 (Chang Pu)
Radix Polygalae 远志 (Yuan Zhi)
Dens Draconis 龙齿 (Long Chi)
Poria 茯苓 (Fu Ling)
Poria cum Ligno Hospite 茯神 (Fu Shen)

Radix Codonopsis Pilosulae	党参	(Dang Shen)
Cinnabaris	朱砂	(Zhu Sha)

50. *Gui Pi Tang* 归脾汤

Radix Codonopsis Pilosulae	党参	(Dang Shen)
Radix Astragali seu Hedysari	黄芪	(Huang Qi)
Rhizoma Atractylodis Macrocephalae	白术	(Bai Zhu)
Radix Glycyrrhizae Praeparata	炙甘草	(Zhi Gan Cao)
Poria	茯苓	(Fu Ling)
Radix Polygalae	远志	(Yuan Zhi)
Semen Ziziphi Spinosae	枣仁	(Zao Ren)
Arillus Longan	元肉	(Yuan Rou)
Radix Angelicae Sinensis	当归	(Dang Gui)
Radix Aucklandiae	木香	(Mu Xiang)
Rhizoma Zingiberis Recens	生姜	(Sheng Jiang)
Fructus Ziziphi Jujubae	红枣	(Hong Zao)

51. *Jia Wei Ting Li Da Zao Xie Fei Tang* 加味葶苈大枣泻肺汤

Semen Lepidii seu Descurainiae	葶苈子	(Ting Li Zi)
Fructus Ziziphi Jujubae	大枣	(Da Zao)
Rhizoma Acorus Calamus	水菖蒲	(Shui Chang Pu)

52. *Tiao Ying Yin* 调营饮

Rhizoma Zedoariae	莪术	(E Zhu)
Rhizoma Ligustici Chuanxiong	川芎	(Chuan Xiong)

Radix Angelicae Sinensis	当归	(*Dang Gui*)
Rhizoma Corydalis	元胡	(*Yuan Hu*)
Radix Paeoniae Rubra	赤芍药	(*Chi Shao Yao*)
Herba Dianthi	瞿麦	(*Qu Mai*)
Radix et Rhizoma Rhei	大黄	(*Da Huang*)
Semen Arecae	槟榔	(*Bing Lang*)
Pericarpium Citri Reticulatae	陈皮	(*Chen Pi*)
Pericarpium Arecae	大腹皮	(*Da Fu Pi*)
Semen Lepidii seu Descurainiae	葶苈子	(*Ting Li Zi*)
Poria Rubra	赤茯苓	(*Chi Fu Ling*)
Cortex Mori Radicis	桑白皮	(*Sang Bai Pi*)
Herba Asari	细辛	(*Xi Xin*)
Cortex Cinnamomi	官桂	(*Guan Gui*)
Radix Glycyrrhizae	甘草	(*Gan Cao*)

53. *Ma Xing Tang* 麻杏汤

Herba Ephedrae	麻黄	(*Ma Huang*)
Semen Armeniacae	杏仁	(*Xing Ren*)
Folium Perillae	苏叶	(*Su Ye*)
Radix Peucedani	前胡	(*Qian Hu*)
Radix Glycyrrhizae	甘草	(*Gan Cao*)

54. *Xiao Qing Long Tang* 小青龙汤

Herba Ephedrae	麻黄	(*Ma Huang*)
Ramulus Cinnamomi	桂枝	(*Gui Zhi*)
Herba Asari	细辛	(*Xi Xin*)
Radix Paeoniae Alba	芍药	(*Shao Yao*)
Rhizoma Zingiberis	干姜	(*Gan Jiang*)

Rhizoma Pinelliae 清半夏(Qing Ban Xia)
Fructus Schisandrae 五味子(Wu Wei Zi)
Radix Glycyrrhizae Praeparata 炙甘草(Zhi Gan Cao)

55. *Xiao Ban Xia Tang* 小半夏汤
Rhizoma Pinelliae 半夏 (Ban Xia)
Rhizoma Zingiberis Recens 生姜 (Sheng Jiang)

56. *Yin Qiao Bai Hu Tang* 银翘白虎汤
Gypsum Fibrosum 生石膏(Sheng Shi Gao)
Rhizoma Anemarrhenae 知母 (Zhi Mu)
Semen Oryzae Sativae 粳米 (Geng Mi)
Radix Glycyrrhizae 甘草 (Gan Cao)
Flos Lonicerae 银花 (Yin Hua)
Fructus Forsythiae 连翘 (Lian Qiao)

57. *Zhu Ye Shi Gao Tang* 竹叶石膏汤
Herba Lophatheri 淡竹叶(Dan Zhu Ye)
Gypsum Fibrosum 生石膏(Sheng Shi Gao)
Rhizoma Pinelliae 半夏 (Ban Xia)
Radix Ginseng 人参 (Ren Shen)
Radix Ophiopogonis 麦冬 (Mai Dong)
Radix Glycyrrhizae 甘草 (Gan Cao)
Semen Oryzae Sativae 粳米 (Geng Mi)

58. *Yang Yin Qing Fei Tang* 养阴清肺汤
Radix Rehmanniae 大生地(Da Sheng Di)
Radix Scrophulariae 玄参 (Xuan Shen)
Radix Ophiopogonis 麦冬 (Mai Dong)

stir-heated Radix Paeoniae Alba	炒白芍 (*Chao Shao Yao*)
Bulbus Fritillariae Cirrhosae	贝母 (*Bei Mu*)
Cortex Moutan Radicis	丹皮 (*Dan Pi*)
Herba Menthae	薄荷 (*Bo He*)
Radix Glycyrrhizae	生甘草 (*Sheng Gan Cao*)

59. *Du Shen Tang* 独参汤

| Radix Ginseng | 人参 (*Ren Shen*) |

60. *Hui Yang Fan Ben Tang* 回阳返本汤

Radix Ginseng	人参 (*Ren Shen*)
Radix Ophiopogonis	麦冬 (*Mai Dong*)
Eructus Schisandrae	五味子 (*Wu Wei Zi*)
Radix Aconiti Praeparata	熟附子 (*Shu Fu Zi*)
Rhizoma Zingiberis	干姜 (*Gan Jiang*)
Radix Glycyrrhizae Praeparata	炙甘草 (*Zhi Gan Cao*)

61. *Shen Fu Zhu She Ye Shen Fu* Injection 参附注射液

| Radix Ginseng | 人参 (*Ren Shen*) |
| Radix Aconiti Praeparata | 附子 (*Fu Zi*) |

62. *Wei Jing Yu Xing Tang* 苇茎鱼腥汤

Reed Stem	苇茎 (*Wei Jing*)
Herba Houttuyniae	鱼腥草 (*Yu Xing Cao*)
Semen Coicis	生薏苡米 (*Sheng Yi Mi*)
Semen Benincasae	冬瓜仁 (*Dong Gua Ren*)
Radix Play	桔梗 (*Jie Geng*)
Herba Patriniae	败酱草 (*Bai Jiang Cao*)

Semen Persicae	桃仁	(Tao Ren)
Flos Lonicerae	银花	(Yin Hua)
Fructus Trichosanthis	瓜蒌	(Gua Lou)
Radix Glycyrrhizae	生甘草	(Sheng Gan Cao)

63. Jia Wei Wei Jing Tang 加味苇茎汤

Reed Stem	苇茎	(Wei Jing)
Rhizoma Phragmitis	芦根	(Lu Gen)
Semen Coicis	生苡米	(Sheng Yi Mi)
Semen Benincasae	冬瓜仁	(Dong Gua Ren)
Radix Platycodi	桔梗	(Jie Geng)
Herba Houttuyniae	鱼腥草	(Yu Xing Cao)
Bulbus Fritillariae Thunbergii	浙贝母	(Zhe Bei Mu)
Flos Lonicerae	银花	(Yin Hua)
Nodus Nelumbinis Rhizomatis	藕节	(Ou Jie)
Semen Persicae	桃仁	(Tao Ren)
Pericarpium Citri Reticulatae	陈皮	(Chen Pi)

64. Xiong Zhi Shi Gao Tang 芎芷石膏汤

Rhizoma Ligustici Chuanxiong	川芎	(Chuan Xiong)
Radix Angelicae Dahuricae	白芷	(Bai Zhi)
Gypsum Fibrosum	石膏	(Shi Gao)
Flos Chrysanthemi	菊花	(Ju Hua)
Rhizoma Ligustici	藁本	(Gao Ben)
Rhizoma seu Radix Notopterygii	羌活	(Qiang Huo)

65. Chuan Xiong Cha Tiao San 川芎茶调汤

Rhizoma Ligustici

Chuanxiong	川芎	(Chuan Xiong)
Herba Schizonepetae	荆芥	(Jing Jie)
Rhizoma seu Radix Notopterygii	羌活	(Qiang Huo)
Radix Angelicae Dahuricae	白芷	(Bai Zhi)
Radix Ledebouriellae	防风	(Fang Feng)
Herba Menthae	薄荷	(Bo He)
Herba Asari	细辛	(Xi Xin)
Radix Glycyrrhizae	甘草	(Gan Cao)

66. *Tong Qiao Zhi Tong Tang* 通窍止痛汤

Rhizoma Ligustici Chuanxiong	川芎	(Chuan Xiong)
Radix Paeoniae Rubra	赤芍	(Chi Shao)
Semen Persicae	桃仁	(Tao Ren)
Flos Carthami	红花	(Hong Hua)
Rhizoma Zingiberis Recens	生姜	(Sheng Jiang)
Bulbus Allii Fistulosi	老葱	(Lao Cong)
Moschus	麝香	(She Xiang)

67. *San Pian Tang* 散偏汤

Rhizoma Ligustici Chuanxiong	川芎	(Chuan Xiong)
Radix Angelicae Dahuricae	白芷	(Bai Zhi)
Radix Paeoniae Alba	白芍	(Bai Shao)
Semen Sinapis Albae	白芥子	(Bai Jie Zi)
Rhizoma Cyperi	香附	(Xiang Fu)
Radix Bupleuri	柴胡	(Chai Hu)
Semen Pruni	郁李仁	(Yu Li Ren)

Radix Glycyrrhizae 甘草 (Gan Cao)

68. *Chuan Xiong Ding Tong Tang* 川芎定痛汤
Rhizoma Ligustici Chuanxiong 川芎 (Chuan Xiong)
Radix Paeoniae Rubra 赤芍 (Chi Shao)
Radix Salviae Miltiorrhizae 丹参 (Dan Shen)
Radix Ledebouriellae 防风 (Fang Feng)
Herba Asari 细辛 (Xi Xin)
Radix Aconiti 川乌 (Chuan Wu)
Semen Sinapis Albae 白芥子 (Bai Jie Zi)
Semen Coicis 生苡米 (Sheng Yi Ren)
Fructus Amomi 砂仁 (Sha Ren)

69. *Qing Shang Juan Tong Tang* 清上蠲痛汤
Radix Ophiopogonis 麦冬 (Mai Dong)
Radix Scutellariae 黄芩 (Huang Qin)
Fructus Viticis 蔓荆子 (Man Jing Zi)
Flos Chrysantheri 菊花 (Ju Hua)
Rhizoma seu Radix
　Notopterygii 羌活 (Qiang Huo)
Radix Ledebouriellae 防风 (Fang Feng)
Rhizoma Atractylodis 苍术 (Cang Zhu)
Radix Angelicae Sinensis 当归 (Dang Gui)
Radix Angelicae Dahuricae 白芷 (Bai Zhi)
Rhizoma Ligustici Chuanxiong 川芎 (Chuan Xiong)
Herba Asari 细辛 (Xi Xin)
Radix Glycyrrhizae 甘草 (Gan Cao)

70. *Quan Xie Gou Teng San* 全蝎钩藤散

Scorpio	全蝎	(*Quan Xie*)
Ramulus Uncariae cum Uncis	钩藤	(*Gou Teng*)
Radix Ginseng Rubra	红参	(*Hong Shen*)
Radix Angelicae Dahuricae	白芷	(*Bai Zhi*)
Radix Aconiti Praeparata	制川乌	(*Zhi Chuan Wu*)

71. *Fu Fang Yang Jiao Chong Ji* 复方羊角冲剂

Cornu Naemorhedus Goral Hardwicke	羊角	(*Yang Jiao*)
Rhizoma Ligustici Chuanxiong	川芎	(*Chuan Xiong*)
Radix Angelicae Dahuricae	白芷	(*Bai Zhi*)
Radix Aconiti Praeparata	制川乌	(*Zhi Chuan Wu*)

72. *Zhi Tong San* 止痛散

Borneolum Syntheticum	冰片	(*Bing Pian*)
Natrii Salfus	芒硝	(*Mang Xiao*)
Moschus	麝香	(*She Xiang*)
Menthol	薄荷冰	(*Bo He Bing*)

73. *Jie Du Tong Luo Yin* 解毒通络饮

Rhizoma Polygoni Cuspidati	虎杖	(*Hu Zhang*)
Herba Bidentis	婆婆针	(*Po Po Zhen*)
Radix Notoginseng	土大黄	(*Tu Da Huang*)
Radix Salviae Miltiorrhizae	丹参	(*Dan Shen*)
Flos Lonicerae	银花	(*Yin Hua*)
Rhizoma Dryopteris	贯众	(*Guan Zhong*)

74. *Huo Xue Qu Tong Tang* 活血驱痛汤

Radix Angelicae Sinensis	当归	(*Dang Gui*)

Radix Salviae Miltiorrhizae 丹参 (Dan Shen)
Flos Carthami 红花 (Hong Hua)
Radix Achyranthis Bidentatae 牛膝 (Niu Xi)
Caulis Spatholobi 鸡血藤 (Ji Xue Teng)
Radix Angelicae Dahuricae 白芷 (Bai Zhi)
Rhizoma Corydalis 元胡 (Yuan Hu)
Radix Astragali seu Hedysari 生黄芪 (Sheng Huang Qi)
Radix Glycyrrhizae 生甘草 (Sheng Gan Cao)
Radix Stephaniae Tetrandrae 汉防己 (Han Fang Ji)
Rhizoma seu Radix
   Notopterygii 羌活 (Qiang Huo)
Resina Olibani 乳香 (Ru Xiang)
Myrrha 没药 (Mo Yao)
stir-heated Semen Ziziphi
   Spinosa 炒枣仁 (Chao Zao Ren)

75. Dang Gui Wei Ling Xian Tang 当归威灵仙汤

Radix Angelicae Sinensis 当归 (Dang Gui)
Radix Clematidis 威灵仙 (Wei Ling Xian)
Rhizoma Ligustici Chuanxiong 川芎 (Chuan Xiong)
Radix Angelicae Dahuricae 白芷 (Bai Zhi)
Radix Stephaniae Tetrandrae 汉防己 (Han Fang Ji)
Rhizoma Atractylodis 苍术 (Cang Zhu)
Rhizoma seu Radix
   Notopterygii 羌活 (Qiang Huo)
Ramulus Cinnamomi 桂枝 (Gui Zhi)
Rhizoma Zingiberis Recens 生姜 (Sheng Jiang)

76. Jie Du Huo Xue Tong Luo Tang 解毒活血通络汤

Radix Isatidis	板兰根	(*Ban Lan Gen*)
Folium Isatidis	大青叶	(*Da Qing Ye*)
Radix Sophorae Subprostratae	山豆根	(*Shan Dou Gen*)
Lasiosphaera seu Calvatia	马勃	(*Ma Bo*)
Radix Platycodi	桔梗	(*Jie Geng*)
Ramulus Cinnamomi	桂枝	(*Gui Zhi*)
Radix Salviae Miltiorrhizae	丹参	(*Dan Shen*)
Radix Paeoniae Rubra	赤芍	(*Chi Shao*)
Radix Angelicae Sinensis	当归	(*Dang Gui*)
Radix Achyranthis Bidentatae	牛膝	(*Niu Xi*)
Retinervus Luffae Fructus	丝瓜络	(*Si Gua Luo*)
Radix Astragali Hedysari	生黄芪	(*Sheng Huang Qi*)
Radix Glycyrrhizae	甘草	(*Gan Cao*)

77. *Bu Qi Yang Xue Chu Wei Tang* 补气养血除痿汤

Radix Astragali seu Hedysari	生黄芪	(*Sheng Huang Qi*)
sugared Radix Ginseng	白糖参	(*Bai Tang Shen*)
Rhizoma Polygonati	黄精	(*Huang Jing*)
Radix Codonopsis Pilosulae	党参	(*Dang Shen*)
Radix Angelicae Sinensis	当归	(*Dang Gui*)
Radix Salviae Miltiorrhizae	丹参	(*Dan Shen*)
Radix Achyranthis Bidentatae	牛膝	(*Niu Xi*)
Radix Ophioponogis	麦冬	(*Mai Dong*)
Radix Dipsaci	川断	(*Chuan Duan*)
Ramulus Cinnamomi	桂枝	(*Gui Zhi*)
Rhizoma Cimicifugae	升麻	(*Sheng Ma*)
Rhizoma Homalomenae	千年健	(*Qian Nian Jian*)
Herba Cistanchis	大云	(*Da Yun*)

Caulis Spatholobi 鸡血藤(*Ji Xue Teng*)
stir-heated Fructus Hordei
    Germinatus 炒麦芽(*Chao Mai Ya*)

78. *Du Huo Ji Sheng Tang* 独活寄生汤
Radix Angelicae Pubescentis 独活 (*Du Huo*)
Radix Ledebouriellae 防风 (*Fang Feng*)
Ramulus Loranthi 桑寄生(*Sang Ji Sheng*)
Radix Gentianae
    Macrophyllae 秦艽 (*Qin Jiao*)
Radix Angelicae Sinensis 当归 (*Dang Gui*)
Radix Paeoniae Alba 芍药 (*Shao Yao*)
Cortex Eucommiae 杜仲 (*Du Zhong*)
Radix Rehmanniae Praeparata 熟地 (*Shu Di*)
Radix Codonopsis Pilosulae 党参 (*Dang Shen*)
Poria 茯苓 (*Fu Ling*)
Radix Achyranthis Bidentatae 牛膝 (*Niu Xi*)
Rhizoma Ligustici
    Chuanxiong 川芎 (*Chuan Xiong*)
Herba Asari 细辛 (*Xi Xin*)
Cortex Cinnamomi 肉桂 (*Rou Gui*)
Radix Glycyrrhizae 甘草 (*Gan Cao*)

79. *Di Tan Tang* 涤痰汤
Rhizoma Pinelliae Praeparata 制半夏(*Zhi Ban Xia*)
Rhizoma Arisaematis
    Praeparata 制南星(*Zhi Nan Xing*)
Pericarpium Citri Reticulatae 陈皮 (*Chen Pi*)
Fructus Citri Aurantii

  Immaturus   枳实 (Zhi Shi)
  Poria   茯苓 (Fu Ling)
  Radix Ginseng   人参 (Ren Shen)
  Rhizoma Acori Graminei 石菖蒲(Shi Chang Pu)
  Caulis Bambusae in Taeniam 竹茹 (Zhu Ru)
  Radix Glycyrrhizae   甘草 (Gan Cao)
  Rhizoma Zingiberis Recens 生姜 (Sheng Jiang)

80. *Zhi Bao Dan* 至宝丹
  Cornu Rhinoceri   乌犀屑(Wu Xi Shao)
  Carapax Eretmochelydis 玳瑁屑(Dai Mao Shao)
  Succinum   琥珀 (Hu Po)
  Cinnabaris   朱砂 (Zhu Sha)
  Realgar   雄黄 (Xiong Huang)
  Borneolum Syntheticum   冰片 (Bing Pian)
  Moschus   麝香 (She Xiang)
  Calculus Bovis   牛黄 (Niu Huang)
  Benzoinum   安息香(An Xi Xiang)

81. *Ping Gan Xie Gan Tang* 平肝泻火汤
  Ramulus Uncariae cum Uncis 钩藤 (Gou Teng)
  Flos Chrysanthemi   菊花 (Ju Hua)
  Spica Prunellae   夏枯草(Xia Ku Cao)
  Cortex Moutan Radicis   丹皮 (Dan Pi)
  Concha Margaritifera Usta 珍珠母(Zhen Zhu Mu)
  Radix Achyranthis Bidentatae 淮牛膝(Huai Niu Xi)
  Radix Paeoniae Rubra   赤芍 (Chi Shao)

82. *Dan Gou Liu Zhi Yin* 丹钩六枝饮

Radix Salviae Miltiorrhizae 丹参 (Dan Shen)
Ramulus Uncariae cum Uncis 钩藤 (Gou Teng)
Herba Siegesbeckiae 豨莶草 (Xi Xian Cao)
Spica Purnellae 夏枯草 (Xia Ku Cao)
Lumbricus 地龙 (Di Long)
Flos Carthami 红花 (Hong Hua)
Ramulus Mori 桑枝 (Sang Zhi)
Ramulus Citri Reticulatae 橘枝 (Ju Zhi)
Ramulus Pini 松枝 (Song Zhi)
Ramulus Persicae 桃枝 (Tao Zhi)
Ramulus Abies 杉枝 (Shan Zhi)
Ramulus Bambosae 竹枝 (Zhu Zhi)
Radix Glycyrrhizae 甘草 (Gan Cao)

83. *Li Xiu Lin Shi Fan Wei Jian Zheng Fan* 李秀林氏犯胃兼证方

Ochra Haematitum 代赭石 (Dai Zhe Shi)
Gypsum Fibrosum 生石膏 (Sheng Shi Gao)
Radix Paeoniae Alba 生白芍 (Sheng Bai Shao)
ginger juice-prepared Caulis
  Bambusae in Taeniam 姜竹茹 (Jiang Zhu Ru)
Exocarpium Citri Grandis 化橘红 (Hua Ju Hong)
Rhizoma Acori Graminei 石菖蒲 (Shi Chang Pu)
Poria 云茯苓 (Yun Fu Ling)
Semen Amomi Cardamomi 白豆蔻 (Bai Dou Kou)
Herba Eupatorii 佩兰 (Pei Lan)

84. *Ping Gan Huo Xue Tang* 平肝活血汤

powder of Cornu Antelopis 羚羊角粉 (Ling Yang Jiao Fen)

Concha Haliotidis	石决明	(Shi Jue Ming)
Ramulus Uncariae cum Uncis	钩藤	(Gou Teng)
Radix Achyranthis Bidentatae	牛膝	(Niu Xi)
Rhizoma Ligustici Chuanxiong	川芎	(Chuan Xiong)
Flos Carthami	红花	(Hong Hua)
Semen Persicae	桃仁	(Tao Ren)
Eupolyphaga seu Steleophaga	䗪虫	(Zhe Chong)

85. *Hua Tan Tong Luo Yin* 化痰通络饮

Rhizoma Pinelliae Praeparata	法半夏	(Fa Ban Xia)
Rhizoma Atractylodis Macrocephalae	生白术	(Sheng Bai Zhu)
Rhizoma Gastrodiae	天麻	(Tian Ma)
Arisaema cum Bile	胆南星	(Dan Nan Xing)
Radix Salviae Miltiorrhizae	丹参	(Dan Shen)
Rhizoma Cyperi	香附	(Xiang Fu)
Radix et Rhizoma Rhei Praeparata	酒军	(Jiu Jun)

86. *Xi Feng Hua Tan Tang* 息风化痰汤

Poria	茯苓	(Fu Ling)
Rhizoma Pinelliae	半夏	(Ban Xia)
Exocarpium Citri Grandis	橘红	(Ju Hong)
Caulis Bambusae in Taeniam	竹茹	(Zhu Ru)
Rhizoma Acori Graminei	石菖蒲	(Shi Chang Pu)
Arisaema cum Bile	胆星	(Dan Xing)
Bombyx Batryticatus	僵蚕	(Jiang Can)
Ramulus Uncariae cum Uncis	钩藤	(Gou Teng)
Scorpio	全蝎	(Quan Xie)

Herba Siegesbeckiae   豨莶草 (*Xi Xian Cao*)
Semen Persicae   桃仁 (*Tao Ren*)
Flos Carthami   红花 (*Hong Hua*)

87. *Tong Fu Hua Tan Yin* 通腑化痰饮
Radix et Rhizoma Rhei   生大黄 (*Sheng Da Huang*)
Natrii Sulfas   芒硝 (*Mang Xiao*)
Fructus Trichosanthis   瓜蒌 (*Gua Lou*)
Arisaema cum Bile   胆南星 (*Dan Nan Xing*)

88. *Da Huang Gua Lou Tang* 大黄瓜蒌汤
Radix et Rhizoma Rhei   大黄 (*Da Huang*)
Fructus Trichosanthis   瓜蒌 (*Gua Lou*)
Eupolyphaga seu Steleophaga   土元 (*Tu Yuan*)
Lumbricus   地龙 (*Di Long*)
Semen Sinapis Albae   白芥子 (*Bai Jie Zi*)

89. *Yi Qi Huo Xue Tang* 益气活血汤
Radix Astragali seu Hedysari   生黄芪 (*Sheng Huang Qi*)
Semen Persicae   桃仁 (*Tao Ren*)
Flos Carthami   红花 (*Hong Hua*)
Radix Paeoniae Rubra   赤芍 (*Chi Shao*)
tail of Radix Angelicae
  Sinensis   归尾 (*Gui Wei*)
Lumbricus   地龙 (*Di Long*)
Rhizoma Ligustici
  Chuanxiong   川芎 (*Chuan Xiong*)

90. *Huang Qi Wu Wu Tang* 黄芪五物汤

Radix Astragali seu Hedysari	黄芪	(*Huang Qi*)
Ramulus Cinnamomi	桂枝	(*Gui Zhi*)
Radix Paeoniae Albae	芍药	(*Shao Yao*)
Rhizoma Zingiberis Recens	生姜	(*Sheng Jiang*)
Fructus Ziziphi Jujubae	大枣	(*Da Zao*)

91. *Xi Xian Zhi Yin Tang* 豨莶至阴汤

Herba Siegesbeckiae Praeparata	制豨莶草	(*Zhi Xi Xian Cao*)
Radix Rehmanniae	干地黄	(*Gan Di Huang*)
salt solution—prepared Rhizoma Anemarrhenae	盐知母	(*Yan Zhi Mu*)
Radix Angelicae Sinensis	当归	(*Dang Gui*)
Fructus Lycii	枸杞子	(*Gou Qi Zi*)
stir—heated Radix Paeoniae Rubra	炒赤芍	(*Chao Chi Shao*)
Plastrum Testudinis	龟板	(*Gui Ban*)
Radix Achyranthis Bidentatae	牛膝	(*Niu Xi*)
Flos Chysanthemi	甘菊花	(*Gan Ju Hua*)
Rhizoma Curcumae	郁金	(*Yu Jin*)
Radix Salviae Miltiorrhizae	丹参	(*Dan Shen*)
Cortex Phellodendri	黄柏	(*Huang Bai*)

92. *Yu Yin Xi Feng Tang* 育阴息风汤

Radix Rehmanniae	生地	(*Sheng Di*)
Radix Scrophulariae	玄参	(*Xuan Shen*)
Fructus Ligustri Lucidi	女贞子	(*Nu Zhen Zi*)
Ramulus Loranthi	桑寄生	(*Sang Ji Sheng*)
Ramulus Uncariae cum Uncis	钩藤	(*Gou Teng*)

Radix Paeoniae Albae	白芍	(Bai Shao)
Radix Salviae Miltiorrhizae	丹参	(Dan Shen)

93. *Jia Jian Ling Jiao Gou Teng Tang* 加减羚角钩藤汤

powder of Cornu Antelopis	羚羊角粉	(Ling Yang Jiao Fen)
Concha Haliotidis	生石决明	(Sheng Shi Jue Ming)
Plastrum Testudinis	生龟板	(Sheng Gui Ban)
Os Draconis	生龙骨	(Sheng Long Gu)
Concha Ostreae	生牡蛎	(Sheng Mu Li)
Radix Paeoniae Albae	白芍	(Bai Shao)
Ramulus Uncariae cum Uncis	钩藤	(Gou Teng)
Radix Achyranthis Bidentatae	牛膝	(Niu Xi)
Radix Rehmanniae	生地	(Sheng Di)
Spica Prunellae	夏枯草	(Xia Ku Cao)
Cortex Moutan Radicis	丹皮	(Dan Pi)

94. *Bao He Wan* 保和丸

stir-heated Fructus Crataegi	炒山楂	(Chao Shan Zha)
Rhizoma Pinelliae	半夏	(Ban Xia)
Poria	茯苓	(Fu Ling)
Pericarpium Citri Reticulatae	陈皮	(Chen Pi)
Semen Raphani	莱菔子	(Lai Fu Zi)
Fructus Forsythiae	连翘	(Lian Qiao)
Massa Fermentata Medicinalis	神曲	(Shen Qu)

95. *Xie Xin Tang* 泻心汤

Radix et Rhizoma Rhei	大黄	(Da Huang)
Rhizoma Coptidis	黄连	(Huang Lian)
Radix Scutellariae	黄芩	(Huang Qin)

96. *Shi Hui San* 十灰散
Herba seu Radix Cirsii
    Japonici   大蓟   (*Da Ji*)
Herba Cephalanoploris   小蓟   (*Xiao Ji*)
Folium Nelumbinis   荷叶   (*He Ye*)
Cacumen Biotae   侧柏叶(*Ce Bai Ye*)
Rhizoma Imperatae   茅根   (*Mao Gen*)
Radix Rubiae   茜草根(*Qian Cao Gen*)
Radix et Rhizoma Rhei   大黄   (*Da Huang*)
Fructus Gardeniae   山栀   (*Shan Zhi*)
Cortex Trachycarpi   棕榈皮(*Zong Lu Pi*)
Cortex Moutan Radicis   牡丹皮(*Mu Dan Pi*)

97. *Long Dan Xie Gan Tang* 龙胆泻肝汤
Radix Gentianae   龙胆草(*Long Dan Cao*)
Radix Scutellariae   黄芩   (*Huang Qin*)
Fructus Gardeniae   山栀   (*Shan Zhi*)
Caulis Aristolochiae
    Manshuriensis   木通   (*Mu Tong*)
Rhizoma Alismatis   泽泻   (*Ze Xie*)
Radix Rehmanniae   生地   (*Sheng Di*)
Radix Bupleuri   柴胡   (*Chai Hu*)
Semen Plantaginis   车前子(*Che Qian Zi*)
Radix Angelicae Sinensis   当归   (*Dang Gui*)
Radix Glycyrrhizae   甘草   (*Gan Cao*)

98. *Zhi Zhu Wan* 枳术丸
stir-heated Fructus Aurantii

Immaturus	炒枳实	(Chao Zhi Shi)
Rhizoma Atractylodis Macrocephalae	白术	(Bai Zhu)

99. *Mu Xiang Bing Lang Wan* 木香槟榔丸

Radix Aucklandiae	木香	(Mu Xiang)
Semen Arecae	槟榔	(Bing Lang)
Pericarpium Citri Reticulatae Viride	青皮	(Qing Pi)
Pericarpium Citri Reticulatae	陈皮	(Chen Pi)
Rhizoma Zedoariae	莪术	(E Zhu)
Rhizoma Coptidis	黄连	(Huang Lian)
Cortex Phellodendri	黄柏	(Huang Bai)
Radix et Rhizoma Rhei	大黄	(Da Huang)
Rhizoma Cyperi Praeparata	制香附	(Zhi Xiang Fu)
Fructus Aurantii	枳壳	(Zhi Ke)
Semen Pharbitidis	牵牛	(Qian Niu)

100. *Gua Di San* 瓜蒂散

Muskmelon Pedicel	瓜蒂	(Gua Di)
Semen Phaseoli	赤小豆	(Chi Xiao Dou)
Semen Sojae Praeparatum	豆豉	(Dou Chi)

101. *Yin Chen Zhu Fu Tang* 茵陈术附汤

Herba Artemisiae Scopariae	茵陈	(Yin Chen)
Rhizoma Atractylodis Macrocephalae	白术	(Bai Zhu)
Radix Aconiti Praeparata	附子	(Fu Zi)
Rhizoma Zingiberis	干姜	(Gan Jiang)

Radix Glycyrrhizae 甘草 (Gan Cao)

102. Su He Xiang Wan 苏合香丸

Styrax Li Quidus 苏合香 (Su He Xiang)
Cinnabaris 朱砂 (Zhu Sha)
Cornu Rhinoceri 犀角 (Xi Jiao)
Moschus 麝香 (She Xiang)
Lignum Santali 檀香 (Tan Xiang)

103. Qian Jin Xi Jiao San 千金犀角散

Cornu Rhinoceri 犀角 (Xi Jiao)
Rhizoma Coptidis 黄连 (Huang Lian)
Rhizoma Cimicifugae 升麻 (Sheng Ma)
Fructus Gardeniae 山栀 (Shan Zhi)
Herba Artemisiae Scopariae 茵陈 (Yin Chen)

104. Wu Mei Wan 乌梅丸

Fructus Mume 乌梅 (Wu Mei)
Herba Asari 细辛 (Xi Xin)
Pericarpium Zanthoxyli 川椒 (Chuan Jiao)
Rhizoma Zingiberis 干姜 (Gan Jiang)
Radix Aconiti Praeparata 附子 (Fu Zi)
Cortex Cinnamomi 肉桂 (Rou Gui)
Rhizoma Coptidis 黄连 (Huang Lian)
Cortex Phellodendri 黄柏 (Huang Bai)
Radix Angelicae Sinensis 当归 (Dang Gui)
Radix Codonopsis Pilosulae 党参 (Dang Shen)

105. Qu Hui Li Dan Tang 驱蛔利胆汤

Herba Artemisiae Scopariae	茵陈	(Yin Chen)
Flos Lonicerae	金银花	(Jin Yin Hua)
Radix Scutellariae	黄芩	(Huang Qin)
Radix Bupleuri	柴胡	(Chai Hu)
Radix et Rhizoma Rhei	大黄	(Da Huang)
Fructus Meliae Toosendan	川楝子	(Chuan Lian Zi)
Fructus Quisqualis	使君子	(Shi Jun Zi)
Semen Arecae	槟榔	(Bing Lang)
Cortex Meliae	苦楝根皮	(Ku Lian Gen Pi)
Radix Glycyrrhizae	甘草	(Gan Cao)
Natrii Sulfas	元明粉	(Yuan Ming Fen)

106. Chai Hu Shu Gan Yin 柴胡疏肝散

Radix Bupleuri	柴胡	(Chai Hu)
Fructus Aurantii	枳壳	(Zhi Ke)
Radix Paeoniae Alba	白芍	(Bai Shao)
Rhizoma Ligustici Chuanxiong	川芎	(Chuan Xiong)
Rhizoma Cyperi Praeparata	制香附	(Zhi Xiang Fu)
Radix Glycyrrhizae	炙甘草	(Zhi Gan Cao)

107. Da Chai Hu Tang 大柴胡汤

Radix Bupleuri	柴胡	(Chai Hu)
Rhizoma Pinelliae	半夏	(Ban Xia)
Radix et Rhizoma Rhei	大黄	(Da Huang)
Radix Scutellariae	黄芩	(Huang Qin)
Radix Paeoniae Alba	芍药	(Shao Yao)
Fructus Aurantii Immaturus	枳实	(Zhi Shi)
Rhizoma Zingiberis Recens	生姜	(Sheng Jiang)

Fructus Ziziphi Jujubae　　　大枣　(*Da Zao*)

108. *Da Xian Xiong Tang* 大陷胸汤
Radix et Rhizoma Rhei　　　大黄　(*Da Huang*)
Natrii Sulfas　　　　　　　　芒硝　(*Mang Xiao*)
Radix Euphorbiae Kansui　　甘遂　(*Gan Sui*)

109. *Lan Wei Qing Hua Tang* 阑尾清化汤
Flos Lonicerae　　　　　　　双花　(*Shuang Hua*)
Herba Taraxaci　　　　　　　蒲公英(*Pu Gong Ying*)
Cortex Moutan Radicis　　　丹皮　(*Dan Pi*)
Radix et Rhizoma Rhei　　　大黄　(*Da Huang*)
Fructus Meliae Toosendan　川楝子(*Chuan Lian Zi*)
Radix Paeoniae Rubra　　　赤芍　(*Chi Shao*)
Semen Persicae　　　　　　　桃仁　(*Tao Ren*)
Radix Glycyrrhizae　　　　　生甘草(*Sheng Gan Cao*)

110. *Lan Wei Hua Yu Tang* 阑尾化瘀汤
Fructus Meliae Toosendan　川楝子(*Chuan Lian Zi*)
Rhizoma Corydalis　　　　　玄胡索(*Xuan Hu Suo*)
Cortex Moutan Radicis　　　牡丹皮(*Mu Dan Pi*)
Semen Persicae　　　　　　　桃仁　(*Tao Ren*)
Radix Aucklandiae　　　　　木香　(*Mu Xiang*)
Flos Lonicerae　　　　　　　双花　(*Shuang Hua*)
Radix et Rhizoma Rhei　　　大黄　(*Da Huang*)

111. *Da Huang Mu Dan Pi Tang* 大黄牡丹皮汤
Radix et Rhizoma Rhei　　　大黄　(*Da Huang*)
Cortex Moutan Radicis　　　丹皮　(*Dan Pi*)

Semen Persicae	桃仁	(Tao Ren)
Semen Benincasae	冬瓜仁	(Dong Gua Ren)
Natrii Sulfas	芒硝	(Mang Xiao)

112. *Lan Wei Jie Du Tang* 阑尾解毒汤

Caulis Sargentodoxae	红藤	(Hong Teng)
Herba Patriniae	败酱	(Bai Jiang)
Flos Lonicerae	金银花	(Jin Yin Hua)
Herba Taraxaci	蒲公英	(Pu Gong Ying)
Semen Benincasae	冬瓜仁	(Dong Gua Ren)
Radix Paeoniae Rubra	赤芍	(Chi Shao)
Radix et Rhizoma Rhei	大黄	(Da Huang)
Radix Aucklandiae	木香	(Mu Xiang)
Radix Scutellariae	黄芩	(Huang Qin)
Semen Persicae	桃仁	(Tao Ren)
Fructus Meliae Toosendan	川楝子	(Chuan Lian Zi)

113. *Xiang Sha Liu Jun Zi Tang* 香砂六君子汤

Radix Aucklandiae	木香	(Mu Xiang)
Fructus Amomi	砂仁	(Sha Ren)
Pericarpium Citri Reticulatae	陈皮	(Chen Pi)
Rhizoma Pinelliae	半夏	(Ban Xia)
Radix Codonopsis Pilosulae	党参	(Dang Shen)
Rhizoma Atractylodis Macrocephalae	白术	(Bai Zhu)
Poria	茯苓	(Fu Ling)
Radix Glycyrrhizae	甘草	(Gan Cao)

114. *Huang Lian Jie Du Tang* 黄连解毒汤

Rhizoma Coptidis	黄连	(*Huang Lian*)
Radix Scutellariae	黄芩	(*Huang Qin*)
Cortex Phellodendri	黄柏	(*Huang Bai*)
Fructus Gardeniae	栀子	(*Zhi Zi*)

115. *Huang Lian Tang* 黄连汤

Rhizoma Coptidis	黄连	(*Huang Lian*)
Rhizoma Zingiberis	干姜	(*Gan Jiang*)
Ramulus Cinnamomi	桂枝	(*Gui Zhi*)
Rhizoma Pinelliae	半夏	(*Ban Xia*)
Codonopsis Pilosulae	党参	(*Dang Shen*)
Radix Glycyrrhizae	炙甘草	(*Zhi Gan Cao*)
Fructus Ziziphi Jujubae	大枣	(*Da Zao*)

116. *Shen Ling Bai Zhu San* 参苓白术散

Radix Ginseng	人参	(*Ren Shen*)
Poria	茯苓	(*Fu Ling*)
Rhizoma Atractylodis Macrocephalae	白术	(*Bai Zhu*)
Radix Platycodi	桔梗	(*Jie Geng*)
Rhizoma Dioscoreae	山药	(*Shan Yao*)
Radix Glycyrrhizae	甘草	(*Gan Cao*)
Semen Dolichoris Album	白扁豆	(*Bai Bian Dou*)
Semen Nelumbinis	莲子仁	(*Lian Zi Ren*)
Fructus Amomi	砂仁	(*Sha Ren*)
Semen Coicis	薏苡仁	(*Yi Yi Ren*)

117. *Qing Wen Bai Du Yin* 清瘟败毒饮

Gypsum Fibrosum	生石膏	(*Sheng Shi Gao*)

Radix Rehmanniae	生地	(*Sheng Di*)
Cornu Rhinoceri	犀角	(*Xi Jiao*)
Rhizoma Coptidis	川连	(*Chuan Lian*)
Fructus Gardeniae	栀子	(*Zhi Zi*)
Radix Platycodi	桔梗	(*Jie Geng*)
Radix Scutellariae	黄芩	(*Huang Qin*)
Rhizoma Anemarrhenae	知母	(*Zhi Mu*)
Radix Paeoniae	赤芍	(*Chi Shao*)
Radix Scrophulariae	玄参	(*Xuan Shen*)
Fructus Forsythiae	连翘	(*Lian Qiao*)
Radix Glycyrrhizae	甘草	(*Gan Cao*)
Cortex Moutan Radicis	丹皮	(*Dan Pi*)
Herba Lophatheri	竹叶	(*Zhu Ye*)

118. *Liu Wei Di Huang Wan* 六味地黄丸

Radix Rehmanniae Praeparata	熟地黄	(*Shu Di Huang*)
Rhizoma Dioscoreae	山药	(*Shan Yao*)
Poria	茯苓	(*Fu Ling*)
Cortex Moutan Radicis	丹皮	(*Dan Pi*)
Rhizoma Alismatis	泽泻	(*Ze Xie*)
Fructus Corni	山茱萸	(*Shan Zhu Yu*)

119. *Suo Quan Wan* 缩泉丸

Radix Linderae	乌药	(*Wu Yao*)
Fructus Alpiniae Oxyphyllae	益智仁	(*Yi Zhi Ren*)
Rhizoma Dioscoreae	山药	(*Shan Yao*)

120. *Du Qi Wan* 都气丸

Radix Rehmanniae Praeparata	熟地黄	(*Shu Di Huang*)

Rhizoma Dioscoreae	山药	(Shan Yao)
Poria	茯苓	(Fu Ling)
Cortex Moutan Radicis	丹皮	(Dan Pi)
Rhizoma Alismatis	泽泻	(Ze Xie)
Fructus Corni	山茱萸	(Shan Zhu Yu)
Fructus Schisandrae	五味子	(Wu Wei Zi)

121. Da Huang Fu Zi Tang 大黄附子汤

Radix et Rhizoma Rhei	大黄	(Da Huang)
Radix Aconiti Praeparata	附子	(Fu Zi)
Herba Asari	细辛	(Xi Xin)

122. Feng Sui Dan 封髓丹

Radix Asparagi	天冬	(Tian Dong)
Radix Rehmanniae Praeparata	熟地	(Shu Di)
Radix Ginseng	人参	(Ren Shen)
Cortex Phellodendri	黄柏	(Huang Bai)
Fructus Amomi	砂仁	(Sha Ren)
Radix Glycyrrhizae	甘草	(Gan Cao)

123. Zhi Bai Di Huang Tang 知柏地黄汤

Radix Rehmanniae Praeparata	熟地	(Shu Di)
Rhizoma Dioscoreae	山药	(Shan Yao)
Fructus Corni	山萸肉	(Shan Yu Rou)
Rhizoma Alismatis	泽泻	(Ze Xie)
Cortex Moutan Radicis	丹皮	(Dan Pi)
Poria	茯苓	(Fu Ling)
Cortex Phellodendri	黄柏	(Huang Bai)
Rhizoma Anemarrhenae	知母	(Zhi Mu)

124. Shi Wei San 石苇散

Folium Pyrrosiae	石苇	(Shi Wei)
Fructus Malvae Vertillatae	冬葵	(Dong Kui)
Herba Dianthi	瞿麦	(Qu Mai)
Talcum	滑石	(Hua Shi)
Semen Plantaginis	车前子	(Che Qian Zi)

125. Dai Di Dang Wan 代抵当丸

Radix et Rhizoma Rhei	大黄	(Da Huang)
tail of Radix Angelicae Sinensis	归尾	(Gui Wei)
Radix Rehmanniae	生地	(Sheng Di)
Squama Manitis	山甲片	(Shan Jia Pian)
Natrii Sulfas	芒硝	(Mang Xiao)
Semen Persicae	桃仁	(Tao Ren)
Cortex Cinnamomi	肉桂	(Rou Gui)

126. Ji Sheng Shen Qi Wan 济生肾气丸

Radix Rehmanniae	地黄	(Di Huang)
Rhizoma Dioscoreae	山药	(Shan Yao)
Fructus Corni	山茱萸	(Shan Zhu Yu)
Cortex Moutan Radicis	丹皮	(Dan Pi)
Poria	云苓	(Yun Ling)
Rhizoma Alismatis	泽泻	(Ze Xie)
Radix Aconiti Praeparata	炮附子	(Pao Fu Zi)
Ramulus Cinnamomi	桂枝	(Gui Zhi)
Radix Achyranthis Bidentatae	牛膝	(Niu Xi)
Semen Plantaginis	车前子	(Che Qian Zi)

127. Niu Bang Jie Ji Tang 牛蒡解肌汤

Fructus Arctii	牛蒡子 (Niu Bang Zi)
Herba Menthae	薄荷　(Bo He)
Herba Schizonepetae	荆芥　(Jing Jie)
Fructus Forsythiae	连翘　(Lian Qiao)
Fructus Gardeniae	山栀子 (Shan Zhi Zi)
Cortex Moutan Radicis	丹皮　(Dan Pi)
Herba Dendrobii	石斛　(Shi Hu)
Radix Scrophulariae	玄参　(Xuan Shen)
Spica Prunellae	夏枯草　(Xia Ku Cao)

128. Jin Huang San 金黄散

Radix Trichosanthis	天花粉 (Tian Hua Fen)
Cortex Phellodendri	黄柏　(Huang Bai)
Radix et Rhizoma Rhei	大黄　(Da Huang)
Rhizoma Curcumae Longae	姜黄　(Jiang Huang)
Radix Angelicae Dahuricae	白芷　(Bai Zhi)
Cortex Magnoliae Officinalis	厚朴　(Hou Po)
Pericarpium Citri Reticulatae	陈皮　(Chen Pi)
Radix Glycyrrhizae	甘草　(Gan Cao)
Rhizoma Atractylodis	苍术　(Cang Zhu)
Rhizoma Arisaematis	天南星 (Tian Nan Xing)

129. Yi Guan Jian 一贯煎

Radix Adenophorae	沙参　(Sha Shen)
Radix Ophiopogonis	麦冬　(Mai Dong)
Radix Angelicae Sinensis	当归　(Dang Gui)
Radix Rehmanniae	地黄　(Di Huang)

Fructus Lycii 枸杞子(Gou Qi Zi)
Fructus Meliae Teesendan 川楝子(Chuan Lian Zi)

130. Di Huang Yin Zi 地黄饮子
Radix Ginseng 人参 (Ren Shen)
Radix Rehmanniae 生地黄(Shen Di Huang)
Radix Rehmanniae Praeparata 熟地黄(Shu Di Huang)
Radix Astragali seu Hedysari 黄芪 (Huang Qi)
Radix Asparagi 天冬 (Tian Dong)
Radix Ophiopogonis 麦冬 (Mai Dong)
Rhizoma Alismatis 泽泻 (Ze Xie)
Herba Dendrobii 石斛 (Shi Hu)
Folium Eriobotryae 枇杷叶(Pi Pa Ye)
Fructus Aurantii 枳壳 (Zhi Ke)
Radix Glycyrrhizae 甘草 (Gan Cao)

131. Wen Dan Tang 温胆汤
Rhizoma Pinelliae 半夏 (Ban Xia)
Exocarpium Citri Grandis 橘红 (Ju Hong)
Poria 茯苓 (Fu Ling)
Radix Glycyrrhizae Praeparata 炙甘草 (Zhi Gan Cao)
Caulis Bambusae in Taeniam 竹茹 (Zhu Ru)
Fructus Aurantii Immaturus 枳实 (Zhi Shi)
Rhizoma Zingiberis Recens 生姜 (Sheng Jiang)
Fructus Ziziphi Jujubae 大枣 (Da Zao)

132. Ba Zhen Tang 八珍汤
Radix Ginseng 人参 (Ren Shen)

Rhizoma Atractylodis Macrocephalae	白术	(*Bai Zhu*)
Poria	茯苓	(*Fu Ling*)
Radix Glycyrrhizae	甘草	(*Gan Cao*)
Radix Angelicae Sinensis	当归	(*Dang Gui*)
Rhizoma Ligustici Chuanxiong	川芎	(*Chuan Xiong*)
Radix Rehmanniae Praeparata	熟地	(*Shu Di*)
Radix Paeoniae Alba	白芍	(*Bai Shao*)

133. *Si Jun Zi Tang* 四君子汤

Radix Ginseng	人参	(*Ren Shen*)
Rhizoma Atractylodis Macrocephalae	白术	(*Bai Zhu*)
Poria	茯苓	(*Fu Ling*)
Radix Glycyrrhizae	甘草	(*Gan Cao*)

134. *You Gui Wan* 右归丸

Radix Rehmanniae Praeparata	熟地	(*Shu Di*)
Rhizoma Dioscoreae	山药	(*Shan Yao*)
Fructus Corni	山茱萸	(*Shan Zhu Yu*)
Radix Angelicae Sinensis	当归	(*Dang Gui*)
Fructus Lycii	枸杞子	(*Gou Qi Zi*)
Colla Cornu Cervi	鹿角胶	(*Lu Jiao Jiao*)
Cortex Eucommiae	杜仲	(*Du Zhong*)
Semen Cuscutae	菟丝子	(*Tu Si Zi*)

135. *Gui Shao Di Huang Tang* 归芍地黄汤

Radix Rehmanniae Praeparata	熟地	(*Shu Di*)
Rhizoma Dioscoreae	山药	(*Shan Yao*)

Fructus Corni 山萸肉(Shan Yu Rou)
Cortex Moutan Radicis 丹皮 (Dan Pi)
Rhizoma Alismatis 泽泻 (Ze Xie)
Radix Angelicae Sinensis 当归 (Dang Gui)
Radix Paeoniae Alba 白芍 (Bai Shao)
Poria 云苓 (Yun Ling)

136. *Jin Kui Shen Qi Wan* 金匮肾气丸
Radix Rehmanniae Praeparata 熟地 (Shu Di)
Fructus Corni 山茱萸(Shan Zhu Yu)
Rhizoma Dioscoreae 山药 (Shan Yao)
Poria 云苓 (Yun Ling)
Cortex Moutan Radicis 丹皮 (Dan Pi)
Rhizoma Alismatis 泽泻 (Ze Xie)
Ramulus Cinnamomi 桂心 (Gui Xin)
Radix Aconiti Praeparata 附子 (Fu Zi)

137. *Yang Xin Tang* 养心汤
Radix Codonopsis Pilosulae 党参 (Dang Shen)
Radix Angelicae Sinensis 当归 (Dang Gui)
Poria 茯苓 (Fu Ling)
Poria cum Ligno Hospite 茯神 (Fu Shen)
Semen Biotae 柏子仁 (Bai Zi Ren)
Radix Astragali seu Hedysari
 Praeparata 炙黄芪(Zhi Huang Qi)
Semen Ziziphi Spinosae 酸枣仁(Suan Zao Ren)
Radix Polygalae 远志 (Yuan Zhi)
Rhizoma Ligustici
 Chuanxiong 川芎 (Chuan Xiong)

Cortex Cinnamomi 肉桂 (*Rou Gui*)
Fructus Schisandrae 五味子 (*Wu Wei Zi*)
fermented Rhizoma Pinelliae 半夏曲 (*Ban Xia Qu*)
Radix Glycyrrhizae Praeparata 炙甘草 (*Zhi Gan Cao*)

138. *Si Wu Tang* 四物汤
Radix Rehmanniae Praeparata 熟地 (*Shu Di*)
Radix Angelicae Sinensis 当归 (*Dang Gui*)
Radix Paeoniae Alba 白芍 (*Bai Shao*)
Rhizoma Ligustici Chuanxiong 川芎 (*Chuan Xiong*)

139. *Zi Yin Jiang Huo Tang* 滋阴降火汤
Radix Paeoniae Alba 白芍 (*Bai Shao*)
Radix Angelicae Sinensis 当归 (*Dang Gui*)
Radix Rehmanniae Praeparata 熟地黄 (*Shu Di Huang*)
Radix Ophiopogonis 麦冬 (*Mai Dong*)
Rhizoma Atractylodis
    Macrocephalae 白术 (*Bai Zhu*)
Radix Rehmanniae 生地 (*Sheng Di*)
Pericarpium Citri Reticulatae 陈皮 (*Chen Pi*)
Rhizoma Anemarrhenae 知母 (*Zhi Mu*)
Cortex Phellodendri 黄柏 (*Huang Bai*)
Rhizoma Zingiberis Recens 生姜 (*Sheng Jiang*)
Fructus Ziziphi Jujubae 大枣 (*Da Zao*)

140. *Xiao ji Yin Zi* 小蓟饮子
Radix Rehmanniae 生地 (*Sheng Di*)
Caulis Aristolochiae
    Manshuriensis 木通 (*Mu Tong*)

tip of Radix Glycyrrhizae	生甘草梢	(*Sheng Gan Cao Shao*)
Herba Lophatheri	竹叶	(*Zhu Ye*)
Herba Cephalanoploris	小蓟	(*Xiao Ji*)
Nodus NelumBinis	藕节	(*Ou Jie*)
Pollen Typhae	蒲黄	(*Pu Huang*)
Radix Angelicae Sinensis	当归	(*Dang Gui*)
Talcum	滑石	(*Hua Shi*)
Fructus Gardeniae	山栀	(*Shan Zhi*)

141. *Bu Zhong Yi Qi Tang* 补中益气汤

Radix Astragali seu Hedysari	黄芪	(*Huang Qi*)
Radix Ginseng	人参	(*Ren Shen*)
Radix Angelicae Radicis	当归	(*Dang Gui*)
Pericarpium Citri Reticulatae	橘皮	(*Ju Pi*)
Rhizoma Cimicifugae	升麻	(*Sheng Ma*)
Radix Bupleuri	柴胡	(*Chai Hu*)
Rhizoma Atractylodis Macrocephalae	白术	(*Bai Zhu*)
Radix Glycyrrhizae	甘草	(*Gan Cao*)

142. *Wu Bi Shan Yao Wan* 无比山药丸

Rhizoma Dioscoreae	山药	(*Shan Yao*)
Herba Cistanchis	肉苁蓉	(*Rou Cong Rong*)
Radix Rehmanniae Praeparata	熟地黄	(*Shu Di Huang*)
Fructus Corni	山茱萸	(*Shan Zhu Yu*)
Poria cum Ligno Hospite	茯神	(*Fu Shen*)
Semen Cuscutae	菟丝子	(*Tu Si Zi*)
Semen Fructus Schisandrae	五味子	(*Wu Wei Zi*)

Halloysitum Rubrum 赤石脂 (Chi Shi Zhi)
Radix Morindae Officinalis 巴戟天 (Ba Ji Tian)
Rhizoma Alismatis 泽泻 (Ze Xie)
Cortex Eucommiae 杜仲 (Du Zhong)
Radix Achyranthis Bidentatae 牛膝 (Niu Xi)

143. *Qing Hao Bie Jia Tang* 青蒿鳖甲汤
Herba Artemisiae Chinghao 青蒿 (Qing Hao)
Carapax Trionycis 鳖甲 (Bie Jia)
Radix Rehmanniae 细生地 (Xi Sheng Di)
Rhizoma Anemarrhenae 知母 (Zhi Mu)
Cortex Moutan Radicis 丹皮 (Dan Pi)

144. *Jing Fang Bai Du San* 荆防败毒散
Radix Ginseng 人参 (Ren Shen)
Rhizoma seu Radix
 Notopterygii 羌活 (Qing Huo)
Radix Angelicae Pubescentis 独活 (Du Huo)
Radix Bupleuri 柴胡 (Chai Hu)
Radix Peucedani 前胡 (Qian Hu)
Rhizoma Ligustici Chuanxiong 川芎 (Chuan Xiong)
Fructus Aurantii 枳壳 (Zhi Ke)
Radix Platycodi 桔梗 (Jie Geng)
Herba Menthae 薄荷 (Bo He)
Herba Schizonepetae 荆芥 (Jing Jie)
Radix Ledebouriellae 防风 (Fang Feng)
Poria 茯苓 (Fu Ling)
Radix Glycyrrhizae 甘草 (Gan Cao)

145. *Jia Jian Fu Mai Tang* 加减复脉汤

Radix Glycyrrhizae Praeparata	炙甘草	(Zhi Gan Cao)
Radix Rehmanniae	生地	(Sheng Di)
Radix Paeoniae Alba	白芍	(Bai Shao)
Radix Ophiopogonis	麦冬	(Mai Dong)
Colla Corii Asini	阿胶	(E Jiao)
Fructus Cannabis	麻仁	(Ma Ren)

146. *Ting Li Da Zao Xie Fei Tang* 葶苈大枣泻肺汤

Semen Lepidii seu Descurainiae	葶苈子	(Ting Li Zi)
Fructus Ziziphi Jujubae	大枣	(Da Zao)

147. *Ba Xian Chang Shou Wan* 八仙长寿丸

Radix Ophiopogonis	麦冬	(Mai Dong)
Fructus Schisandrae	五味子	(Wu Wei Zi)
Radix Rehmanniae Praeparata	熟黄	(Shu Huang)
Fructus Corni	山茱萸	(Shan Zhu Yu)
Rhizoma Dioscoreae	干山药	(Gan Shan Yao)
Rhizoma Alismatis	泽泻	(Ze Xie)
Poria	茯苓	(Fu Ling)
Cortex Moutan Radicis	丹皮	(Dan Pi)

148. *Wei Ling Tang* 胃苓汤

Rhizoma Atractylodis	苍术	(Cang Zhu)
Cortex Magnoliae Officinalis	厚朴	(Hou Po)
Pericarpium Citri Reticulatae	陈皮	(Chen Pi)
Radix Glycyrrhizae	甘草	(Gan Cao)
Rhizoma Zingiberis	生姜	(Sheng Jiang)

Fructus Ziziphi Jujubae 大枣 (*Da Zao*)
Cortex Cinnamomi 肉桂 (*Rou Gui*)
Rhizoma Atractylodis
　Macrocephalae 白术 (*Bai Zhu*)
Rhizoma Alismatis 泽泻 (*Ze Xie*)
Poria 茯苓 (*Fu Ling*)
Polyporus Umbellatus 猪苓 (*Zhu Ling*)

149. *Lu Dou Gan Cao Jie Du Tang* 绿豆甘草解毒汤
Semen Phaseoli Radiati 绿豆 (*Lu Dou*)
Radix Glycyrrhizae 甘草 (*Gan Cao*)
Radix Salviae Miltiorrhizae 丹参 (*Dan Shen*)
Fructus Forsythiae 连翘 (*Lian Qiao*)
Herba Dendrobii 石斛 (*Shi Hu*)
Radix et Rhizoma Rhei 大黄 (*Da Huang*)

150. *Dang Gui Si Ni Tang* 当归四逆汤
Radix Angelicae Sinensis 当归 (*Dang Gui*)
Ramulus Cinnamomi 桂枝 (*Gui Zhi*)
Rhizoma Dioscoreae 芍药 (*Shao Yao*)
Herba Asari 细辛 (*Xi Xin*)
Radix Glycyrrhizae Praeparata 炙甘草 (*Zhi Gan Cao*)
Medulla Tetrapanacis 通草 (*Tong Cao*)
Fructus Ziziphi Jujubae 大枣 (*Da Zao*)

151. *Shi Quan Da Bu Tang* 十全大补汤
Radix Rehmanniae Praeparata 熟地黄 (*Shu Di Huang*)
Radix Paeoniae Alba 白芍 (*Bao Shao*)
Radix Angelicae Sinensis 当归 (*Dang Gui*)

Rhizoma Ligustici Chuanxiong	川芎	(*Chuan Xiong*)
Radix Ginseng	人参	(*Ren Shen*)
Rhizoma Atractylodis Macrocephalae	白术	(*Bai Zhu*)
Poria	茯苓	(*Fu Ling*)
Radix Glycyrrhizae Praeparata	炙甘草	(*Zhi Gan Cao*)
Radix Astragali seu Hedysari	黄芪	(*Huang Qi*)
Cortex Cinnamomi	肉桂	(*Rou Gui*)

152. *Tiao Gan Tang* 调肝汤

Rhizoma Dioscoreae	山药	(*Shan Yao*)
Colla Corii Asini	阿胶	(*E Jiao*)
Radix Angelicae Sinensis	当归	(*Dang Gui*)
Radix Paeoniae Alba	白芍	(*Bai Shao*)
Fructus Corni	山茱萸	(*Shan Zhu Yu*)
Radix Morindae Officinalis	巴戟天	(*Ba Ji Tian*)
Radix Glycyrrhizae	甘草	(*Gan Cao*)

153. *Yin Qiao Hong Teng Tang* 银翘红藤汤

Flos Lonicerae	金银花	(*Jin Yin Hua*)
Fructus Forsythiae	连翘	(*Lian Qiao*)
Caulis Sargentodoxae	红藤	(*Hong Teng*)
Herba Patriniae	败酱草	(*Bai Jiang Cao*)
Semen Coicis	苡仁	(*Yi Ren*)
Cortex Moutan Radicis	丹皮	(*Dan Pi*)
Fructus Gardeniae	栀子	(*Zhi Zi*)
Radix Paeoniae Rubra	赤芍	(*Chi Shao*)
Semen Persicae	桃仁	(*Tao Ren*)
Rhizoma Corydalis	元胡	(*Yuan Hu*)

Fructus Meliae Toosendan　　川楝子(*Chuan Lian Zi*)
Resina Olibani　　乳香　(*Ru Xiang*)
Resina Commiphorae
　　Myrrhae　　没药　(*Mo Yao*)

154.*Tao Hong Si Wu Tang* 桃红四物汤
Radix Angelicae Sinensis　　当归　(*Dang Gui*)
Rhizoma Ligustici Chuanxiong　川芎　(*Chuan Xiong*)
Radix Paeoniae Alba　　白芍　(*Bai Shao*)
Radix Rehmanniae Praeparata　熟地　(*Shu Di*)
Semen Persicae　　桃仁　(*Tao Ren*)
Flos Carthami　　红花　(*Hong Hua*)

155.*Shi Xiao San* 失笑散
Faeces Trogopterorum　　五灵脂(*Wu Ling Zhi*)
Pollen Typhae　　蒲黄　(*Pu Huang*)

156.*Qing Re Gu Jing Tang* 清热固经汤
Plastrum Testudinis
　　Praeparata　　炙龟板　(*Zhi Gui Ban*)
Concha Ostreae　　牡蛎　(*Mu Li*)
Colla Corii Asini　　阿胶　(*E Jiao*)
Radix Rehmanniae　　大生地(*Da Sheng Di*)
Cortex Lycii Radicis　　地骨皮(*Di Gu Pi*)
stir-heated Fructus Gardeniae　焦山栀(*Jiao Shan Zhi*)
Radix Scutellariae　　黄芩　(*Huang Qin*)
Radix Sanguisorbae　　地榆　(*Di Yu*)
carbonized Cortex Trachycarpi　棕榈炭(*Zong Lu Tan*)
Nodus Nelumbinis Rhizomatis　藕节　(*Ou Jie*)

Radix Glycyrrhizae 甘草 (Gan Cao)

157. Ping Gan Kai Yu Zhi Xue Tang 平肝开郁止血汤

Radix Bupleuri 柴胡 (Chai Hu)
Radix Angelicae Sinensis 当归 (Dang Gui)
stir-heated Radix
    Paeoniae Alba 炒白芍(Chao Bai Shao)
stir-heated Rhizoma Atracty-
    lodis Macrocephalae 炒白术(Chao Bai Zhu)
Cortex Moutan Radicis 丹皮 (Dan Pi)
Radix Rehmanniae 生地 (Sheng Di)
vinegar-prepared Rhizoma
    Cyperi 醋香附(Cu Xiang Fu)
Poria 茯苓 (Fu Ling)
carbonized Cortex Trachycarpi 棕榈炭(Zong Lu Tan)
carbonized Spica Schizonepetae 芥穗炭(Jie Sui Tan)
Herba Menthae 薄荷 (Bo He)
powder of Radix Notoginseng 三七粉(San Qi Fen)
Radix Glycyrrhizae 甘草 (Gan Cao)

158. Tiao Jing Sheng Yang Chu Shi Tang 调经升阳除湿汤

Rhizoma Atractylodis 苍术 (Cang Zhu)
Radix Astragali seu Hedysari
    Praeparata 炙黄芪(Zhi Huang Qi)
Radix Bupleuri 柴胡 (Chai Hu)
Radix Angelicae Sinensis 当归 (Dang Gui)
Rhizoma Cimicifugae 升麻 (Sheng Ma)
Rhizoma seu Radix
    Notopterygii 羌活 (Qiang Huo)

Radix Angelicae Pubescentis	独活	(Du Huo)
Radix Ledebouriellae	防风	(Fang Feng)
Rhizoma Ligustici	藁本	(Gao Ben)
Fructus Viticis	蔓荆子	(Man Jing Zi)
Radix Glycyrrhizae Praeparata	炙甘草	(Zhi Gan Cao)

159. *Gu Ben Zhi Beng Tang* 固本止崩汤

Radix Ginseng	人参	(Ren Shen)
Radix Astragali seu Hedysari	黄芪	(Huang Qi)
Rhizoma Atractylodis Macrocephalae	白术	(Bai Zhu)
Radix Rehmanniae Praeparata	熟地	(Shu Di)
Radix Angelicae Sinensis	当归	(Dang Gui)
baked Rhizoma Zingiberis	炮姜	(Pao Jiang)

160. *San Ao Tang* 三拗汤

Herba Ephedrae	麻黄	(Ma Huang)
Semen Armeniacae Amarum	杏仁	(Xing Ren)
Radix Glycyrrhizae	甘草	(Gan Cao)

161. *Liu Jun Zi Tang* 六君子汤

Radix Ginseng	人参	(Ren Shen)
Radix Glycyrrhizae Praeparata	炙甘草	(Zhi Gan Cao)
Poria	茯苓	(Fu Ling)
Rhizoma Atractylodis Macrocephalae	白术	(Bai Zhu)
Pericarpium Citri Reticulatae	陈皮	(Chen Pi)
Rhizoma Pinelliae Praeparata	炙半夏	(Zhi Ban Xia)

162. Ge Gen Huang Qin Huang Lian Tang 葛根黄芩黄连汤

Radix Puerariae	葛根	(Ge Gen)
Radix Scutellariae	黄芩	(Huang Qin)
Rhizoma Coptidis	黄连	(Huang Lian)
Radix Glycyrrhizae Praeparata	炙甘草	(Zhi Gan Cao)

163. Cang Er San 苍耳散

Radix Angelicae Dahuricae	白芷	(Bai Zhi)
Herba Menthae	薄荷	(Bo He)
Flos Magnoliae	辛夷	(Xin Yi)
Fructus Xanthii	苍耳子	(Can Er Zi)

164. Huang Qin Hua Shi Tang 黄芩滑石汤

Radix Scutellariae	黄芩	(Huang Qin)
Talcum	滑石	(Hua Shi)
Caulis Aristolochiae Manshuriensis	木通	(Mu Tong)
Fructus Forsythiae	连翘	(Lian Qiao)
Poria	茯苓	(Fu Ling)
Polyporus Umbellatus	猪苓	(Zhu Ling)
Pericarpium Arecae	大腹皮	(Da Fu Pi)
Semen Amomi Gardamomi	白蔻	(Bai Kou)
Rhizoma Acori Graminei	石菖蒲	(Shi Chang Pu)
Herba Agastachis	藿香	(Huo Xiang)

165. Shu Feng Qing Re Tang 疏风清热汤

Herba Schizonepetae	荆芥	(Jing Jie)
Radix Ledebouriellae	防风	(Fang Feng)

Flos Lonicerae	银花	(*Yin Hua*)
Forsythiae	连翘	(*Lian Qiao*)
Radix Scutellariae	黄芩	(*Huang Qin*)
Radix Paeoniae	赤芍	(*Chi Shao*)
Radix Scrophulariae	元参	(*Yuan Shen*)
Bulbus Fritillariae Thunbergii	浙贝母	(*Zhe Bei Mu*)
stir-heated Fructus Arctii	炒牛子	(*Chao Niu Zi*)
Radix Trichosanthis	花粉	(*Hua Fen*)
Radix Platycodi	桔梗	(*Jie Geng*)
Cortex Mori Radicis	桑白皮	(*Sang Bai Pi*)
Radix Glycyrrhizae	甘草	(*Gan Cao*)

166. Qing Yan Li Ge Tang 清咽利膈汤

Fructus Forsythiae	连翘	(*Lian Qiao*)
Fructus Gardeniae	生栀仁	(*Sheng Zhi Ren*)
Radix Scutellariae	黄芩	(*Huang Qin*)
Herba Menthae	薄荷	(*Bo He*)
Radix Ledebouriellae	防风	(*Fang Feng*)
Herba Schizonepetae	荆芥	(*Jing Jie*)
Natrii Sulfas	玄明粉	(*Xuan Ming Fen*)
Radix Platycodi	桔梗	(*Jie Geng*)
Flos Lonicerae	金银花	(*Jin Yin Hua*)
Radix Scrophulariae	玄参	(*Xuan Shen*)
Radix et Rhizoma Rhei	大黄	(*Da Huang*)
Radix Glycyrrhizae	甘草	(*Gan Cao*)
Rhizoma Coptidis	黄连	(*Huang Lian*)

167. Qi Ju Di Huang Wan 杞菊地黄丸

Fructus Lycii	枸杞子	(*Gou Qi Zi*)

Flos Chrysanthemi	菊花	(Ju Hua)
Radix Rehmanniae Praeparata	熟地黄	(Shu Di Huang)
Fructus Corni	山茱萸	(Shan Zhu Yu)
Rhizoma Dioscoreae	山药	(Shan Yao)
Rhizoma Alismatis)	泽泻	(Ze Xie)
Cortex Moutan Radicis	丹皮	(Dan Pi)
Poria	茯苓	(Fu Ling)

168. Tian Ma Gou Teng Tang 天麻钩藤汤

Rhizoma Gastrodiae	天麻	(Tian Ma)
Ramulus Uncariae cum Uncis	钩藤	(Gou Teng)
Ramulus Loranthi	桑寄生	(Sang Ji Sheng)
Radix Cyathulae	牛膝	(Niu Xi)
Cortex Eucommiae	杜仲	(Du Zhong)
Herba Leonuri	益母草	(Yi Mu Cao)
Poria	云苓	(Yun Ling)
Fructus Gardeniae	山栀	(Shan Zhi)
Radix Scutellariae	黄芩	(Huang Qin)
Concha Haliotidis	石决明	(Shi Jue Ming)
Caulis Polygoni Multiflori	夜交藤	(Ye Jiao Teng)

169. Ban Xia Bai Zhu Tian Ma Tang 半夏白术天麻汤

Rhizoma Pinelliae	半夏	(Ban Xia)
Rhizoma Atractylodis Macrocephalae	白术	(Bai Zhu)
Rhizoma Gastrodiae	天麻	(Tian Ma)
Pericarpium Citri Reticulatae	陈皮	(Chen Pi)
Poria	茯苓	(Fu Ling)
Radix Glycyrrhizae	甘草	(Gan Cao)

Rhizoma Zingiberis Recens 生姜 (Sheng Jiang)
Fructus Ziziphi Jujubae 大枣 (Da Zao)
Fructus Viticis 蔓荆子 (Man Jing Zi)

170. *Chu Shi Wei Ling Tang* 除湿胃苓汤
Radix Ledebouriellae 防风 (Fang Feng)
Rhizoma Atractylodis 苍术 (Cang Zhu)
Rhizoma Atractylodis
   Macrocephalae 白术 (Bai Zhu)
Poria Rubra 赤苓 (Chi Ling)
Pericarpium Citri Reticulatae 陈皮 (Chen Pi)
Cortex Magnoliae Officinalis 厚朴 (Hou Po)
Polyporus Umbellatus 猪苓 (Zhu Ling)
Fructus Gardeniae 山栀 (Shan Zhi)
Rhizoma Alismatis 泽泻 (Ze Xie)
Caulis Aristolochiae
   Manshuriensis 木通 (Mu Tong)
Talcum 滑石 (Hua Shi)
Radix Glycyrrhizae 甘草 (Gan Cao)
Medulla Junci 灯芯 (Deng Xin)
Cortex Cinnamomi 肉桂 (Rou Gui)

171. *Huo Xue San Yu Tang* 活血散瘀汤
tail of Radix Angelicae Sinensis 当归尾 (Dang Gui Wei)
Radix Paeoniae Rubra 赤芍 (Chi Shao)
Semen Persicae 桃仁 (Tao Ren)
Radix et Rhizoma Rhei 大黄 (Da Huang)
Rhizoma Ligustici
   Chuanxiong 川芎 (Chuan Xiong)

Lignum Sappan	苏木	(*Su Mu*)
Cortex Moutan Radicis	丹皮	(*Dan Pi*)
Fructus Aurantii	枳壳	(*Zhi Ke*)
Fructus Trichosanthis	瓜蒌仁	(*Gua Lou Ren*)
Semen Arecae	槟榔	(*Bing Lang*)

172. *Wu Wei Xiao Du Yin* 五味消毒饮

Flos Lonicerae	金银花	(*Jin Yin Hua*)
Flos Chrysanthemi	野菊花	(*Ye Ju Hua*)
Herba Taraxaci	蒲公英	(*Pu Gong Ying*)
Herba Violae	紫花地丁	(*Zi Hua Di Ding*)
Radix Semiaquilegiae	紫背天葵	(*Zi Bei Tian Tui*)

173. *Yi Wei Tang* 益胃汤

Radix Adenophorae	沙参	(*Sha Shen*)
Radix Ophiopogonis	麦冬	(*Mai Dong*)
Radix Rehmanniae	生地黄	(*Sheng Di Huang*)
Rhizoma Polygonati Odorati	玉竹	(*Yu Zhu*)
Crystal sugar	冰糖	(*Bing Tang*)

174. *Cong Chi Tang* 葱豉汤

Semen Sojae Praeparatum	淡豆豉	(*Dan Dou Chi*)
Bulbus Allii Fistulosi	葱白	(*Cong Bai*)

175. *Gua Di San* 瓜蒂散

Musk-mellon Pedicel	瓜蒂	(*Gua Di*)
Semen Phaseoli	赤小豆	(*Chi Xiao Dou*)

19 急症学

学 友 会

序

　　《英汉实用中医药大全》即将问世，吾为之高兴。

　　歧黄之道，历经沧桑，永盛不衰。吾中华民族之强盛，由之。世界医学之丰富和发展，亦由之。然而，世界民族之差异，国别之不同，语言之障碍，使中医中药的传播和交流受到了严重束缚。当前，世界各国人民学习、研究、运用中医药的热潮方兴未艾。为使吾中华民族优秀文化遗产之一的歧黄之道走向世界，光大其业，为世界人民造福，徐象才君集省内外精英于一堂，主持编译了《英汉实用中医药大全》。是书之问世将使海内外同道欢呼雀跃。

　　世界医学发展之日，当是歧黄之道光大之时。

　　吾欣然序之。

中华人民共和国卫生部副部长
兼国家中医药管理局局长
世界针灸学会联合会主席
中国科学技术协会委员
中华全国中医学会副会长
中国针灸学会会长

胡熙明
1989 年 12 月

序

中华民族有同疾病长期作斗争的光辉历程，故而有自己的传统医学——中国医药学。中国医药学有一套完整的从理论到实践的独特科学体系。几千年来，它不但被完好地保存下来，而且得到了发扬光大。它具有疗效显著、副作用小等优点，是人们防病治病，强身健体的有效工具。

任何一个国家在医学进步中所取得的成就，都是人类共同的财富，是没有国界的。医学成果的交流比任何其他科学成果的交流都应进行得更及时，更准确。我从事中医工作30多年来，一直盼望着有朝一日中国医药学能全面走向世界，为全人类解除病痛疾苦做出其应有的贡献。但由于用外语表达中医难度较大，中国医药学对外传播的速度一直不能令人满意。

山东中医学院的徐象才老师发起并主持了大型系列丛书《英汉实用中医药大全》的编译工作。这个工作是一项巨大工程，是一种大型科研活动，是一个大胆的尝试，是一件新事物。对徐象才老师及与其合作的全体编译者夜以继日地长期工作所付出的艰苦劳动，克服重重困难所表现出的坚韧不拔的毅力，以及因此而取得的重大成绩，我甚为敬佩。作为一个中医界的领导者，对他们的工作给予全力支持是我应尽的责任。

我相信《英汉实用中医药大全》无疑会在中国医学史和世界科学技术史上找到它应有的位置。

中华全国中医学会常务理事
山东省卫生厅副厅长

张奇文
1990年3月

出 版 前 言

  中国医药学是我中华民族优秀文化遗产之一，建国以来由于党和国家对待中医药采取了正确的政策，使中医药理论宝库不断得到了发掘整理，取得了巨大的成绩。当前，世界各国人民对中国医药学的学习和研究热潮日益高涨，为促进这一热潮更加蓬勃的发展，为使中国医药学能更好地为全人类解除病痛服务，就必须促进中医中药在世界范围内的传播和交流，而要使这一传播和交流进行得更及时、更准确，就必须首先排除语言障碍。因此，编译一套英汉对照的中医药基本知识的书籍，供国内外学习、研究中医药时使用，已成为国内外医药学界和医药学教育界许多人士的迫切需要。

  多年来，在卫生部门的号召下，在"中医英语表达研究"方面，已经作出了一些可喜的成绩。本书《英汉实用中医药大全》的编辑出版就是在调查上述研究工作的历史和现状的基础上，继续对中医药英语表达作较系统、较全面的研究，以适应中国医药学对外传播交流的需要。

  这部"大全"的版本为英汉对照，共有21个分册，一个分册介绍论述中国医药学的一个分科。在编者上注意了中医药汉文稿的编写特色，在内容上注意了科学性、实用性、全面性和简明易读。汉文稿的执笔撰写者主要是有20年以上实践经验的教授、副教授、主任医师和副主任医师。各分册汉文稿撰写成后，均经各学科专家逐一审订。各分册英文主译、主审主要是国内既懂中医又懂英语的权威人士，还有许多中医院校的英语教师及医药卫生部门的专业翻译人员。英译稿脱稿后，经过了复审、终审，有些译稿还召开全国22所院校和单位人员参加的英译稿统稿定稿研讨会，对英译稿进行细致的研讨和推敲，对如何较全面、较系统、较准确地用英语表达中国医药学进行了探讨，从而推动整个译文达到较高水平，因此，这部"大全"可供中医院校高年级学生作为泛读教材使用。

这部"大全"的编纂得到了国家教育委员会、国家中医药管理局、山东省教育委员会、山东省卫生厅等各部门有关领导的支持。在国家教委高等教育司的指导下，成立了《英汉实用中医药大全》编译领导委员会。还得到了全国许多中医院校和中药生产厂家领导的支持。

希望这部"大全"的出版，对中医院校加强中医英语教学，对国内卫生界培养外向型中医药人才，以及在推动世界各国人民对中医药的学习和研究方面，都将产生良好的影响。

高等教育出版社

1990年3月

前　言

　　《英汉实用中医药大全》是一部以中医基本理论为基础，以中医临床为重点，较为全面系统、简明扼要、易读实用的中级英汉学术性著作。它的主要读者是：中医药院校高年级学生和中青年教师，中医院的中青年医生和中医药科研单位的科研人员，从事中医对外函授工作的人员和出国讲学或行医的中医人员，西学中人员，来华学习中医的外国留学生和各类进修人员。

　　由于中国医药学为我中华民族之独有，因此，英译便成了本《大全》编译工作的重点。为确保译文能准确表达中医的确切含义，我们邀集熟悉中医的英语人员、医学专业翻译人员、懂英语的中医药人员乃至医古文人员于一堂，共同翻译、共同对译文进行研讨推敲的集体翻译法，这样，就把众人之长融进了译文质量之中。然而，即使这样，也难确保译文都能尽如人意。汉文稿虽反映了中国医药学的精髓和概貌，但也难能十全十美。我衷心地盼望读者能提出批评和建议，以便《大全》再版时修改。

　　参加本《大全》编、译、审工作的人员达200余名，他们来自全国28个单位，其中有山东、北京、上海、天津、南京、浙江、安徽、河南、湖北、广西、贵阳、甘肃、成都、山西、长春等15所中医学院，还有中国中医研究院，山东省中医药研究所等中医药科研单位。

　　山东省教育委员会把本《大全》的编译列入了科研计划并拨发了科研经费，山东省卫生厅和一些中药生产厂家也给了很大支持，济南中药厂的资助为编译工作的开端提供了条件。

　　本《大全》的编译成功是全体编译审者集体劳动的结晶，是各有关单位主管领导支持的结果。在《大全》各分册即将陆续出版之际，我诚挚地感谢全体编译审者的真诚合作，感谢许多专家、教授、各级领导和生产厂家的热情支持。

愿本《大全》的出版能在培养通晓英语的中医人才和使中医早日全面走向世界方面起到我所期望的作用。

主编　徐象才

于山东中医学院

1990年3月

目 录

说明 …………………………………………………………… 367
1 危重征象的急救 …………………………………………… 369
    1.1 高热 …………………………………………………… 369
    1.2 休克 …………………………………………………… 377
    1.3 昏迷 …………………………………………………… 379
2 循环系统疾病 ……………………………………………… 384
    2.1 心绞痛 ………………………………………………… 384
    2.2 急性心肌梗塞 ………………………………………… 386
    2.3 急性肺源性心脏病 …………………………………… 390
    2.4 急性感染性心内膜炎 ………………………………… 391
    2.5 病毒性心肌炎 ………………………………………… 393
    2.6 心律失常 ……………………………………………… 396
    2.7 心力衰竭 ……………………………………………… 399
3 呼吸系统疾病 ……………………………………………… 401
    3.1 急性支气管炎 ………………………………………… 401
    3.2 大叶肺炎 ……………………………………………… 403
    3.3 急性吸入性肺脓肿 …………………………………… 405
4 神经系统及精神疾病 ……………………………………… 407
    4.1 三叉神经痛 …………………………………………… 407
    4.2 偏头痛 ………………………………………………… 408
    4.3 急性多发性神经炎 …………………………………… 410
    4.4 急性感染性多发神经根炎 …………………………… 412
    4.5 高血压性脑出血 ……………………………………… 413
    4.6 动脉硬化性脑梗塞 …………………………………… 416

5 消化系统疾病 …… 422
5.1 急性胃肠炎 …… 422
5.2 胃十二指肠溃疡出血 …… 423
5.3 急性胃扩张 …… 425
5.4 肝性昏迷 …… 427
5.5 胆道蛔虫病 …… 429
5.6 急性胆囊炎与胆石症 …… 430
5.7 急性胰腺炎 …… 432
5.8 急性阑尾炎 …… 434
5.9 胃十二指肠溃疡病穿孔 …… 435
5.10 急性腹膜炎 …… 437
5.11 伪膜性肠炎 …… 439

6 泌尿系统疾病 …… 441
6.1 急性肾功能衰竭 …… 441
6.2 尿毒症 …… 443
6.3 急性泌尿系统感染 …… 445
6.4 急性尿潴留 …… 446

7 内分泌及新陈代谢系统疾病 …… 449
7.1 急性化脓性甲状腺炎 …… 449
7.2 急性肾上腺皮质机能减退症 …… 450
7.3 自发性低血糖症 …… 451

8 造血系统疾病 …… 453
8.1 再生障碍性贫血 …… 453
8.2 急性失血性贫血 …… 455
8.3 过敏性紫癜 …… 456

9 结缔组织及变态反应性疾病 …… 459
9.1 系统性红斑狼疮 …… 459
9.2 变应性亚急性败血症 …… 460

10 传染性疾病 …… 462

10.1	流行性感冒	462
10.2	流行性脑脊髓膜炎	463
10.3	流行性乙型脑炎	465
10.4	流行性出血热	467
10.5	急性重症肝炎	471

11 中毒及理化损伤性疾病 ………… 474

11.1	酒精中毒	474
11.2	乌头类药物中毒	475
11.3	马钱子中毒	476
11.4	中暑	477

12 妇科常见急症 ………… 479

12.1	痛经	479
12.2	急性盆腔炎	481
12.3	功能性子宫出血	482

13 儿科常见急症 ………… 486

13.1	新生儿肺炎	486
13.2	新生儿败血症	488
13.3	中毒性痢疾	489

14 耳鼻喉科常见急症 ………… 491

14.1	急性鼻窦炎	491
14.2	急性扁桃体炎	493
14.3	急性化脓性中耳炎	494
14.4	美尼尔氏病	496

15 皮肤科常见急症 ………… 499

15.1	带状疱疹	499
15.2	急性丹毒	501

16 中医急症主要治疗措施 ………… 503

16.1	催吐导泻法	503
16.2	针刺按摩法	504

16.3	静脉输液法	507
16.4	雾化吸入法	508
16.5	肛肠纳药法	508

《英汉实用中医药大全》（书目） ………………………… 509

说　　明

《急症学》是《英汉实用中医药大全》中的第19分册。

中医对急症的诊治历史悠久,内容广泛,经验丰富,理论独特。

本分册共有危重征象的急救、循环系统疾病、呼吸系统疾病、神经系统及精神疾病、消化系统疾病、泌尿系统疾病、内分泌及新陈代谢系统疾病、造血系统疾病、结缔组织及变态反应性疾病、传染性疾病、中毒及理化损伤性疾病、妇科常见急症、儿科常见急症、耳鼻喉科常见急症、皮肤科常见急症和中医急症主要治疗措施等16章。全册采用西医病名,但内容的编写体系则是按中医的理论辨证论治。对每个病症均在"病名"之后列出"病因病机"、"诊断要点"、"辨证施治"等项分别进行论述;所用药物剂量,除第13章儿科常见急症外,均为成人剂量。

在本分册所涉及的中药材中,有些来自濒临灭绝的珍稀动物。由于这些动物已明令禁捕禁杀,这些药材的来源也就因之而继绝。无论在医院里还是在制药厂里,它们均已被具有相同药效的其他药材替代,本分册仍将其涉及,只是为方便学习探讨。

本分册汉文稿经中华医学会副主任委员王永炎教授审阅。

编者

1 危重征象的急救

西医所说危重征象,相当于中医的危重证候,常见如下几种。

1.1 高热

高热是在大量急重病中反映不同病机的临床急危证候,在中医文献中属"发热"、"壮热"、"潮热"、"寒热"等范畴。

病因病机

外感高热:多系感受六淫之邪或疫毒之气而起,以正邪相争为其病机。

内伤高热:多因劳倦过度、饮食失调、情志抑郁、瘀血内停、湿热滞留等而起,使脏腑功能失调或气血津液内耗,致阳热偏盛为其病机。

诊断要点

1. 体温骤然升高,大多在39℃以上,或由低热骤然转成高热。具有起病急、体温高、病程短、传变快的发病特点。

2. 临床以烦渴、身热、脉数为主要特征。

3. 外感高热病初有恶寒或恶风,兼见其他外感证候;内伤高热则不伴恶寒,而见于其他内伤证候。

辨证施治

1. 表证期

(1) 表寒证

主症:恶寒,发热,无汗,头痛项强,身痛,腰痛,骨节疼痛,舌苔薄白,脉象浮紧。

治法:辛温解表。

方药：荆防解表汤（1）。

荆芥9克，防风12克，羌活9克，白芷9克，葱白1寸，生姜3片，甘草6克。水煎服。

(2) 表热证

主症：发热，微恶风寒，无汗或少汗，头痛，鼻塞，咳嗽，口微渴，咽红肿或疼痛。舌边舌尖红，苔薄白或薄黄，脉象浮数。

治法：辛凉解表。

方药：银翘散（2）加减。

金银花30克，连翘12克，桔梗9克，薄荷9克，炒杏仁9克，荆芥穗9克，淡竹叶9克，甘草3克。水煎服，日1～2剂。

加减：发热甚者可加板兰根、黄芩、生石膏。

头痛明显者加桑叶12克、菊花12克。

咳嗽频者加枇杷叶12克、桑白皮12克、川贝母12克。

咽喉红肿疼痛加炒牛子12克，山豆根12克。

(3) 表湿证

主症：恶寒，身热不扬或午后热势加重，头痛如裹，肢体困重，汗出而粘，胸脘痞闷，纳呆不知饥，口渴不欲饮，饮则恶心。舌苔白腻或黄腻，脉象濡数。

治法：芳香透表，清热利湿。

方药：藿香正气散（3）加减。

藿香12克，紫苏9克，白术12克，大腹皮12克，茯苓9克，半夏曲9克，陈皮9克，厚朴9克，桔梗9克，连翘12克，薄荷9克，车前子12克（包），甘草3克。水煎服。

(4) 肺燥证

主症：身热，口渴，鼻唇干燥，干咳无痰，甚者痰中带血。舌质红，苔薄黄乏津，脉象浮滑而数。

治法：辛凉润肺，清热生津。
方药：桑杏汤（4）合沙参麦门冬汤（5）加减。
桑叶12克，炒杏仁9克，沙参24克，麦冬15克，贝母9克，栀子9克，玉竹12克，花粉12克，梨皮9克。水煎服。
(5) 暑热证
主症：发热恶寒，无汗或汗出，头痛头重，心烦，口渴不欲饮，胃脘痞满，恶心呕吐，小便短赤。舌质红，苔白腻，脉象濡数或浮濡。
治法：清暑除湿，解表退热。
方药：新加香薷饮（6）加减。
香薷9克，鲜扁豆花24克，厚朴12克，金银花30克，连翘12克，藿香12克，佩兰叶9克，苡米24克，六一散15克。水煎服。

2. 表里证期
(1) 半表半里证
主症：寒热往来，口苦咽干、目眩，胸胁苦满，心烦喜呕，纳呆。舌苔白滑，脉弦。
治法：和解表里。
方药：小柴胡汤（7）加减。
柴胡30克，黄芩15克，半夏9克，人参9克，生姜3片，大枣3枚，甘草3克。水煎服。
(2) 表寒里热证
主症：高热、咳喘、气急、口渴、咳吐黄白粘稠痰，甚或带血，或吐铁锈色痰，咳则胸痛。舌苔黄白相兼，脉浮数。
治法：清热宣肺。
方药：麻杏石甘汤（8）加减。
麻黄6克，炒杏仁9克，生石膏30克，甘草6克，桑白皮12克，枇杷叶12克，黄芩9克，鱼腥草18克，桔梗9克，地骨皮18克。水煎服。

加减：若咳痰带血可加生地15克，丹皮12克。
(3) 表里俱热证
主症：发热、咽痛、口渴、便秘。舌质红，苔黄，脉数。
治法：透热清里。
方药：凉膈散（9）加减。
大黄6～9克，黄芩9克，栀子9克，连翘15克，薄荷9克，芒硝6克，甘草3克。水煎服。

3 里证期
(1) 气分热炽证
主症：壮热，烦渴，汗出，或伴有口臭、牙龈肿痛、腐烂或出血。舌质红，苔黄燥，脉洪大。
治法：清热生津。
方药：白虎汤（10）加减。
生石膏60～120克，知母30克，花粉30克，太子参30克，柴胡30～60克，板兰根45～60克，蚤休15克，生地30克，甘草6克。水煎服，日1～2剂。
加减：热毒盛者可加金银花30克、连翘15克。
神昏谵语、大便燥结者加大黄10克、芒硝6克。
高热发斑者加犀角粉3克（冲）元参30克、丹皮12克。
口臭、牙龈肿甚者加黄连10克、栀子12克。
(2) 热结肠胃证
主症：高热，午后尤甚，大便秘结或下利清水，腹疼拒按、痞满不舒，甚则烦躁谵语，或神志不清，循衣摸床。舌红，苔黄燥或焦黄起芒刺，脉弦数或沉实有力。
治法：通腑泻热。
方药：大承气汤（11）加减。
大黄9克（后入），芒硝9克，枳实12克，厚朴12克。水煎服。

加减：如热结阴亏，大便燥结不得下行者加生地、麦冬、元参各 30 克。

伴胸膈烦热，口舌生疮者加栀子 12 克、黄芩 15 克、连翘 12 克、薄荷 10 克、竹叶 10 克。

(3) 热毒壅肺证

主症：身热、汗出、咳喘、咯痰、口渴、气促。舌质红，苔黄，脉象滑数。

治法：清肺化痰，排脓解毒。

方药：千金苇茎汤（12）加减。

鲜苇茎 30 克，苡仁 30 克，冬瓜仁 12 克，桃仁 9 克，桔梗 24 克，甘草 6 克，金银花 30 克，红藤 30 克，鱼腥草 24 克，黄芩 9 克、桑白皮 12 克，瓜蒌 24 克，公英 24 克，地丁 18 克。水煎服。

加减：咳而喘满，咯痰量多不得卧者加葶苈子 12 克。

咯血者酌加丹皮 12 克、山栀 12 克、白芨 15 克、藕节 30 克、三七粉 3 克（冲）。

胸痛呼吸不利者可加香附、郁金各 10 克。

烦渴重者可加花粉、知母各 15 克。

(4) 湿热困脾证

主症：身热不扬午后为重，胸闷，脘痞，嗜卧。舌苔腻，脉濡数。

治法：宣化湿热。

方药：三仁汤（13）加减。

薏苡仁 30 克，炒杏仁 9 克，白蔻仁 6 克，滑石 24 克，厚朴 9 克，半夏 6 克，白通草 9 克，竹叶 9 克。水煎服。

(5) 肝胆湿热证

主症：发热口渴，口苦胁痛，脘腹痞胀，恶心欲吐或吐黄水，小便短赤。舌质红，苔黄腻，脉弦滑数。

治法：疏利肝胆，清利湿热

方药：蒿芩清胆汤（14）加减。

青蒿24克，黄芩12克，陈皮9克，竹茹9克，半夏9克，赤茯苓12克，枳壳9克，黄连6克，佩兰6克，菖蒲9克，碧玉散10克。水煎服。

(6) 膀胱湿热证

主症：发热，小便频急涩痛，淋沥不畅，腰痛。舌质红，苔黄腻或薄黄，脉滑数。

治法：清热利湿，通利下焦。

方药：八正散（15）加减。

金银花30克，连翘15克，车前草30克，萹蓄15克，瞿麦18克，木通9克，生栀子9克，滑石24克，甘草梢9克，灯心3克，大黄6克（后入）。水煎服，日1～2剂。

加减：小便赤者可加白茅根、小蓟、坤草各30克。

小便混浊者可加石苇、萆薢各24克。

(7) 湿热痢疾证

主症：发热、腹疼，下痢赤白，里急后重，或暴注下迫，肛门灼热，恶心呕逆。舌质红，苔腻，脉滑数。

治法：清热解毒，燥湿止痢。

方药：白头翁汤（16）加味。

白头翁24克，黄连9克，黄柏9克，秦皮12克，木香9克，槟榔9克，芍药30克，甘草10克，山楂15克。水煎服。

(8) 气营两燔证

主症：壮热，口渴，汗出，不眠，神昏。舌质绛，苔黄而干，脉数。

治法：清气凉营。

方药：玉女煎（17）加减。

生石膏90-120克，生地30克，麦冬30克，知母30克，黄芩15克，栀子9克，黄连9克，板兰根30克，苇根30克，白蔻9克。水煎服。

加减：有出血者加犀角粉3克（冲）、赤芍、丹皮各12克。
神昏者加羚羊角粉6克（冲）、菖蒲、郁金各10克。
心力衰竭者加西洋参15克。

(9) 邪热入营证

主症：发热，夜间为甚，心烦不寐，口渴不欲饮，重者可有神昏谵语，斑疹隐隐。舌质红，少苔乏津，脉细数。

治法：清营泻热。

方药：清营汤（18）加减。

犀角10克（或水牛角30克，先煎），生地黄15克，元参15克，麦冬30克，丹参15克，黄连9克，金银花30克，连翘15克，竹叶心9克。水煎服。

(10) 邪热入心证

主症：高热炽盛，神昏谵语，或昏愦不语，躁扰不宁。舌质绛红，苔黄燥，脉细数。

治法：清心开窍。

方药：清宫汤（19）加减。

犀角10克（或水牛角30克，先煎），元参15克，麦冬30克，莲子心9克，连翘心9克，竹叶心9克，菖蒲12克。水煎服。

加减：若患者高热神昏伴口噤目张，两手握固，出现高热闭证，立即给安宫牛黄丸（20）或牛黄清心丸（21）每次1丸，每日3～4次，鼻饲。亦可加用清开灵20～40毫升或醒脑静10～20毫升加入5%葡萄糖液中静脉点滴，每日1～2次。

若痰涎壅盛者，可用竹沥水1支，日服3次。可加用人工牛黄粉3克冲服。亦可用皂角粉1克或麝香粉0.01克吹鼻，或用猴枣散0.1克吹鼻。

(11) 热极生风证（亦为高热痉证）

主症：高热烦渴，头痛目眩，抽搐颈强，两目上翻，牙关紧闭，甚则角弓反张，神昏厥逆。舌质绛，苔黄，脉弦数而急。

治法：清热凉肝，养血熄风。

方药：羚羊钩藤汤（22）加减。

羚羊角6克（先煎），钩藤15克，桑叶12克，菊花15克，生地15克，白芍12克，川贝9克，茯苓12克，竹茹9克，甘草6克。水煎报，日1~2剂。

亦可用清开灵40~60毫升加入5%葡萄糖液500毫升中静脉点滴，每日1~2次。

另可加用止痉散1.5克，每日1~2次，冲服。或用安宫牛黄丸1丸，每日3~4次。

(12) 阴虚风动证

主症：低热，手足蠕动，口舌干燥，目陷睛迷。舌绛少苔，脉象细数无力。

治法：滋阴养血，潜阳熄风。

方药：大定风珠（23）加减。

生地黄24克，麦冬30克，白芍12克，生牡蛎30克，生鳖甲18克，生龟板18克，炙甘草9克，麻仁6克，五味子6克，阿胶9克（烊化），鸡子黄2个。水煎服。

(13) 血热发斑证（亦为高热出血证）

主症：壮热谵妄，躁扰发狂，斑疹透露，或有吐、咯、衄、便、溲血等。舌绛，苔黄燥，脉数。

治法：清热凉血，解毒化斑。

方药：化斑汤（24）合犀角地黄汤（25）加减。

犀角6~10克（先煎），生石膏30~90克，知母15~30克，生地黄30克，元参30克，丹皮12克，白芍12克，粳米9克，生甘草6克。水煎服。

针剂：清开灵注射液30~40毫升加入5%葡萄糖液100毫升静滴，每日1~2次。

(14) 阴虚阳脱证（亦为高热脱证）

主症：神昏，咽燥目陷，躁动不安，四肢厥逆，冷汗出或汗出如油。舌质红，无苔，脉微欲绝。

治法：益气滋阴，回阳固脱。

方药：生脉散（26）合四逆汤（27）加减。

人参9~30克，麦冬30克，五味子6克，附子9克，干姜6克，甘草6克。水煎服。

1.2 休克

各种原因引起的休克，据其临床表现，均属于中医"厥脱"的范畴。故对各类休克，中医均按厥脱来辨证论治。

病因病机

本病多由六淫邪毒内侵而陷入营血、剧痛惊恐、亡津失血耗精、过敏、中毒、久病等致耗气伤阴，损及五脏，使气血运行障碍，阴阳失调，气机逆乱而发为厥脱。

诊断要点

1. 本病具起病急，病情转变快的发病特点。

2. 本病为多种疾病的变证，故有原发病因及证候。

3. 临床以四肢厥冷，面色苍白或潮红或发绀，时出冷汗，烦躁不安或神情淡漠，气息微弱或气促息粗。脉沉细或微细欲绝或不能触及为特点。

4. 血压下降：一般收缩压低于80毫米汞柱，脉压差小于20毫米汞柱。原有高血压者，收缩压可低于平时血压的三分之一或降低30毫米汞柱。

辨证施治

1. 热毒炽盛

主症：发热不寒，手足厥冷，烦躁口渴，谵妄，胸腹灼热，小便短赤，大便秘结且便下腐臭。舌质红，苔黄燥，脉沉细数。

治法：泄热解毒。

方药：人参白虎汤（28）加味。

人参15~30克，石膏30~90克，知母15克，粳米15克，甘草6克，黄芩15克，黄连9克，蒲公英30～60克。水煎服。

加减：大便秘结、腑气不通者可加用大黄6克、芒硝6克、枳实、厚朴各12克。

痰壅气滞、喉间痰鸣者可用半夏、陈皮、南星、枳实各12克等酌情加入。

针剂：清气解毒针300～400毫升静脉点滴，日1～2次。

清开灵注射液20～40毫升加入5%葡萄糖液250毫升中静脉点滴，日1～2次。

2. 气阴两亏

主症：神萎倦怠，四肢欠温，口渴汗出，气息微弱。舌质红或淡红，苔薄白，脉细数。

治法：益气养阴。

方药：生脉散（26）加味。

人参15克，黄芪24克，麦冬30克，五味子9克。水煎服。

针剂：参麦注射液50-100毫升加入5%葡萄糖液250～500毫升静脉点滴，持续至病情好转。或参麦注射液20毫升加入25%葡萄糖液20毫升静脉注射，每隔10～15分钟一次，连用3～5次，待血压回升或稳定后，再以50～100毫升加入增液针或养阴针500毫升中静脉点滴。

3. 阳气暴脱

主症：四肢厥冷，体温不升或过低，大汗淋漓，气促息微。舌质淡，脉微弱欲绝或不能触及。

治法：回阳救脱。

方药：参附汤（29）或四逆汤（27）加减。

红参15～30克，附子15～30克（先入），干姜9克，炙甘草9克。水煎服。

4. 真阴耗竭

主症：神志恍惚，面色潮红，发热烦躁，惊恐心悸，口渴喜饮，尿少色黄，肢厥不温。舌光剥干枯无苔，脉虚数或结代。

治法：养阴增液固脱。

方药：三甲复脉汤（30）加减。

牡蛎30克，鳖甲30克（先入），龟板30克（先入），生地30克，麦冬30克，山萸肉15克，五味子9克，炙甘草9克。水煎服。

5. 心气不足

主症：怔忡不安，气促息弱，气短乏力。舌质淡，苔薄白，脉细而促或结代。

治法：补养心气。

方药：炙甘草汤（31）加减。

炙甘草9～15克，人参9～30克，桂枝9克，麦冬30克，龙骨15克（先入），牡蛎30克（先入）。水煎服。

6. 气滞血瘀

主症：唇甲青紫，皮肤瘀斑，腹胀。舌质暗紫，脉沉细而涩。

治法：活血化瘀，理气救逆。

方药：血府逐瘀汤（32）加减。

柴胡9克，枳实9克，青皮9克，白芍9克，川芎9克，红花6克，丹参15克，制大黄9克，甘草6克。水煎服。

针剂：丹参注射液20～30毫升加入5%葡萄糖液250毫升静脉点滴。

1.3 昏迷

昏迷是内科常见急症之一，据其临床表现，当属中医"神昏"、"昏蒙"、"昏厥"、"昏愦"的范畴。

病因病机

昏迷乃心脑受扰而发，清窍闭塞或失养所致。故不论外感时疫热毒内攻，或风痰瘀血上扰清阳而致清窍闭塞；还是内伤痰病，气血亏耗，阴阳衰竭不相维系，以致清窍失养，神无所倚均可导致昏迷。

诊断要点

1. 本病起病急暴。
2. 临床以神志不清，呼之不应，昏不知人为特征。
3. 昏迷系多种疾病发展演变而成的危重证候，故多兼有原发病证候。

辨证施治

1. 实证昏迷

(1) 热陷心营

主症：高热神昏，烦躁谵语，或昏迷不醒，呼之不应。兼见斑疹衄血，抽搐时作，或角弓反张。舌质红绛，苔黄少津，脉滑数或细数。

治法：清心开窍，泄热救阴。

方药：清营汤（18）加减。

犀角10克（切片先煎），生地24克，元参18克，竹叶心3克，莲子心3克，连翘15克，麦冬30克，金银花30～60克，黄连6克。水煎服。

加减：昏迷深重者可加菖蒲、郁金各12克并送服（或鼻饲）安宫牛黄丸或至宝丹，每次1丸，每日4～6次。

若抽搐者可加羚羊角粉6克（冲）钩藤30克、地龙12克。并送服（或鼻饲）紫雪丹，每次1丸，每日3～4次。

(2) 喘促痰蒙

主症：神志呆痴，时昏时醒，语言错乱或意识朦胧。兼胸闷、恶心、咳逆喘促，痰涎壅盛，身热不扬，午后热甚。舌苔白腻或黄厚垢浊，脉濡滑而数。

治法：豁痰开窍，清热化湿平喘。

方药：菖蒲郁金汤（33）合三子养亲汤（34）加减。

石菖蒲12克，郁金12克，炒山栀12克，连翘15克，竹叶

9克，牛子12克，姜半夏9克，茯苓12克，陈皮9克，白芥子6克，苏子9克，莱菔子12克，丹皮12克，菊花12克。滑石15克。水煎服。同时，鲜竹沥15～30毫升冲服，每日2～3次。

加减：若昏迷深重可加服苏合香丸，每次1丸，每日3～4次。

(3) 腑实燥结

主症：神昏谵语、燥扰不宁，高热或日晡潮热，腹满胀疼，大便秘结。舌红苔黄燥或起芒刺，脉沉实而数。

治法：攻积泄热。

方药：大承气汤（11）。

大黄9～30克（后入），芒硝6～9克（烊化），枳实12克，厚朴12克。水煎服。

加减：若口渴多饮可加生石膏30～60克，知母15克。

若谵语、狂躁重者可加用紫雪丹0.15～0.3克，每日2～3次。

(4) 湿热急黄

主症：神昏谵语或昏迷不省或昏而时醒，烦躁不安，黄疸日深，斑疹衄血或腹胀如鼓。舌绛，苔腻，脉弦数。

治法：利湿泄热，凉血开窍。

方药：茵陈蒿汤（35）加减。

茵陈30克，炒山栀子12克，水牛角15克（锉末冲服），大黄9克，生地15克，丹皮12克，元参15克，菖蒲12克，石斛12克。水煎服。

亦可同时加服神犀丹，每次3克，每日3～4次。

针剂：醒脑静10～20毫升加入5%葡萄糖液250毫升静脉点滴，日1～2次。

(5) 瘀热阻窍

主症：神昏谵语或狂躁，高热夜甚，口渴喜饮，少腹满痛，大便秘结，唇甲青紫。舌质深绛或紫暗，脉弦数。

治法：清热通瘀开窍。

方药：犀地清络饮（36）加减。

犀角15～30克（切片先煎），生地30克，赤芍15克，丹皮12克，连翘30克，桃仁12克，菖蒲12克，大黄6克，琥珀粉2克（冲）。水煎服。

加减：热甚神昏可加紫雪丹0.15～0.3克，日2～3次；或加安宫牛黄丸1丸，日2～3次。

若心火炽盛者可加黄连、栀子各10克。

若吐衄可加侧柏叶、旱莲草、白茅根各30克。

(6) 阳亢中风

主症：突然昏仆，不省人事，牙关紧闭，口噤不开，两手握固，二便闭，肢体强痉偏瘫。兼见颜面潮红，呼吸气粗，鼾声时作。舌质红，苔黄而少津，脉弦滑而数。

治法：镇肝潜阳，熄风开窍。

方药：羚羊钩藤汤（22）加减。

羚羊角3克（锉末冲服），钩藤15克，生地15克，丹皮12克，龟板18克，夏枯草30克，石决明30克，白芍12克，柴胡9克，薄荷9克，菊花15克，牡蛎30克（先煎）。水煎服。

加减：痰壅神昏者加胆星12克、瓜蒌30克、天竺黄12克、鲜竹沥30克、生姜汁12克。

伴抽搐者加全虫10克、蜈蚣3条僵蚕12克等。

便秘口臭，腹胀者可加大黄10克、芒硝6克、枳实12克。

针剂：清开灵注射液20～40毫升加入5%葡萄糖液250毫升静脉点滴，日1～2次。

(7) 夏令中暑

主症：头晕头痛，胸闷身热，面色潮红，继则卒仆，手足厥冷，不省人事，或有神昏谵妄。舌红而干，脉洪数。

治法：解暑益气，清心开窍。

方药：首先用万氏牛黄清心丸或紫雪丹1丸，以凉开水调服，继则用加味白虎汤（10）。

生石膏30～60克，知母12克，甘草3克，石斛15克，太子参30克，荷梗10克，黄连6克，麦冬30克，竹叶12克，西瓜翠衣30克。水煎服。

加减：若伴四肢抽搐可加羚羊角5克（冲）、钩藤30克、全虫10克。

2. 虚证昏迷

(1) 亡阴神昏

主症：神昏汗出，面红身热，唇舌干红，脉虚数。

治法：救阴敛阳。

方药：生脉散（26）加味。

红参6～30克（另煎），麦冬30克，五味子6克，山萸肉15克，黄精30克，龙骨30克，牡蛎30克。水煎服。

加减：热痰上泛，堵塞窍道，舌强不语者加贝母、竹沥、胆星、天竺黄。

(2) 亡阳神昏

主症：面色苍白，昏愦不语，目合口开，呼吸微弱，手撒肢厥，大汗淋漓，二便失禁。唇舌淡润或口唇青紫，脉微欲绝。

治法：回阳救逆。

方药：参附汤（29）加味。

红参12克，制附片30克，龙骨30克，牡蛎30克，菖蒲12克。水煎急服。

加减：若阴寒内盛者亦可加干姜10克、炙甘草10克。

(曹晓岚)

2 循环系统疾病

西医所称循环系统疾患,相当于中医心脉系所出现的病症。临床常见急症有如下几种。

2.1 心绞痛

西医的心绞痛,属中医的"心痛"、"胸痛"、"心痹"、"胸中痛"的范畴。

病因病机

凡由饮食不节、情志不遂、劳倦内伤、寒邪内侵等原因,导致脏腑功能失调、气机不畅、心脉瘀阻、心失所养,而发心痛。

诊断要点

1. 年龄在40岁以上。
2. 胸中闷痛,痛引左侧肩背或左臂内侧。
3. 常于劳累、饱食、受寒、情绪激动后突然发作,历时3～6分钟,休息或用药后缓解。
4. 心电图示慢性冠状动脉供血不足或急性心肌缺血。

辨证施治

1. 心血瘀阻

主症:胸部刺痛。固定不移,入夜更甚,伴有胸闷气短,心悸不宁,舌质紫暗,脉象沉涩。

治法:活血化瘀,通络止痛。

方药:血府逐瘀汤(32)加减。

当归12克,赤芍12克,川芎10克,桃仁10克,红花10克,柴胡10克,郁金10克,炒枳壳10克,炒元胡10克,水煎服。

2. 痰浊闭阻

主症：胸闷如窒，痛引肩背，形体肥胖，气短喘促，咳嗽痰多，舌体肥胖，舌苔浊腻，脉象弦滑。

治法：宣痹泻浊，豁痰通阳。

方药：瓜蒌薤白半夏汤（37）加味。

瓜蒌24克，薤白15克，半夏12克，炒枳实10克，干姜6克，白蔻仁10克，陈皮10克，水煎服。

加减：兼有瘀血者，加服复方丹参片4片，日3次。

3. 阴寒凝滞

主症：胸痛彻背，感寒痛甚，伴有胸闷气短，心悸不安，面色苍白，四肢厥冷，舌苔白滑，脉象沉细。

治法：辛温通阳，开痹散寒。

方药：瓜蒌薤白白酒汤（38）加味：

瓜蒌24克，薤白15克，制附子12克，桂枝10克，炒枳实10克，炙甘草6克，檀香6克，白酒10克为引，水煎服。

加减：若兼咳吐痰涎者，加陈皮、茯苓、炒杏仁各12克。

若疼痛剧烈而不解者，加服苏合香丸1丸，日2次。

4. 气阴两虚

主症：胸闷隐痛，时作时止，心悸气短，倦怠懒言，头晕目眩，舌质偏红，脉细弱无力，或兼结代。

治法：益气养阴，活血通络。

方药：生脉散（26）加味。

人参12克，麦冬24克，五味子10克，白术12克，茯苓12克，黄精15克，丹参15克，炙甘草6克，水煎服。

加减：若胸闷痛明显者，可加郁金10克，参三七6克。

胸痛剧烈不止者，加服复方丹参注射液2毫升加入5%葡萄糖溶液20毫升静脉推注。

5. 心肾阴虚

主症：胸中闷痛，头晕耳鸣，心悸盗汗，心烦不寐，腰膝酸软，舌质红或有紫斑，脉象细数。

治法：滋养心肾，通脉安神。

方药：左归饮（39）加减。

熟地15克，山萸肉12克，枸杞子15克，淮山药12克，茯苓12克，丹皮12克，丹参20克，麦冬15克，炒枣仁20克，水煎服。

加减：若心烦不寐明显者，加栀子10克，炙百合24克。

若头晕目眩，头面烘热明显者，加钩藤30克，生石决明30克，制首乌30克，怀牛膝15克。

6. 气虚血瘀

主症：胸闷气短，胸中刺痛，活动加剧，伴有倦怠乏力，心悸不安，舌质淡红有瘀斑，脉象细弱而涩。

治法：益气养心，活血安神。

方药：补阳还五汤（40）加减。

生黄芪60克，党参15克，当归尾6克，桃仁6克，红花6克，川芎6克，炒枣仁20克，炙甘草6克，水煎服。

加减：若胸痛明显，加服冠心Ⅱ号片6片，口服，日3次；或七厘散1克，口服，日2至3次。

2.2 急性心肌梗塞

西医的急性心肌梗塞，属中医的"真心病"、"卒心痛"、"厥心痛"等。

病因病机

胸阳与心脉因痰浊、瘀血、气滞、寒凝所阻滞，心脉失养，心体损伤而突然发病。

诊断要点

1. 在胸痛时发时止的基础上突然出现胸中闷痛，怔忡不安，大汗淋漓，四肢厥冷，脉微欲绝。

2. 多由劳累过度、七情过极、气候突变、酒食过量而诱发。

3. 多数病人在 40 岁以上。
4. 心电图示急性心肌梗塞图的演变规律。
5. 实验室检查：CPK、GOT、LOH、HBDH 升高。

辨证施治

急性期（发病 1 周以内）

1. 心气欲绝

主症：突然胸痛，继则四肢厥冷，昏不知人，伴有面色苍白，身冷汗出，舌质淡，苔白滑，脉微细。

治法：补气通阳，固脱醒神。

方药：生脉散（26）加味。

人参 30 克，黄芪 30 克，麦冬 30 克，桂枝 12 克，五味子 9 克，麝香 0.15 克（绢包煎），水煎服，日 1 至 2 剂。

补气固脱散（自拟方）：人参 6 份，上肉桂 1.5 份，麝香 0.05 份，照比例配制，为极细末，每服 9 克，或鼻饲，日 2 至 3 次。

针刺内关、鸠尾穴。强刺激，不留针。

益心丸 1 丸口服，或鼻饲。

2. 心阳欲脱

主症：胸痛欲死，冷汗淋漓，畏寒肢冷，面色苍白，精神萎靡，唇舌青紫，脉微欲绝。

治法：回阳救逆，敛阴固脱。

方药：人参桂枝汤（41）加减。

人参 15 克，桂枝 12 克，熟附子 12 克，山萸肉 15 克，干姜 10 克，炙甘草 10 克，水煎服。

加减：汗出不止者，加龙骨 30 克，牡蛎 30 克以敛汗固脱。耳针或压豆取心、交感、皮质下。

3. 水气凌心

主症：心痛暴作，憋气欲死，咳吐泡沫痰涎，形寒肢冷，精神萎靡，喘息自汗，怔忡不安，面浮肢肿，舌体胖边有齿痕，脉沉迟微弱。

治法：温阳化饮，开痹醒神。

方药：温阳化饮汤（自拟方）。

桂枝15克，茯苓30克，干姜12克，葶苈子30克，水菖蒲24克，郁金10克，炙麻黄6克，水煎服。

加减：若浮肿心悸明显者，加熟附子12克，泽泻、车前子各24克。

咳嗽吐痰明显者，加瓜蒌24克，前胡12克，清半夏12克，细辛6克。

苏冰滴丸2至4丸，口服或舌下含化，日2至3次。

宽胸丸每服1丸，日3次。

耳针心、皮质下、肾上腺、肺等穴。

缓解期

1. 气滞血瘀

主症：胸痛如刺，痛有定处，伴有胸闷憋气，两胁胀满，心烦易怒，少寐多梦，舌边红有瘀斑，脉弦。

治法：理气活血，通络安神。

方药：血府逐瘀汤（32）加减。

柴胡、郁金各12克，丹参30克，炒枳实10克，川芎10克，丹皮10克，玫瑰花12克，红花10克，香附8克，水煎服。

加减：若阵痛剧烈，痛引右胁，加炒元胡12克，川楝子10克。

心烦不寐明显，加寸冬24克，琥珀粉、朱砂粉各1克（冲服）。

2. 痰浊痹阻（参看心绞痛节）

3. 寒凝心脉（参看心绞痛节）

4. 痰食阻滞

主症：胸中闷痛，脘腹胀满，恶心纳呆，大便秘结，心悸少寐，舌质红苔淡黄厚腻，脉象弦滑。

治法：通腑和胃，化浊通络。

方药：半夏泻心汤（42）加减。

清半夏 12 克，炒枳实 10 克，黄连 6 克，干姜 3 克，酒军 6 克，降香 10 克，瓜蒌 24 克，丹参 24 克，生山楂 24 克，水煎服。

加减：若脘腹纳呆明显者，加砂仁 10 克，鸡内金粉 3 克（冲服）。

5. 心气虚弱

主症：胸中隐痛阵作，有空虚感，气短乏力，怔忡不安，动则诸症加剧，舌质淡红，舌苔薄白，脉象细弱。

治法：补益心气，通络安神。

方药：保丹饮（自拟方）。

人参 12 克，生黄芪 30 克，丹参 24 克，当归 12 克，桃仁 6 克，桂枝 6 克，檀香 6 克，炙甘草 3 克，水煎服。

加减：兼阴虚者，加麦冬 21 克，五味子 6 克。

出现阳虚症状者，加熟附子 6 克。

6. 心阳不振

主症：胸中闷痛，怔忡不安，怯寒肢冷，自汗短气，动则加剧，舌质淡红，舌苔白滑，脉象沉迟细弱，或有结、代。

治法：温阳益气，通脉化饮。

方药：人参四逆汤（43）加减。

人参 12 克，熟附子 10 克，茯苓 24 克，白术 12 克，甘草 3 克，水煎服。

加减：若水饮上泛，恶心呕吐，或吐痰涎者，加姜半夏 10 克，葶苈子 15 克，陈皮 10 克。

心悸不安明显者，加龙骨 30 克，牡蛎 30 克。（本节与心阳欲脱互参）。

7. 气阴两虚（参看心绞痛节）
8. 心肾阴亏（参看心绞痛节）

2.3 急性肺源性心脏病

西医的急性肺源性心脏病，属中医"咳喘"、"咯血"、"水肿"等范畴。

病因病机

久病咳喘，正气耗伤，痰瘀内阻，气机逆乱，水饮泛滥，射肺凌心。

诊断要点

1. 久患咳喘，突然加剧，息促不安，不能平卧。
2. 近来咳吐痰液量多或出现脘腹胀满，下肢浮肿。
3. 剧咳、咯血、憋闷欲死。

辨证施治

1. 痰饮内停，射肺凌心

主症：咳喘气短，咳痰量多，呈泡沫状，倚息不得卧，心悸不安，下肢浮肿，舌体胖质暗红，脉象滑数。

治法：温阳化痰，开肺宁心。

方药：苓桂术甘汤（44）加味。

茯苓30克，桂枝12克，生白术24克，葶苈子18克，水菖蒲24克，桑白皮15克，前胡12克，桃仁6克，甘草3克，车前子24克（包煎），水煎服。

加减：吐痰粘稠者，加瓜蒌24克，桔梗10克。

大便干结者，加生大黄6克，炒莱菔子15克。

针刺肺俞、尺泽、阴陵泉透阳陵泉、水分、复溜、足三里等穴。

2. 心肾阳虚，痰瘀内阻

主症：心悸气短，颜面及下肢浮肿，腰膝酸软，咳喘胸闷，脘腹胀满，恶心纳呆，唇舌青紫，舌苔黄腻，脉象沉细弱。

治法：益气温阳，化痰消瘀。
方药：真武汤（45）加味。

熟附子 12 克，茯苓 30 克，生白术 30 克，赤白芍各 12 克，生姜 10 克，葶苈子 15 克，北五加皮 15 克，桃仁 6 克，泽兰 10 克，炒莱菔子 15 克，生山楂 16 克，水煎服。

加减：脘腹胀满，右胁症块明显者，加三棱 6 克，莪术 6 克，鸡内金 12 克。

纳呆明显者，加砂仁 10 克，神曲 12 克。

参附针每次 4 毫升，肌肉注射，每日 2 次。

3. 痰瘀阻络，络破血溢

主症：剧烈咳嗽，胸痛向颈间放射，咳出紫黑色血痰，胸闷憋气，喘促不安，唇舌青紫，脉象细涩。

治法：清肺化痰，祛瘀止血。

方药：犀角地黄汤（25）加味。

犀角粉 3 克（冲），生地 30 克，丹皮 12 克，赤芍 12 克，大黄粉 15 克（冲服），白芨粉 15 克（冲服），浙贝 12 克，桑白皮 15 克。水煎服。

生大黄粉 6～12 克，每日 3 次吞服。

三七粉 6 克，血余炭 6 克，浙贝 6 克，花蕊石 24 克，共研细末，分四次用开水冲服。

云南白药每次 0.5 克，每日 3 次，温开水送服。

2.4 急性感染性心内膜炎

西医的急性感染性心内膜炎，属中医温热病范畴。

病因病机

毒邪外袭入里化热，伤及营血，诸症丛生。

诊断要点

1. 有表证病史。
2. 呈不规则持续性发热，或呈低热。

3. 常伴五脏损伤以及皮肤瘀点或瘀斑。

辨证施治

1. 风热在表

主症：发热微恶风寒，头痛、鼻塞，流黄涕，咽痛，口微渴，舌质红，苔薄黄，脉象浮数。

治法：辛凉解表，清热败毒。

方药：银翘散（2）。

银花30克，连翘15克，野菊花12克，荆芥10克，薄荷10克，牛蒡子12克，鲜芦根30克，黄芩12克，炒杏仁12克，生甘草3克。水煎服。

犀羚解毒片24片，为末冲服，日2～3次。

2. 肺胃热盛

主症：高热，喘咳，吐黄脓痰，口渴喜冷饮，溲赤，便结，牙龈肿痛，舌红苔黄，脉象洪数。

治法：清热泻火，通便平喘。

方药：白虎汤（10）合泻白散（46）。

生石膏30克，知母12克，桑白皮15克，地骨皮12克，炒杏仁12克，玄参30克，大黄8克，竹叶10克。水煎服。

3. 热入营血

主症：发热夜甚，口渴不欲饮，躁动不安，神昏谵语，肤起斑疹，或吐血、衄血，溺血，舌质红绛，苔黄燥，脉细数。

治法：清热解毒，泄热凉血。

方药：犀角地黄汤（25）加减。

犀角粉3克（冲），生地30克，丹皮12克，赤芍12克，金银花30克，羚羊角粉6克（冲），茅根45克，竹叶卷心6克。水煎服。

加减：衄血、便血明显者，加仙鹤草30克，三七粉3克（冲）。

咯血明显者，加生大黄粉，白芨粉各15克（冲）。

神昏便秘者，加紫雪散 3 克（冲），日 2～3 次，或用安宫牛黄丸 1 丸，或用至宝丹 1 丸，口服，日 1～2 次。

清开灵注射液 90 毫升，加入 10% 葡萄糖 500 毫升中静滴，每日 1～2 次。

4. 热极动风

主症：高热神昏，四肢抽搐，颈项强直，甚至角弓反张，牙关紧闭，或一侧肢体抽搐，日发数次，终至半身瘫痪，舌质红绛，脉象弦数。

治法：清热凉肝，熄风镇痉。

方药：羚羊钩藤汤（22）加减。

羚羊角粉 5 克（冲），玳瑁粉 3 克（冲），生石决明 30 克，钩藤 30 克，牛膝 15 克，生地 30 克，生栀子 12 克，青黛 6 克，全蝎 12 克，地龙 12 克，杭芍 15 克，菊花 12 克，桑叶 12 克，茯神 15 克。水煎服。

加减：抽搐明显者，加服止痉散 1.5 克，每日 1～2 次，冲服。

高热不退者，加服安宫牛黄丸 1 丸，每日 3 次，口服，或用牛黄清心丸，或用紫雪散等，或针刺十宣放血。

2.5 病毒性心肌炎

西医的病毒性心肌炎，属于中医"心痹"、"心痛"、"心悸"的范畴。

病因病机

外邪不解，内含于心，心脉痹阻，心肌失养，心主不明，而生诸症。

诊断要点

1. 近两周内有感冒病史，突然出现怔忡不安，胸闷气短等症状。

2. 患者常有发热胸痛等症状。

3. 多有结、代脉象。
辨证施治
1. 心脉瘀阻

主症：胸痛如刺，固定不移，持续不止，夜间尤甚，伴有心悸不定，舌质紫暗或有瘀斑，脉象沉涩。

治法：活血化瘀，通络安神。

方药：血府逐瘀汤（32）加味。

当归12克，川芎10克，赤芍10克，丹皮12克，桃仁6克，红花6克，玫瑰花12克，炒枳壳10克，柴胡10克，甘草3克，水煎服。

加减：心悸明显者，去炒枳壳，加甘松12克，麦冬24克。

胸痛甚者，加炒元胡12克，降香10克。

挟痰者，加瓜蒌18克，半夏10克。

针刺内关、心俞、膻中穴。强刺激，不留针。

2. 寒凝心络

主症：胸痛彻背，如锥如刺，遇寒更剧，伴有胸闷气短，心悸不宁，面色苍白，四肢欠温，舌苔白滑，脉象沉迟。

治法：温经散寒，通络止痛。

方药：麻黄附子细辛汤（47）加味。

炙麻黄12克，熟附子10克，细辛6克，当归15克，川芎12克，檀香10克，炙甘草12克，桂枝6克。水煎服。

加减：剧痛不止者，加服苏合香丸1丸，口服，日3次，或用冠心苏合丸。

阴寒极盛者，加肉桂6克。

3. 阴虚火旺

主症：心悸盗汗，低热心烦，不寐健忘，胸中隐痛，头晕耳鸣，舌红少津，脉象细数。

治法：滋阴清火，养心安神。

方药：天王补心丹（48）加味。

生地 20 克，玄参 21 克，天冬 15 克，麦冬 15 克，丹参 20 克，西洋参 10 克，苦参 10 克，茯苓 15 克，炒枣仁 30 克，夜交藤 30 克，炙远志 10 克。水煎服。

加减：若胸中闷痛明显者，加川芎 12 克，郁金 12 克。

心悸眩晕明显者，加制首乌 30 克，生石决明 30 克，女贞子 18 克，钩藤 30 克。

针刺内关，神门。中刺激，不留针。

参麦针 20 毫升加入 5% 葡萄糖液 20 毫升中静注，日 3 次。

苦参碱注射液，每次 2 毫升，肌肉注射，日 2～3 次。或用苦参浸膏片和丹参片各 3 片，口服，日 2 次。

4. 心阳不振

主症：心悸气短，动则加剧，伴有面色苍白，倦怠神疲，畏寒自汗，舌淡苔白，脉沉微细。

治法：益气通阳，通络安神。

方药：参附汤（29）加味。

人参 12 克，熟附子 10 克，桂枝 10 克，炙甘草 12 克，茯苓 20 克，生姜 10 克，大枣 6 枚（劈）。水煎服。

加减：若四肢厥冷，脉微欲绝者，用红参 24 克，加山萸肉 24 克，龙骨 30 克，牡蛎 30 克。

浮肿心悸者，加汉防己 12 克，车前子 24 克。

5. 气阴双亏

主症：胸闷隐痛，心悸不宁，时作时止，午后加剧，伴有面色少华，倦怠懒言，头晕目眩，遇劳更甚，舌质红，脉细弱而结代。

治法：益气养阴，定悸安神。

方药：炙甘草汤（31）加味。

炙甘草 12 克，桂枝 12 克，人参 10 克，柏子仁 15 克，生地 30 克，麦冬 24 克，阿胶 12 克（烊化），炒枣仁 30 克，丹参 21 克，生姜 8 克，大枣 30 枚（劈）。水煎服。

2.6 心律失常

西医的心律失常,属于中医的心悸怔忡病证。

病因病机

心虚胆怯,心血不足,心阳虚衰,阴虚火旺,水饮内停,瘀血阻络等,不论是心神失养,还是邪气干扰,均能导致心神不宁而出现心悸怔忡。

诊断要点

1. 病人自觉心中悸动,惊悸不定,甚则不能自主,脉数、迟、结、代、涩、促等。
2. 每因情志过激或劳累过度而诱发。
3. 常与失眠、健忘、眩晕、耳鸣等同时并见。
4. 心电图示心律失常。

辨证施治

1. 气滞血瘀

主症:心悸不安,胸痛如刺,固定不移,遇怒加剧,舌质紫暗,或有瘀斑,脉象沉涩。

治法:行气活血,通络止痛。

方药:血府逐瘀汤(32)加减。

当归12克,川芎10克,桃仁12克,红花12克,柴胡12克,炒枳壳10克,甘松12克,丹参21克,玫瑰花10克。水煎服。

加减:失眠明显者,加夜交藤30克。

怔忡不宁者,加琥珀粉3克(冲)。

胸痛甚者,加服冠心苏合丸1丸,口服,日2次。

复方丹参注射液8毫升加入50%葡萄糖250毫升静脉滴注。

2. 痰浊阻络

主症:心悸胸闷,咳嗽痰多,胸痛彻背,头晕目眩,形体肥胖，舌质淡红，舌苔白滑，脉滑或结代。

治法：温阳化痰，开痹化痰。

方药：瓜蒌薤白半夏汤（37）加味。

瓜蒌24克，薤白12克，清半夏12克，炒枳实12克，厚朴10克，白蔻仁10克，橘红12克，茯苓12克。水煎服。

瓜蒌片每服4片，日3次。

苏冰滴丸，每服2丸，日2次。

3. 水饮凌心

主症：心悸眩晕，胸脘满闷，泛恶欲呕，形寒肢冷，下肢浮肿，小便短少，渴不欲饮，舌苔白滑，脉象弦滑。

治法：振奋心阳，化气行水。

方药：苓桂术甘汤（44）加味。

茯苓30克，桂枝12克，生白术15克，甘草3克，汉防己12克，葶苈子15克，前胡12克，姜半夏12克。水煎服。

加减：若恶心呕吐明显，加陈皮10克，生姜10克。

下肢浮肿，小便不利明显者，加熟附子12克，车前子24克（布包煎）。

4. 心虚胆怯

主症：心悸，善惊易恐，少寐多梦，舌质淡红苔薄白，脉象动数。

治法：镇惊安神，养心安志。

方药：安神定志丸（49）加味。

人参10克，炙甘草10克，炒枣仁30克，五味子6克，茯神20克，炙远志10克，生龙齿30克，琥珀粉、硃砂粉各1克（冲服），磁石15克。水煎服。

加减：若惊悸而烦者，去五味子，生龙齿，加清半夏10克，橘红10克，川连6克。

5. 心血不足

主症：心悸头晕，面色无华，倦怠懒言，舌质淡红，脉象细弱或结代。

治法：补血安神，益气养心。

方药：归脾汤（50）加减。

人参10克，炙黄芪12克，茯苓12克，白术12克，炙甘草10克，当归10克，龙眼肉12克，砂仁6克，木香6克，炒枣仁24克，炙远志10克，干地黄12克，水煎服。

加减：若脉结代明显者，用炙甘草汤加减。

若心烦不安，心阴受损者，加麦冬24克，五味子10克。

6. 心肾阴虚

主症：心悸失眠，五心烦热，头晕目眩，耳鸣腰痠，舌红少苔，脉象细数，或结代。

治法：滋阴降火，养心安神。

方药：天王补心丹（48）加减。

生地30克，麦冬20克，天冬20克，当归12克，制首乌24克，丹参21克，炒枣仁30克，柏子仁12克，茯神12克，硃砂粉、琥珀粉各1克（冲服），童参10克，玄参15克。水煎服。

加减：心烦热明显者，加服知柏地黄丸1丸，口服，日2次。

7. 心肾阳虚

主症：心悸气短，动则加剧，面色苍白，形寒肢冷，腰膝痠软，尿少浮肿，舌淡苔白，脉象沉弱。

治法：温补心肾，安神止悸。

方药：参附汤（29）合苓桂术甘汤（44）加减。

人参10克，熟附子10克，茯苓24克，桂枝10克，白术10克，甘草3克，干姜6克，龙骨15克，牡蛎15克。水煎服。

加减：若病情严重，怔忡不安，喘不得卧，人参、附子各用

30 克。

参附针每次 20 毫升加入 50%葡萄糖液 30 毫升缓慢静脉注射，日 3 次。

2.7 心力衰竭

西医的心力衰竭，属中医的心悸怔忡、喘咳、痰饮、水肿等范畴。

病因病机

病延日久，内损五脏，心衰则悸，肺损则咳，肾伤则喘，脾亏则肿。

诊断要点

1. 主要表现为呼吸困难，咳喘不宁，咯血，紫绀，腹胀水肿。

2. 有心肺等脏慢性病史。

辨证施治

1. 心肾阳虚，饮凌心肺

主症：悸忡不宁，喘促胸闷，咳吐泡沫痰涎，不能平卧，动则诸症加剧，自汗倦怠，形寒肢冷，精神萎靡，舌质淡暗，舌苔白滑，脉沉细微。

治法：温阳化饮，泻肺醒神。

方药：参附汤（29）合加味葶苈大枣泻肺汤（51）。

人参 30 克，熟附子 24 克，桂枝 15 克，茯苓 30 克，葶苈子 18 克，水菖蒲 12 克，大枣 10 枚。水煎服。

加减：颜面浮肿者，加车前子 24 克（包煎）。

唇舌青紫者，加丹参 24 克，桃仁 12 克。

2. 痰瘀内阻，浊气上逆

主症：心悸气短，胸闷憋气，脘腹胀闷，右胁坚满，恶心欲呕，下肢浮肿，精神困倦，怯寒懒动，小便短少，大便溏泻，舌质暗红，舌苔白腻，脉缓而涩。

治法：消瘀化痰，行气醒脾。

方药：调营饮（52）加减。

当归尾 10 克，川芎 10 克，赤芍 12 克，莪术 12 克，大黄 6 克，厚朴 10 克，茯苓 30 克，姜半夏 10 克，葶苈子 12 克，大腹皮 15 克，车前子 18 克，桂枝 6 克，生姜三片，大枣 10 枚。水煎服。

加减：若恶心呕吐明显者，加陈皮 10 克，姜竹茹 10 克。

下肢浮肿明显者，加汉防己 12 克，泽兰 12 克。

（邵念方）

3 呼吸系统疾病

西医所称呼吸系统疾病，相当于中医呼吸系所出现的病症。临床常见急症有如下几种。

3.1 急性支气管炎

可见于中医的"感冒"、"咳嗽"、"喘逆"等病症。

病因病机

邪气犯肺，清肃失令，肺气上逆而咳嗽气喘。

诊断要点

1. 一般有鼻塞、流涕、咽痛、畏寒、发热等上呼吸道感染症状。

2. 咳嗽为主要症状。开始痰少色白，以后变黄，可有血丝。

3. 肺部可闻及呼吸音粗糙，并有干、湿罗音。

4. X线检查大多正常或肺纹理增粗。

辨证施治

1. 风寒犯肺

主症：咳嗽声重，气急，咽痒，痰稀薄色白，常伴发热、畏寒、头痛。苔薄白，脉浮或浮紧。

治法：疏散风寒，宣肺止咳。

方药：麻杏汤（53）加味。

麻黄9克，苏叶9克，荆芥12克，杏仁9克，甘草6克。加水煎两次，共取药汁300毫升左右，早晚2次温服。

加减：痰多、苔白腻者加陈皮9克，茯苓12克，厚朴9克。

痰多、喘咳甚、苔厚腻、脉滑者为寒饮郁肺,可用小青龙汤(54)加减:麻黄20克,桂枝12克,干姜20克,细辛3克,五味子3克,大枣20克,甘草20克,法半夏30克,生石膏120克。水煎两次,取汁约300毫升,早晚二次温服。

2. 风热犯肺

主症:咳嗽剧频,气粗咽痛,痰黄稠粘,发热口渴,苔黄,脉浮数。

治法:疏散风热,清肺止咳。

方药:银翘散(2)合麻杏石甘汤(8)加减。

金银花30克,连翘15克,炙麻黄9克,生石膏45克(先入),杏仁10克,桔梗12克,牛蒡子12克,芦根30克,桑白皮12克,黄芩15克,甘草6克。水煎2次,取药汁约300毫升,早晚分2次服用。

加减:口干渴,痰难出,加沙参30克,麦冬15克。

痰黄稠者,加鱼腥草30克。

身热重者,加知母12克,生石膏用至60克。

3. 热痰蕴肺

主症:咳喘甚,痰黄稠,身热重,苔黄腻,脉滑数,口干渴,大便干。

治法:清热泻火,化痰止咳。

方药:麻杏石甘汤(8)合小半夏汤(55)加减。

炙麻黄12克,生石膏30克,杏仁9克,瓜蒌30克,黄连9克,黄芩12克,鱼腥草30克,法半夏9克,甘草6克。水煎2次,取汁300毫升,早晚分2次温服。

加减:痰黄腥臭,加苇茎30克,生苡米30克,桃仁12克,冬瓜仁30克,蚤休20克,以清热解毒,化痰排脓。

大便干者,加大黄9克。

3.2 大叶肺炎

大叶肺炎大致属于中医"咳嗽"的范畴。

病因病机

风热疫邪由口鼻侵犯肺脏，痰热壅滞，肺失肃降而咳嗽胸痛。

诊断要点

1. 多因风（寒、热、疫毒），邪外袭，过度劳累，精神创伤以及病后体弱所致。

2. 有寒战、发热、咳嗽、胸痛、痰黄或铁锈样痰。

3. 肺部可闻及干湿罗音，叩诊浊音。X线检查实变期为大片状密度均匀增高的阴影。化验血白细胞总数 $10000/mm^3$ 以上，嗜中性粒细胞70%以上，核左移。

4. 病情严重者可出现神志模糊，甚至昏迷，伴有血压下降，出冷汗，面色苍白等周围循环衰竭征象。

辨证施治

1. 温邪袭肺，肺卫不宣

主症：寒战发热（体温突然升至38℃以上），头痛，体痛，胸痛，咳嗽较剧，咳吐粘痰。苔薄白或薄黄少津，脉浮数。

治法：疏散风热，宣肺止咳。

方药：银翘散（2）加减。

银花30克，连翘12克，鱼腥草30克，薄荷9克，杏仁9克，桔梗9克，芦根30克，竹叶3克，牛蒡子12克，瓜蒌24克，青黛6克（冲），甘草6克。水煎2次，共取汁约400毫升，分2～3次温服。

加减：身热甚者，加生石膏30—45克。

高热神昏者，加羚羊粉1.5克（冲）。

2. 肺胃热盛，气血两燔。

主症：高热（体温39℃以上），咳嗽剧烈，呼吸气粗，胸痛如焚，咯吐大量铁锈色痰或脓血痰，出汗多，口干渴，面色红，大便秘结。苔黄燥，脉洪大或洪数。

治法：清解肺胃热邪，凉血解毒。

方药：银翘白虎汤（56）加减。

银花30克，连翘12克，大青叶30克，败酱草30克，鱼腥草30克，知母12克，生石膏45克，桑白皮30克，鲜生地30克，桃仁15克，生苡米30克。水煎2次，取汁300毫升，早晚分2次温服。

加减：胸痛咳剧者，加浙贝12克，，郁金12克，瓜蒌皮15克。

大便秘结，数日不行者，加大黄12克，芒硝9~12克（冲），瓜蒌30克。

3. 肺胃阴伤

主症：高热渐降，或低热不退，咳吐血痰减少，心烦口渴。舌红少苔、脉细数。

治法：滋阴养胃，润肺清热。

方药：竹叶石膏汤（57）合养阴清肺汤（58）。

生石膏30克，麦冬30克，太子参30克，竹叶3克，丹皮9克，生地15克，玄参15克，白芍12克，川贝12克。水煎2次，取汁300毫升，早晚2次温服。

加减：大便干结者，加大黄9克，麻仁30克。

4. 变证

(1) 肺胃热盛，气阴两伤。

主症：高热，咳嗽，胸痛，气粗，咯吐血痰或铁锈色痰，出汗多，口烦渴。舌红苔黄燥，脉扰数。

治法：清泄肺胃热邪，益气育阴。

方药：银翘白虎汤（56）加减，或生脉散（26）加减。

银花30克，连翘15克，鱼腥草30克，生石膏45克，知母12克，西洋参15克（或太子参30克），麦冬30克，生桑皮15

克。水煎 2 次，早晚 2 次温服。

(2) 亡阳欲脱

主症：高热骤降（体温 36℃以下），冷汗淋漓，面色苍白，血压骤降（收缩压低于 80 毫米汞柱），小便减少，脉象微细欲绝或急促无力。

治法：回阳救逆固脱。

方药：①独参汤 (59)。

大力参 30 克。水煎 2 次，取汁 200 毫升，温服。

②参附汤 (29) 加味。

人参 15 克，熟附子 30 克，吴茱萸 15 克，五味子 9 克。水煎 2 次，取汁 200 毫升温服。

③回阳返本汤 (60)。

人参 12 克，麦冬 24 克，五味子 9 克，熟附子 15 克，干姜 12 克，炙甘草 6 克。水煎服。

亦可酌情选用：参附注射液 (61) 或生脉注射液 10 毫升，加入 500 毫升 10%的葡萄糖注射液中，静脉点滴，每日 1～2 次。

3.3 急性吸入性肺脓肿

急性吸入性肺脓肿，属于中医"肺痈"的范畴。

病因病机

热毒蕴肺，毒瘀互结，败腐成痈。

诊断要点

1. 身热不已，咳喘胸痛，痰黄腥臭。苔黄，质红，脉洪滑或数。

2. 化验血：白血球计数及中性粒细胞明显升高。

3. 胸部 X 线检查，早期表现为肺胞型浸润性炎症改变，脓成后病灶内可见有液平的空洞。

辨证施治

1. 毒热蕴结

主症：壮热不已，咳喘胸痛，痰黄稠粘，苔黄腻，质红暗，脉滑数或洪数。

治法：清解热毒，化痰散瘀。

方药：苇茎鱼腥汤（62）加减。

苇茎30克，鱼腥草30克，桔梗9克，败酱草24克，桃仁15克，冬瓜仁30克，银花30克，生苡米30克，瓜蒌18克，生甘草12克。水煎2次，取汁300毫升，早晚2次温服。

加减：身热甚者，加生石膏45克。

大便不通者，加大黄6～10克。

2. 脓毒内溃

主症：痰多黄稠，腥臭难闻，身热胸痛。苔黄腻，脉滑数。

治法：解毒排脓。

方药：加味苇茎汤（63）。

芦根30克，生苡米45克，冬瓜仁30克，桔梗12克，鱼腥草30克，浙贝母15克，银花30克，藕节12克，桃仁12克，陈皮9克。水煎2次，取汁约400毫升，分早、中、晚三次饭后温服，每日1剂。

加减：身热甚加生石膏45克，知母9克。

脓痰带血者，加三七粉6克，分2次温开水冲服。

大便干者，加大黄6～10克。

壮热神昏者，加紫雪丹2克冲服，日2次。或加安宫牛黄丸（20）1丸，日2次温开水送服。

（卢尚岭）

4 神经系统及精神疾病

4.1 三叉神经痛

本病大致属于中医"头风头痛"、"偏头痛"的范畴。

病因病机

风邪外袭，滞而不除，血行不畅，不通则痛，风邪善行而数变，故乍痛乍止，发作突然。

诊断要点

1. 面部阵发烧灼，有闪电样剧痛，常伴面肌抽搐、流泪，一般只持续数秒钟即可自行消失，间歇时可无任何不适。

2. 疼痛多从面部动作引起，可涉及上下唇、鼻翼、硬腭等。

3. 一般无明显神经系统阳性体征。

4. 注意与牙痛、鼻窦炎、舌咽神经痛鉴别。

辨证施治

1. 外感风热

主症：头面一侧阵发疼痛如刺如烧，常伴身热、口渴、咽痛、苔薄黄、脉弦数。

治法：疏散风热，活络止痛。

方药：芎芷石膏汤（64）加减。

川芎 20 克，生石膏 30 克，菊花 9 克，薄荷 9 克，地龙 9 克，牛蒡子 12 克，白芷 12 克，芦根 30 克。水煎服。

配合针刺听宫、下关、合谷、四白、鱼腰等穴，效果更著。

2. 外感风寒

主症：疼痛剧烈，畏寒发热，苔薄白，脉弦紧。

治法：散寒止痛。

方药：川芎茶调散（65）加减。

川芎 30 克，白芷 12 克，羌活 12 克，防风 12 克，细辛 6 克，荆芥 9 克，薄荷 6 克，制草乌 9 克（先煎 1 小时），甘草 6 克。水煎服。

加减：疼痛剧或服上方后效果不著者，可加蜈蚣 6 条、全蝎 6 克共焙焦，研细末，温开水冲服。

配合针刺头维、列缺、颊车、下关、合谷等穴，效果更著。

3. 瘀血阻络

主症：痛如锥刺，痛处固定，反复发作，耳鸣耳聋，舌质暗，或有瘀斑，脉细涩。

治法：活血化瘀止痛。

方药：通窍止痛汤（66）加减。

川芎 30 克，赤芍 15 克，桃仁 15 克，红花 9 克，生姜 3 片，老葱 3 寸，麝香 0.2 克（冲）。水煎服。

4.2 偏头痛

偏头痛属于中医"脑风"、"首风"、"头风头痛"等的范畴。

病因病机

先天不足，后天失宜，脏腑受损，阴阳相失，复感外邪，或情志相激，或劳累过度等，气血逆乱，髓海失养，而头痛骤发，时痛时止。

诊断要点

1. 常与遗传有关，可反复发作，首次多始自儿童或少年。

2. 多以感受时邪或疲劳、紧张、情绪激动、睡眠欠佳、月经来潮等为发作诱因。

3. 发作前常伴有嗜睡、精神不振或过分舒适，视物模糊，畏光，可出现盲点、偏盲，眼球胀痛，或肢体感觉异常，或动作障碍等症状。

4. 多一侧，有时也有两侧额、颞、眼眶等处反复发生剧烈跳痛或钻痛，如锥刺，似火烧，痛不能忍。疼痛可持续几分钟或1～2天。有时1日发作数次，有时亦可数月乃至数年后才再度发作。

5. 常伴有恶心，呕吐，腹胀，腹泻，多汗流泪，面色苍白，皮肤青紫水肿等症状。

辨证施治

1. 风痰瘀血痹阻络脉

主症：头痛时作，遇寒风冷气尤剧，口不渴，苔薄白，脉浮弦而紧。

治法：祛风、化痰、活血止痛。

方药：①散偏汤（67）加减。

川芎30克，白芷15克，白芍15克，白芥子9克，香附6克，柴胡3克，郁李仁3克，甘草3克。水煎2次，取汁300毫升，早晚2次温服，每日1剂。

②川芎定痛汤（68）加减。

川芎30克，赤芍15克，丹参30克，防风12克，细辛6克，川乌6克，白芥子15克，生苡米30克，砂仁9克。水煎服。

痛止后预防再复发可用下方：

川芎15克，当归、红花、白芷、白蒺藜、菊花各9克，钩藤6克，珍珠母30克。水煎服。

2. 风火上扰

主症：突然头痛如裂如刺，面红汗出，口渴心烦。舌红、苔薄黄，脉弦数。

治法：疏散风热，活络定痛。

方药：清上蠲痛汤（69）。

麦冬15克，黄芩12克，蔓荆子12克，菊花12克，羌活、防风、苍术、当归、白芷各9克，川芎15克，细辛6克，甘草3克。水煎服。

加减：左边头痛，加红花9克，柴胡9克，龙胆草12克，生地黄9克，用法同上。

右边头痛，加天麻12克，清半夏12克，山楂12克，枳实12克。用法同上。

头顶痛，加藁本9克，大黄6克，荆芥9克。

风入脑髓作痛，鼻流浊涕或通气不畅，加苍耳子12克，木瓜9克，荆芥9克。用法同上。

气血不足头痛，加黄芪15克，人参9克（或太子参30克），赤白芍各12克，生熟地各12克，用法同上。或用全蝎钩藤散（70）加减（全蝎6只，钩藤9克，红参6克，白芷9克，制川乌6克，共焙干，研极细末，分9包）1包，新开水250毫升冲，加盖闷20分钟，去渣温服，每日3次。

若是在月经期发作者，可用羊角冲剂（71）。

羊角18克（或羚羊角3克），川芎6克，白芷9克，制川乌6克，共研细末，分2包，1包放茶杯内，加新开水250毫升，加盖，半小时后，去渣，温服。早晚各1次。10天为1疗程。

另外，亦可用止痛散（72）。

冰片3克，芒硝6克，麝香0.5克，薄荷冰2克，共研细末，装瓶盖紧备用。用法：取药粉0.3克，用新纱布包，塞鼻孔内，右侧头痛塞左侧鼻孔，左侧头痛塞右侧鼻孔。可立时止痛，适用于各种正偏头痛。

4.3 急性多发性神经炎

多发性神经炎又名周围神经炎或多发性末梢神经炎，其临床表现多属于中医"痿证"的范畴。

病因病机

多系热毒侵犯，耗伤津血，筋脉失养，肢体痿弱不用。

诊断要点

1. 起病急，多系突然发病。

2. 肢体有对称性远端针刺、虫行、烧灼等异常感觉，肌肉压痛。

3. 对称性肢体远端肌力减退，肌张力降低，腱反射降低或消失，急性期后远端肌肉可见萎缩。

4. 肢体远端皮肤发汗、光滑、干燥或出汗，或无汗。

辨证施治

1. 热毒蕴结，筋脉失养

主症：发热、口渴，肢体远端刺疼，或烧灼样疼痛，小便黄，苔薄黄，脉数。

治法：清热解毒，活血通络。

方药：解毒通络饮（73）加减。

虎杖15克，婆婆针15克，土大黄15克，丹参15克，银花60克，贯仲30克。水煎服。

加减：湿热重者，加土茯苓30克，泽泻15克。

疼痛明显者，延胡索15克，天仙藤6克。

心烦，口干，舌红少苔者，加生苡米30克，沙参30克，天花粉15克，麦冬15克。

2. 瘀血阻络

主症：肢体远端麻木疼痛较甚，夜间痉痛明显，舌质暗红，脉弦涩。

治法：活血通络止痛。

方药：①活血驱痛汤（74）加减。

当归12克，丹参20克，红花12克，牛膝12克，鸡血藤30克，白芷9克，延胡索9克，生黄芪30克，生甘草12克，汉防己12克，羌活9克，乳没各6克，炒枣仁15克。水煎服。

②当归威灵仙汤（75）加减。

当归12克，威灵仙12克，川芎12克，白芷9克，汉防己9克，苍术6克，羌活6克，桂枝6克，生姜2片。水煎服。

4.4 急性感染性多发性神经根炎

本病是一种特殊的多发性神经炎，一般属于中医"痿证"的范畴。

病因病机

风热疫毒侵犯，郁结蕴蒸，耗伤津血，筋脉失于濡养，四肢痿弱不用。

诊断要点

1. 发病前 2～3 周多有上呼吸道感染或腹泻史。
2. 起病急，自觉肢体麻木，检查可无明显体征，或有带手套、穿袜子样肢体远端感觉减退、消失或过敏。
3. 四肢和躯干肌无力是其主要症状，多在数日内造成四肢对称性瘫痪，一周内达到高峰。近端较远端为重，重者可有呼吸困难。
4. 四肢肌张力降低，腱反射消失，肌肉压痛明显。
5. 脑脊液细胞蛋白分离，即细胞数正常或略高而蛋白明显增高。

辨证施治

1. 热毒壅盛，血脉不畅

主症：突然自觉肢体麻木，伴有身热、口渴、小便黄、大便干，舌红苔黄，脉数。

治法：清热解毒，活血通络。

方药：解毒活血通络汤（76）加减。

板兰根 30 克，大青叶 15 克，山豆根 15 克，马勃 9 克，桔梗 6 克，桂枝 6 克，丹参 30 克，赤芍 15 克，当归 15 克，牛膝 12 克，丝瓜络 18 克，生黄芪 15 克，甘草 9 克。水煎服。

加减：热毒甚者，可用解毒通络饮（73）加减。

虎杖 15 克，婆婆针 15 克，土大黄 15 克，丹参 15 克，银花藤 60 克，贯仲 30 克。水煎服。

疼痛甚者，加元胡15克，威灵仙15克。

口干、心烦、舌红、脉数者，加沙参30克，麦冬15克，天花粉12克。

2. 气血亏虚，筋脉失养

主症：多见于本病的慢性恢复期，四肢痿弱，体倦乏力，食欲不振或正常，二便调，脉细弱，苔薄白。

治法：益气养血活络。

方药：①补气养血除痿汤（77）加减。

生黄芪60克，白糖参10克，黄精12克，党参15克，桂枝6克，炒麦芽18克，当归12克，丹参18克，牛膝15克，麦冬15克，川断12克，升麻6克，千年健15克，大云15克，鸡血藤18克，每日1剂，水煎服。

②独活寄生汤（78）加减。

独活6克，桑寄生12克，秦艽6克，防风6克，细辛3克，当归12克，杭芍12克，川芎10克，熟地12克，杜仲9克，牛膝9克，红参6克。每日1剂，水煎服。

针灸及穴位注射：上肢取穴手三里合谷，配穴为肩髃、肩髎、曲池。下肢取穴肾俞，大肠俞、环跳，配穴为足三里、阳陵泉。隔日1次，10次为1疗程，一般以温针效果较好。穴位药物注射可用维生素B_1、维生素B_{12}或当归注射液作相应穴位注射，或复方丹参注射液，2毫升肌肉注射，每日1次。

4.5 高血压性脑出血

高血压性脑出血的临床表现，多属于中医"卒中风"中脏腑的范畴。

病因病机

多由于先天不足，后天失养，阴精暗耗，阴阳失调，阴虚于下，阳亢于上，复有所触，如：六淫侵犯，情志过激，劳逸过度，饥饱失宜等等，致气血升降逆乱，血不循经，血溢脉外，而卒然昏仆,发为中风。

诊断要点

1. 起病急,多于活动中卒然昏仆。

2. 多发于40岁以上的中老年人。

3. 多素有头痛、眩晕或高血压病史。发病前常有病情加重、短暂失语、肢体发麻、眼前发花等先兆;发病时常有感受外邪、情绪激动、酗酒暴食和劳累等诱因。

4. 以卒然昏仆,不省人事,半身不遂,口眼㖞斜,语言蹇涩或不语等为主要临床表现。

5. 脑脊液多为血性,脑CT可见有出血灶。

辨证施治

1. 风火上扰清窍

主症:平素多有头痛眩晕,情志相激或酗酒、劳累等,病势突变,神识恍惚或迷蒙,半身不遂,肢体强硬拘急,便干便秘。舌质红绛,舌苔黄腻而干,脉弦滑大数。

治法:熄风泻火,化痰开窍。

方药:①安宫牛黄丸(20)化服或鼻饲,第1~3天每6小时1次,每次1丸。第4天~第7天,改为每日2次,每次1丸。用法同上。

②安宫牛黄散1.6克。冲服或鼻饲,每日1~2次。

③清开灵40~60毫升加入250毫升5%葡萄糖液内,静脉滴注,每日1次。

④验方:蝎尾3克,天竹黄5克,羚羊角粉3克,生珍珠0.5克,上药共细末为1日量,分3~4次冲服,或鼻饲。

⑤验方:羚羊角粉4.5克(冲),生石决明30克(先下),生龟板30克(先下),生龙骨30克(先下),生牡蛎30克(先下),白芍30克,勾藤15克(后下),牛膝15克,生地30克,丹皮9克,夏枯草15克。水煎2次,取汁300毫升,分早中晚3次灌服或鼻饲。

⑥白药 0.5 克，口服或鼻饲，最初 3～4 天每 4 小时 1 次，第 5～10 天 6 小时 1 次，第 11～14 天，每日 3 次。

2. 痰湿蒙塞心神

主症：素体阳虚湿痰内蕴，发病后神昏，半身不遂，肢体松懈瘫软不温，甚则四肢逆冷，面白唇暗，痰涎壅盛。舌质暗淡，舌苔白腻，脉沉滑或沉缓。

治法：化痰开窍。

方药：涤痰汤（79）加减。

制南星 12 克，制半夏 12 克，枳实 9 克，陈皮 9 克，石菖蒲 6 克，茯苓 9 克，生姜 3 片，人参 3 克，甘草 3 克，竹沥汁 30 克（冲）。水煎 2 次，取汁 250 毫升，分 3 次灌服或鼻饲。每 1～2 剂。

加减：病情严重或用上药效果不好者，可上药汁送服苏合香丸 1 丸或猴枣散 3 克，每日 1～2 次。

3. 痰热内闭心窍

主症：起病急骤，神昏或昏愦，鼻鼾，痰鸣，半身不遂，肢体强硬拘急，项强身热，躁扰不宁，甚则手足逆冷，频繁抽搐，偶见呕血。舌质红绛，舌苔褐黄干腻，脉弦滑数。

治法：清热化痰，熄风开窍。

方药：菖蒲郁金汤（33）加减。

菖蒲 12 克，郁金 12 克，栀子 9 克，竹沥汁 30 毫升（冲），牡丹皮 12 克，竹叶 3 克，连翘 9 克，玉枢丹 3 克（冲）胆星 6 克，天竹黄 6 克，羚羊角粉 1.5 克（冲）。水煎 2 次，取汁 200 毫升，分数次灌服或鼻饲。日 1 剂。

加减：昏迷重者，加服至宝丹（80）或安宫牛黄丸（20）1 丸，每日 1～2 次。或蝎尾 3 克，天竺黄 5 克，羚羊角粉 3 克，生珍珠 0.5 克，上药共细末，分 3～4 次冲服或鼻饲，6 至 8 小时 1 次。

4. 元气败脱，心神散乱

主症：突然神昏、昏愦，肢体瘫软，手撒肢冷汗多，重则周身湿冷，二便自遗，舌痿，舌质紫暗，苔白腻，脉沉缓或沉微。

治法：益气回阳，扶正固脱。

方药：参附汤（29）加味。

人参15克，熟附子15克，麦冬30克。水煎2次取汁约150~200毫升，分3~4次灌服或鼻饲。

加减：呃逆不止者，加赭石30克，大黄6克。服法同上。

汗出淋漓不止者，加龙骨20克，牡蛎20克，山萸肉12克，五味子6克。水煎分3~4次鼻饲。

呕血者，加白药0.5~1克（冲）。

抽搐者，加全蝎尾3克，羚羊角粉1.5克，共细末，分2次冲服。

4.6 动脉硬化性脑梗塞

本病属于中医"中风病"的范畴。

病因病机

先天不足或后天失宜，损伤脏腑，耗伤阴精，阴阳相失，复受某邪所犯，如酗酒、劳累、情志过极等，使气血升降失调发生突变而逆乱，痰、火、瘀血闭阻脉络，而发为中风病。

诊断要点

1. 发病较急，常在夜间休息时发病。
2. 多发于40岁以上的中老年人。
3. 发病前常有短暂的头晕，肢麻、口斜、舌强等先兆症状。
4. 常以情志不舒，劳累过度，酗酒、暴食等为诱因。
5. 以神志昏蒙，半身不遂，偏身麻木，口舌歪斜，舌强言蹇或不语等为主症。
6. 瘫侧脉象多较健侧弦滑，苔多白腻或黄燥。

辨证施治

1. 肝阳暴亢，风火上扰

主症：半身不遂，口舌歪斜，舌强言蹇或不语，偏身麻木，头痛头晕，面红目赤，口苦咽干，心烦易怒，尿赤便干，舌质红或红绛，舌苔薄黄，脉弦有力。

治法：平肝泻火，活血通络。

方药：①平肝泻火汤（81）。

钩藤30克，菊花10克，夏枯草15克，丹皮15克，珍珠母30克，怀牛膝20克，赤芍10克。水煎2次，取汁300毫升，分早晚2次温服。

②丹钩六枝饮（82）加减。

丹参30～60克，钩藤15～30克（后入），豨莶草12～24克，夏枯草12～24克，地龙9克，红花6克，桑枝15克，橘枝15克，松枝15克，桃枝15克，杉枝15克，竹枝15克，甘草3克。水煎2次，取汁400毫升，分早晚2次温服。

痰涎壅盛者加瓜蒌15克，莱菔子20克。

神昏加郁金9克，菖蒲9克。

血压持续不降加赭石30克，牛膝20克。

③李秀林氏犯胃兼证方（83）。

代赭石30克，生石膏30克，生白芍15克，姜竹茹20克，化橘红9克，石菖蒲9克，云茯苓30克，白豆蔻3克，佩兰20克。水煎服。

呕血便血者，酌加三七粉3～6克（冲），或云南白药0.5克，每日4次冲服。

④平肝活血汤（84）。

羚羊角粉1.5～3克（冲），生石决明30克，钩藤（后下）15～30克，牛膝9～15克，川芎9克，红花9～15克，桃仁9克，土元9克。水煎服。

便秘腹满者酌加大黄9克，瓜蒌30克，服法同上。

失语苔厚腻者，加胆星9克，菖蒲9克。

抽搐加全蝎粉3～6克分2次冲服。

烦躁甚加寒水石30克，用法同上。

2. 风痰瘀血，痹阻脉络

主症：半身不遂，口舌㖞斜，舌强言蹇或不语，偏身麻木，头晕目眩，舌质暗淡，舌苔薄白或白腻，脉弦滑。

治法：熄风化痰，活血通络。

方药：①化痰通络饮（85）。

法半夏10克，生白术10克，明天麻10克，胆星6克，丹参30克，香附15克，酒军5克。水煎服。

②熄风化痰汤（86）加减。

茯苓21克，清半夏12克，橘红12克，竹茹12克，石菖蒲12克，制南星12克，僵蚕9克，钩藤20克（后入），全蝎9克，豨莶草30克，桃仁12克，红花9克。水煎服。

3. 痰热腑实，风痰上扰

主症：半身不遂，口舌㖞斜，舌强言蹇或不语，偏身麻木，腹胀便干便秘，头晕目眩，咯痰或痰多。舌质暗红或暗淡，苔黄或黄腻，脉弦滑而大。

治法：通腑化痰。

方药：①通腑化痰饮（87）。

生大黄10克，瓜蒌30克，胆星6克，芒硝10克（冲）。水煎服。

②大黄瓜蒌汤（88）。

大黄6～12克，瓜蒌15～30克，地龙9克，土元9克，白瓜子9克。水煎服。

4. 气虚血瘀

主症：半身不遂，口舌㖞斜，言语蹇涩或不语，偏身麻木，面色㿠白，气短乏力，口流涎，自汗出，心悸便溏，手足肿胀。舌质暗淡，舌苔薄白或白腻，脉沉细、细缓或细弦。

治法：益气活血。

方药：①益气活血汤 (89)。

生黄芪 30 克，桃仁 10 克，红花 10 克，赤芍 20 克，归尾 10 克，地龙 10 克，川芎 5 克。水煎服。

②补阳还五汤 (40) 加减。

黄芪 15～30 克，川芎 6～9 克，当归 9～12 克，赤芍 9～12 克，地龙 9～12 克，桃仁 9～12 克，牛膝 15 克，丹参 15～30 克。水煎服。

③黄芪五物汤 (90)。

黄芪 60 克，白芍 15 克，桂枝 24 克，生姜 9 克，大枣五枚。水煎服。

加减：偏枯在左侧者加当归 30 克；在下肢者加牛膝 9 克；筋不舒者加木瓜 15 克；腿软加虎骨 9 克；脉迟小者加附子片 15 克。

5. 阴虚风动

主症：半身不遂，口舌歪斜，舌强言蹇或不语，偏身麻木，烦躁失眠，眩晕耳鸣，手足心热。舌质红绛或暗红，少苔或无苔，脉细弦或细弦数。

治法：育阴熄风。

方药：①豨莶至阴汤 (91)。

制豨莶草 50 克，干地黄 15 克，盐知母 20 克，当归 15 克，枸杞子 15 克，炒赤芍 25 克，龟板 10 克，牛膝 10 克，甘菊花 15 克，郁金 15 克，丹参 15 克，黄柏 5 克。水煎服。

②育阴熄风汤 (92)。

生地 20 克，玄参 15 克，女贞子 15 克，桑寄生 30 克，钩藤 30 克（后入），白芍 20 克，丹参 15 克。水煎服。

6. 痰火内闭心窍

主症：发病急骤，神识迷蒙或昏愦，鼻鼾痰鸣，半身不遂肢体强疼拘急，项强身热，躁扰不宁，肢体抽搐。舌红绛，苔褐黄干腻，脉弦滑数。

治法：清热化痰开窍。

方药：①清开灵注射液 40～60 毫升，加入 300 毫升 5%葡萄糖液内，静脉滴注，每日 1 次。

②醒脑静注射液 20 毫升，加入 300 毫升 5%葡萄糖注射液内，静脉滴注，每日 1 次。

③加减羚角钩藤汤 (93)。

羚羊角粉 4.5 克（冲），生石决明 30 克（先下），生龟板（先下）30 克，生龙骨 30 克（先下），生牡蛎 30 克（先下），白芍 30 克，钩藤 15 克（后入），牛膝 15 克，生地 30 克，夏枯草 15 克，丹皮 9 克。水煎 2 次，取汁 300 毫升，分 3～4 次灌服或鼻饲，每日 1 剂。

④至宝丹 (80) 或安宫牛黄丸 (20) 或安宫牛黄散 2.5 克。灌服或鼻饲，6 小时 1 次，连服 3～5 天。

7. 痰湿蒙塞心神

主症：素体多是阳虚湿痰内蕴，病发神昏，半身不遂肢体松懈瘫软不温，甚则四肢逆冷，面白唇暗，痰涎壅盛。舌质淡暗，舌苔白腻，脉沉滑或沉缓。

治法：熄风化痰开窍。

方药：①制南星 12 克，制半夏 12 克，枳实 9 克，陈皮 9 克，石菖蒲 6 克，茯苓 9 克，生姜 6 克，人参 3 克，甘草 3 克，竹沥汁 30 克（冲）。水煎 2 次，取汁 300 毫升，送服苏合香丸 1 丸，灌服或鼻饲，每日 1～2 剂。

②病情严重或用上方效果不著者，以上方药汁送服猴枣 0.3 克，用法同上。

8. 元气脱败，心神散乱

主症：突然神昏，昏愦，肢体瘫软，手撒肢冷汗多，重则周身湿冷，二便自遗。舌痿，舌质紫暗，苔白腻，脉沉缓，沉微。

治法：补益元气，回阳固脱。

方药：参附汤 (29)。

人参 30 克,附子 9 克。水煎 2 次,取汁 200 毫升,频频灌服或鼻饲。

(卢尚岭)

5 消化系统疾病

消化系统急症，主要介绍胃、肠、肝、胆、胰等脏器的疾病，常见急腹症的非手术治疗亦为本章的主要内容。这些疾病分别属于中医学中"急性腹痛"、"肠结"、"关格"、"胁痛"、"黄疸"、"呕吐"、"肠风"等范畴，主要有热郁、寒积、气滞、血瘀、食滞、虫积等因素所致。本章以"急治其标，缓治其本"，"六腑以通为用"为基本的治疗原则。

5.1 急性胃肠炎

急性胃肠炎，是以呕吐、腹泄，脘腹疼痛为主症的疾病，属于中医学中"呕吐"、"泄泻"、"胃脘痛"的范畴。

病因病机

多因暴饮暴食，进食不洁以及感受风、寒暑，湿之邪，侵犯脾胃，脾不健运，胃失和降所致。若吐泻太过，则津气耗伤，可出现虚脱现象。

诊断要点

1. 本病有饮食不当史，常发生于夏秋季节，起病急，突发性呕吐，腹泻，脘腹疼痛，或兼有外感表症等。

2. 呕吐腹泻剧烈时，可出现目眶凹陷，皮肤松弛，心烦口干，尿少等脱水症状。

3. 上腹部和脐周围有压痛，肠鸣音亢进，血液白细胞计数，及嗜中性粒细胞升高。

4. 本病应与霍乱、菌痢等相鉴别。

辨证施治

1. 外邪犯胃

主症：突然呕吐，胸脘满闷，腹痛肠鸣，泄泻清稀，兼有恶寒发热，头身疼痛，舌苔白或白腻，脉濡。

治法：解表疏邪，芳香化浊。

方药：藿香正气散（3）加减。

藿香9克，紫苏9克，厚朴9克，半夏9克，陈皮9克，云苓21克，大腹皮9克，炒白术12克，桔梗6克，白芷6克，甘草3克，生姜、大枣，适量为引，水煎服。

加减：夏秋之际湿热伤中。传化失常者，可应用葛根黄芩黄连汤（162）随证加减。

热重者加连翘15克，地锦9克。

湿重者加苍术9克，白蔻9克。

2. 饮食停积

主症：呕吐酸腐，脘腹胀满，腹痛肠鸣，泻下粪便，臭如败卵，泻后痛减，或泻而不畅嗳气纳呆，舌苔垢浊。脉滑实，或沉弦。

治法：消食导滞。

方药：保和丸（94）加减。

神曲20克，焦山楂21克，莱菔子12克，云苓5克，陈皮12克，半夏12克，连翘15克，砂仁9克，木香9克，枳壳10克，炒麦芽30克。水煎服。

加减：若腹痛胀甚，大便泻下不畅者，加大黄9克，枳实12克，槟榔9克等。

若有积滞化热者，加黄连9克。

若呕吐，苔黄脉数者，加竹茹9克，白蔻9克。

5.2 胃十二指肠溃疡出血

胃十二指肠溃疡出血，属于中医学中"血症"范畴。

病因病机

多因胃中积热，或肝郁化火、逆乘于胃、阳络损伤所致；亦有劳倦过度、脾胃受伤、气不摄血而致者。

诊断要点

1. 出血前,大部分病人均有典型的溃疡病史,但也有部分病人可无症状,突然出现呕血与黑粪(潜血试验为阳性),而后疼痛往往可减轻或消失。

2. 大多数病人在出血量达 400 毫升以上时,有不同程度的眩晕,舌苔白,冷汗,脉细数,血压下降甚至昏迷等表现。

3. 约半数病人可发生低热,一般在 24 小时内产生,持续一周左右。

4. 在大出血后 6 至 12 小时甚至 20 至 30 小时,出现红细胞、白细胞的降低。

5. 血尿素氮大出血后数小时内可增高,24 至 48 小时内达高峰。

辨证施治

1. **胃热炽盛,灼伤络脉**

主症:胃脘灼热或闷痛,吐血鲜红或紫黯或夹有食物残渣,口臭,便秘或大便色黑,舌红,苔黄腻,脉滑数。

治法:清胃泻火,化瘀止血。

方药:泻心汤(95)合十灰散(96)加减。

大黄 12 克,黄连 9 克,黄芩 9 克,侧柏炭 12 克,大小蓟各 15 克,茜草炭 15 克,丹皮 10 克。水煎服。

加减:若脘胀者加陈皮 9 克。

脘痛者,加香附 9 克。

呕吐者,加竹茹 9 克。

2. **肝火犯胃,迫血外溢。**

主症:突然呕血,鲜红或紫黯,心烦易怒,口干唇燥,舌红少苔,脉弦细数。

治法:清泻肝火,凉血止血。

方药:龙胆泻肝汤(97)加减。

龙胆草 12 克，黄芩 12 克，炒山栀 9 克，生地 30 克，丹皮 12 克，白茅根 30 克，藕节 18 克，墨旱莲 12 克，赤芍 9 克水煎服。

加减：若吐血不止，兼有脘闷、口渴者，可加花蕊石粉 15 至 30 克，三七粉 6 至 9 克，白芨粉 9 至 12 克，紫珠草 18 克研粉调服。

若暴吐血者用犀角地黄汤（25）并以三七粉 9 克，大黄粉 12 克等调服。

若吐血不止者，加侧柏叶 30 克，三七粉 10 克（冲）。

3. 脾气亏虚，气不摄血

主症：吐血，病多久延不愈，时轻时重，心悸气短，面色苍白，倦怠乏力，舌淡，苔薄白，脉细弱。

治法：健脾补气，摄血止血。

方药：归脾汤（50）加味。

黄芪 15 克，党参 15 克，茯神 15 克，龙眼肉 20 克，木香 6 克，远志 9 克，酸枣仁 15 克，甘草 6 克，当归 12 克，白芨 12 克，乌贼骨 30 克，地榆炭 15 克。水煎服。

加减：可按上述列举之收敛药随证加减。

若呕血取足三里、内关、内庭、公孙等穴针刺。耳针可刺皮质下、心、肾上腺，配合肝、脾、神门等，另血余炭，或地榆炭 3 至 9 克研细末口服，每日 3 次，云南白药 0.5 克口服，日 3 至 4 次。白芨粉 3 克，生大黄粉 1.5 克口服，日 2 次，出血量多者，增至 4 至 6 次亦可。

5.3 急性胃扩张

急性胃扩张，属于祖国医学中"宿食"、"伤食"的范畴。

病因病机

多由于内伤疾病恢复期或术后以及暴饮暴食，肠胃受损，以致脾不健运，胃失和降，气机阻滞所致。

诊断要点

1. 多发生于胸腹部大手术之后，此外，感染、损伤、病重恢复期患者，以及过量饮食者。

2. 溢出性呕吐。呕吐物先为棕绿色，以后为咖啡色，病情逐渐加重，重者可出现脱水碱中毒及休克。

3. 上腹部膨胀，可见到毫无蠕动的巨大胃轮廓，局部有压痛。叩诊呈过度回响，有振水音，腹部 X 线平片，可显示扩大的胃腔内有液平等现象。

4. 还可并发急性胃穿孔和急性腹膜炎。

辨证施治

首先禁饮食，胃肠减压，静脉输液以纠正水电解质平衡失调，中药煎剂可以在减压后服用，或应用胃管注入。

1. 脾弱气滞，胃失和降

主症：心下坚满，胀满如盘，呕吐秽浊，舌苔腻，脉弦滑。

治法：导滞健脾，和胃降逆。

方药：枳术汤（98）加减。

枳实 30 克，白术 15 克，川朴 10 克，炒莱菔子 15 克，生甘草 9 克。水煎服。

加减：若伴有胃肠积热者，可加生大黄 9 到 15 克（后入）。

如应用上方疗效不著者，可用精制马钱子粉 0.6 克，每次 0.3 克，日 2 次口服。

2. 积滞内停，脾不健运

主症：胸腹痞满、呕吐酸腐，嗳气厌食，大便秘结，舌苔黄腻，脉沉实。

治法：消食导滞，行气健脾。

方药：木香槟榔丸（99）加减。

木香 6 至 9 克，槟榔 9 克，青皮、陈皮各 9 克，莪术 6 克，黄连 6 克，黄柏 6 克，大黄 6 克，制香附、枳壳、牵牛、各 9 克。水煎服。

加减：若腹部胀满，硬痛拒按，痞满燥实俱在者，可应用大承气汤（11）。

若痰涎宿食，填塞上脘，胸中痞硬烦懊不宁者，可用瓜蒂散（100）。

因受寒而致者，可于上方祛苦寒泻下之品加良姜6克，乌药6克。

5.4 肝性昏迷

肝性昏迷，属于中医学中"昏不知人"、"昏蒙"、"昏愦"、"神昏"等病证。

病因病机

由水、火、湿、热之毒，内盛交结，蒙蔽心神，内陷心包所致。痰火内闭，可引动肝风，气阴衰竭者可由闭证转为脱证。

诊断要点

1. 有原发性肝病史。多因消化道大出血，急性感染，腹腔内大量放水，长期用利尿剂，或进食大量含蛋白的食物及某些对肝脏有毒性的药物等诱发。

2. 典型表现：

(1) 前驱期。起病缓慢，可有食减、恶心、呕吐、呃逆、腹泻、尿少、肝臭、扑击性震颤，继而出现精神症状，或欣快，或异常沉默、健忘、智力减退及性格行为改变等。

(2) 昏迷前期。理解力减退，兴奋不安，躁动，语言不清，书写不灵，定向障碍，幻觉，嗜睡等。有的呈木僵状态，扑击性震颤，腱反射亢进，巴氏征阳性。

(3) 昏迷期。嗜睡，呈木僵状态，意识丧失，渐进入深昏迷；严重者可有酸中毒深呼吸，抽搐，高热等。

3. 肝功明显损害，钾、钠、氯，二氧化碳结合力降低，尿素氮升高，氮性昏迷，血氨多在百分之一百微克以上。非氮性昏迷，血氨不高。从昏迷前期到昏迷期，可有异常的脑电图出现，应注意与其他昏迷鉴别。

辨证施治

1. 闭证

(1) 湿浊上蒙

主症：表情呆滞，神志模糊，嗜睡，言语不利，大便不爽，甚者昏迷、臌胀、黄疸、舌苔浊腻，脉沉缓或沉迟。

治法：温阳化湿，泻浊开窍。

方药：茵陈术附汤（101），合菖蒲玉金汤（33）加减。

茵陈24克，白术9克，熟附片6克，干姜6克，大黄9克，石菖蒲6克，郁金10克，云苓6克，远志6克。水煎服。

加减：昏迷者加苏合香丸（102），每次1丸，日服2次。

湿重苔厚者加苍术9克，川朴6克。

痰浊壅盛者加胆星6克，半夏6克。

(2) 痰火内闭

主症：臌胀，或有黄疸，发热烦燥，谵语，意识模糊，甚至昏迷不醒，便干溲赤，舌红苔黄，脉数。

治法：清热泻火，通腑开窍。

方药：千金犀角散（103）合大承气汤（11）加减。

犀角3克，茵陈20克，黄连6克，山栀子6克，紫草20克，板兰根20克，公英20克，石菖蒲9克，大黄9克，川朴6克，枳实6克，元明粉6克（冲）。水煎服。

加减：抽搐者加钩藤12克，石决明20克。

舌红少津者加沙参15克，麦冬9克。

火毒伤络迫血妄行，见齿衄、鼻衄皮下瘀血时，宜加大清热解毒凉血之品，可用犀角粉1.5克冲服，丹皮9克，赤芍9克。

2. 脱症：

(1) 亡阴

主症：神志昏迷，两手抖动，气息低微，汗出肢冷，舌质淡、脉细数。

治法：救阴敛阳，补气固脱。

方药：生脉散（26）加味。

人参 12 克，麦冬 12 克，五味子 6 克，山萸肉 9 克，黄精 9 克，煅龙骨 15 克，煅牡蛎 15 克。水煎服。

加减：舌红少苔者，加生地 12 克，龟板 15 克，阿胶 6 克(烊)。

手足厥冷者，加熟附子 6 克。

(2) 亡阳

主症：神志昏迷，目合口干，手撒肢厥，大汗淋漓，二便自遗，舌淡润，脉微欲绝。

治法：回阳救逆。

方药：参附汤（29）。

人参 10 克，熟附子 10 克，急煎鼻饲。

5.5 胆道蛔虫病

胆道蛔虫病属于祖国医学"蛔厥"的范畴。

病因病机

祖国医学认为，蛔虫寄生于肠内，喜温恶寒，若肠寒不利蛔虫生存，则上移于胆道而诱发本病。

诊断要点

1. 在上腹部、剑突下部突然有阵发性钻顶样绞痛，病人常捧腹曲膝，辗转不安或呻吟不止，少数病人疼痛可向肩部放射。疼痛缓解后如同常人，疼痛时常伴有恶心呕吐、冷热、黄疸等。

2. 即往有便蛔或吐蛔史，大便常规可找到蛔虫卵。本病常因过饥、受寒、发热、妊娠及驱蛔不当诱发。

3. 据病情可进行血生化等项检查以鉴别诊断。

辨证施治

1. 肠寒胃热

主症：腹疼发作突然，时发时止，痛剧则辗转反侧，恶心呕吐，甚则吐蛔，汗出肢冷，舌苔薄白，脉弦紧。

治法：寒热并用，安痛止痛。

方药：乌梅丸（104）加减。

乌梅30克，川椒9克，干姜6克，细辛3克，黄连9克。半夏12克，槟榔10克，甘草3克，水煎分2次服用，或急煎频频而服。

加减：气虚疲惫者加党参9克，当归6克。

　　　寒象重者加桂枝6克，熟附子6克。

　　　热象多者加黄柏9克。

2. 胆经湿热

主症：发冷发热，腹痛拒按，口苦咽干，呕恶吐蛔，甚则面赤身热，烦燥厥逆，舌质红脉弦数。

治法：清热祛湿，利胆祛蛔。

方药：驱蛔利胆汤（105）加减。

茵陈30克，双花40克，黄芩15克，柴胡9克，大黄9克，川楝子9克，使君子15克，槟榔9克，苦楝根9克，甘草6克，元明粉9克（冲）。水煎服。

加减：高热者加连翘15克，山栀9克。

　　　呕吐甚者，加竹茹9克，姜夏9克。

　　　黄疸者，加金钱草30克。

5.6　急性胆囊炎与胆石症

急性胆囊炎和胆石症两种疾病，常同时并存，互为因果，互相影响。它们均属于中医学中"腹痛"、"胁痛"、"黄疸"、"结胸发黄"等范畴。

病因病机

中医学认为："胆为中清之腑"，以通宣下行为顺。凡情志不畅，饮食失节，寒湿不适，虫积等均可引起肝胆气郁，疏泄失职，湿热熏蒸，通调失常，不通则痛，乃致诸症。

诊断要点

1. 本病以痛、吐、热、黄为特点，右上腹及剑突下呈持续性绞痛，或阵发性加剧，疼痛可放射至右肩或左肩及腰背等部。伴有恶心呕吐，寒热往来，出现黄疸。

2. 右上腹胆囊区有明显压痛，深吸气有触痛反应。有时可触及肿大的胆囊，重症常伴有右上腹肌紧张，及反跳痛等局限性腹膜炎征象。

3. 可作腹部B超、腹部平片或胆囊造影等项检查，以协助诊断。

辨证施治

1. 肝胆气滞

主症：胁脘胀痛或绞痛，口苦咽干、食少呕恶，舌苔薄白或微黄，脉弦紧。

治法：疏肝利胆，缓急止痛。

方药：柴胡疏肝散（106）加减。

柴胡9克，白芍15克，枳壳9克，香附9克，半夏9克，郁金9克，元胡9克，木香9克，甘草3克，茵陈12克，水煎服。

加减：若日久结石未排出者，加海金砂18克，鸡内金9克，金钱草15克。

2. 肝胆湿热

主症：腹满胀痛，寒热往来，口苦咽干，烦闷纳呆，目黄身黄，大便秘结，小便黄赤，舌红，苔黄腻或厚，脉弦数。

治法：疏肝利胆、清热利湿。

方药：大柴胡汤（107）加减。

柴胡9克，大黄9克，（后入），枳实9克，黄芩9克，清半夏9克，白芍9克，茵陈15克，郁金9克，木香6克，车前子6克（包），金钱草30克。水煎服。

加减：湿重者加苡米30克，白蔻9克。
　　　　　热重者加板兰根15克，双花15克，连翘15克。
　　　　　便秘者加芒硝9克（冲），川朴12克。
　3. 肝胆实火
　　主症：胁脘胀痛拒按，腹胀而满，寒热往来，口苦咽干，目黄身黄，尿黄浊或赤涩，大便秘结，舌红或降，苔黄燥或有芒刺。脉弦滑数、洪数。
　　治法：清泻肝胆。
　　方药：龙胆泻肝汤（97）合茵陈蒿汤（35）加减。
　　龙胆6克，黄芩9克，柴胡9克，山栀6克，生地6克，当归3克，泽泻6克，郁金9克，生大黄9克（后入），芒硝9克，木香6克，茵陈10克。水煎服。
　　加减：热毒炽盛者，加黄芩9克，白花舌草18克，半枝莲9克等。

5.7　急性胰腺炎

　　急性胰腺炎属于中医学中"胃脘痛"、"膈痛"、"腹痛"、"脾心痛"、"结胸"等病症范畴。
　病因病机
　　多因饮食、情志、虫扰等因素导致肝胆脾胃功能失调，出现疏泄不利，气滞血瘀，生湿化热，实热蕴结为主的证候，若正不胜邪，可发生厥脱等危象。
　诊断要点
　1. 腹痛常在饱餐或酗酒后1至2小时突然起病、疼痛剧烈而持续，可伴有阵发性加剧疼痛如刀割，腹痛大多在中上腹或脐部，常放射到腰背及左肩部，且压痛无自觉疼痛明显。
　2. 起病时伴有恶心、呕吐，以后逐渐减少，呕吐后腹痛并不减轻，可出现发热，腹胀，黄疸，甚至休克。
　3. 化验检查，可见血淀粉酶、尿淀粉酶及腹水淀粉酶升高。

辨证施治

1. 气滞热郁

主症：卒然脘腹疼痛，时有加剧，痛连两胁，恶心呕吐，口苦咽干，寒热往来，腹胀，大便不下，舌苔薄黄，脉弦紧。

治法：疏肝解郁，泻热导滞。

方药：大柴胡汤（107）加减。

柴胡12克，大黄15克，(后入)，黄芩15克，枳实6克，杭芍15克，半夏9克，水煎服。若呕吐较重不能口服者，可用上方煎剂保留灌肠。必要时可胃肠减压，或胃肠减压后，用鼻管注入。

加减：腹痛剧者加元胡9克，川楝子9克。

便秘者加芒硝6克（冲）。

腹胀者加厚朴9克，木香6克。

2. 热盛腑实

症状：腹痛拒按、胀满痞硬、发热不退，口干欲饮，舌质红，苔黄厚腻或燥，脉洪数。

治法：清热解毒、通里攻下。

方药：大承气汤（11）加减。

生大黄12克，芒硝12克（冲），枳实9克，川厚朴15克，黄芩15克，栀子9克，连翘15克，公英15克，红藤15克，败酱草15克。水煎服，日1至2次；或应用保留灌肠法用药。

加减：热与水结者（伴有胸腹水者）可用大陷胸汤（108）。先煮大黄汤，加入芒硝和甘遂末，先服用二分之一，未便时再服。

若有血瘀者，加丹参9克，红花9克，赤芍9克，丹皮9克，桃仁9克。

伴有厥逆者，并用参附汤（29）：人参9克，熟附子9克，急煎服。

5.8 急性阑尾炎

急性阑尾炎属于中医学"肠痈"、"缩脚肠痈"的范畴。

病因病机

主要由于劳倦或情志刺激等因素，致使脏腑功能失调，加之寒温不适，饮食不节，以致气滞血瘀，湿热内生，积于肠中而发病。

诊断要点

1. 初起上腹部或脐周围疼痛，随后转移至右下腹，呈持续性疼痛，可阵发性加重。常伴有恶心、呕吐、发烧、尿黄等。病人站立或行走时，常弯腰，以双手按着右下腹；平卧时右胯关节屈曲。

2. 右下腹阑尾点（马氏点），有明显固定压痛、严重时可有反跳痛。腰大肌有刺激症状，提示阑尾后位，孕妇的压痛点可随妊娠子宫大小面上右上腹推移，幼儿患者若有明显全身症状，右下腹肌紧张时，应首先考虑到急性阑尾炎的可能。有阑尾炎的典型病史，而在右下腹触及疼痛包块者为阑尾脓肿性包块。

3. 血常规检查，白细胞总数及中性粒细胞增高。

辨证施治

1. 热瘀气滞

主症：腹痛、绕脐走窜、渐移痛于右下腹。伴有恶心、呕吐、发热、或便秘、尿赤、舌苔薄白，脉弦或微数。

治法：行气活血、清热解毒。

方药：阑尾化瘀汤（109）。

川楝子15克，元胡9克，丹皮9克，桃仁9克，木香9克，双花15克，大黄9克（后入）。水煎服。

加减：热瘀甚者加红藤30克，丹参9克。

2. 热毒壅结

主症：右下腹痛拒按，或右脚屈而不伸，伸则痛甚，口干身热,便秘尿黄,舌质红,舌苔黄腻或黄燥,脉弦数或滑数。

治法:清热解毒、消肿散结。

方药:阑尾清化汤(110)。

双花30克,公英30克,丹皮15克,大黄15克(后入),川楝子9克,桃仁9克,赤芍12克,生甘草9克。水煎服。

加减:湿热重者加黄连12克,黄芩12克。

湿重加佩兰12克,白蔻15克,藿香梗10克。

便秘者可应用大黄牡丹皮汤(111)。

3. 热毒炽盛

主症:右下腹疼拒按,伴高热,面红目赤唇干口燥,便秘尿赤,舌质红,苔黄燥或黄腻脉弦数。

治法:解毒泻热,化毒导滞。

方药:阑尾解毒汤(112)。

红藤30克,败酱30克,双花30克,蒲公英30克,冬瓜仁15克,赤芍12克,大黄12克(后入)。木香9克,黄芩9克,桃仁9克,川楝子9克。水煎服。

加减:若腹胀甚者可加川朴12克,枳实12克。

腹痛重者加元胡12克。

恶心呕吐加竹茹9克、半夏9克。

常用外敷药:

1. 消炎散。伴有腹膜炎阑尾脓肿时可配合应用。芙蓉叶、大黄各30克,黄芩、黄连、黄柏、泽兰叶各240克,冰片9克,共研细末,用黄酒,或葱酒煎调敷患处,每日两次。

2. 硝黄散。大黄粉与芒硝适量加醋调成糊状外敷。

5.9 胃十二指肠溃疡病穿孔

胃、十二指肠溃疡病穿孔属于中医学中"厥心痛"、"心腹痛"、"胃脘痛"、"厥逆"等病证范畴。

病因病机

多因过伤饱餐、饥饿、情绪激动、过劳等所致,若穿孔大,腹膜炎重,邪盛而正虚,可出现关格、厥脱等危证。

诊断要点

1. 多为突发刀割样上腹部持续疼痛,并由上腹迅速波及右下腹与及全腹;部分病人疼痛向右肩及腰背放射;有些年老体弱患者腹痛不明显,部分患者早期可因剧烈腹痛而休克,后期因细菌性腹膜炎引起休克;半数病人有恶心呕吐,后期因肠麻痹而腹胀、便秘。

2. 腹部早期平坦,晚期腹胀明显,腹式呼吸受限或消失,全腹有压痛及反跳痛,以上腹及右下腹为重,腹壁肌肉紧张,甚至呈板状强直。肝浊音界缩小或消失,肠鸣音减弱或消失。X线检查,绝大多数病人可发现膈下呈气腹征,白细胞数轻度增高。

辨证施治

1. 气血郁闭

主症:突然发作刀割样剧烈腹痛,拒按,或伴有呕吐,舌苔白,脉弦。此型为穿孔后1至2天内属闭合穿孔期。

治法:缓急止痛,促进闭孔。

以针刺为主,取穴上脘、中脘、梁门(双)、天枢(双)、内关(双)、足三里(双)。手法强刺激,有针感后留针30到60分钟,留针期间可用电针刺激,或每15分钟手法捻转1次。每6小时针刺1次,可配合禁饮食,胃肠减压,以纠正水电解质平衡失调。

2. 郁闭化热

主症:腹痛局限,发热,便燥,尿赤,舌苔黄,脉数。此型为穿孔已闭合,合并腹腔积液及感染,属炎症消散期。

治法:清热下实,理气通腑。

方药:凉膈散(9)加减。

大黄9克,芒硝9克(冲),甘草6克,山栀9克,薄荷6

克，黄芩12克，连翘15克，竹叶9克，枳壳9克，木香6克。水煎服。

加减：腹腔感染重者加双花30克，公英30克，红藤15克。

瘀血重者加桃仁9克，红花9克，生蒲黄9克，川芎9克。

腹痛不止者，加元胡9克，川楝子9克，杭芍12克。

3. 脾胃虚弱

主症：面黄体瘦，胸腔痞闷，腹胀，腹泻或便秘，纳食不化，嗳气，吐酸，舌淡薄白，脉沉细弱，此型属修复溃疡期。

治法：健脾和胃。

方药：香砂六君子汤（113）加减。

党参15克，白术9克，云苓12克，陈皮9克，半夏9克，木香9克，砂仁9克，甘草6克，乌贼骨30克，凤凰衣9克。水煎服。

加减：胃寒呃逆者，加吴茱萸9克，干姜9克，乌药9克。

胃疼者加元胡9克，川楝子9克，杭芍12克。

肝郁气滞者加香附9克，佛手9克，槟榔9克。

5.10 急性腹膜炎

急性化脓性腹膜炎，属于中医学中"急性腹痛"的范畴。

病因病机

多因正气不足，邪毒内陷而致湿热积滞，气血凝聚，瘀热内结，壅阻肠腑发为本病。

诊断要点

1. 多剧烈腹痛，呈持续性。疼痛以原发病灶部位最明显，伴有恶心、呕吐及全身中毒症状。

2. 腹式呼吸减弱，腹部压痛，腹肌紧张，反跳痛，腹胀时叩之鼓音，腹腔积液多时可出现移动性浊者，肠鸣音减弱。

3. 可配合肛诊、腹穿、化验、X线透视确诊。

辨证施治

1. 热壅三焦

主症：腹胀，腹痛拒按，大热烦渴，口干咽燥，尿赤，舌苔黄燥，脉弦数，多见于腹内脏器急性感染扩散性感染。

治法：清热解毒。

方药：黄连解毒汤（114）加减。

黄连6克，黄芩9克，栀子9克，黄柏6克，双花30克，连翘15克，公英30克，地丁30克，杭芍18克，甘草6克。水煎服。

加减：伴有黄疸者加茵陈30克，赤小豆30克。

伴有腹水者加云苓30克，猪苓30克，泽泻15克。冬瓜皮30克。

伴有恶心呕吐者，加竹茹12克，清夏9克。

2. 郁闭化热

主症：腹胀痛拒按，发热便燥、尿赤，舌苔黄，脉数。相当于胃肠穿孔闭合后的腹腔感染。

治法：清热解毒，化瘀通腑。

方药：大柴胡汤（107）。

柴胡6克，黄芩9克，清夏9克，枳实9克，白芍12克，大黄9克（后入），双花30克，冬瓜仁15克。水煎服。

加减：伴有口苦胁痛者，加胆草6克，郁金12克，香附9克。

伴有热盛者，可加生石膏30克。知母12克，公英30克。

伴有血瘀重者，加桃仁9克，赤芍12克，皂刺9克，山甲9克。

5.11 伪膜性肠炎

伪膜性肠炎属于中医学中"泄泻"的范畴。

病因病机

体虚久病,正气亏虚,寒湿邪热乘虚而入,致脾胃气机阻滞,运化传导失常,发为本病。

诊断要点

1. 在使用广谱抗生素(特别是林可霉素,氯林可霉素)期间,或停用抗生素短期内突然出现腹泻;或在腹部手术后病情反而恶化,出现腹泻时,应考虑本病。

2. 粪细菌特殊条件下培养,可发现有难辨梭状芽孢杆菌生长。本病可通过检验粪毒素,参考结肠镜及 X 线检查确诊。

辨证施治

1. 上热下寒

主症:暴泻、稀薄多水、腹部胀痛,恶心呕吐,发热,口渴,胸中烦闷,舌苔薄黄,脉沉数。

治法:清上温下,调和脾胃。

方药:黄连汤(115)加减。

黄连 4.5 克,干姜 6 克,肉桂 4.5 克,半夏 6 克,党参 9 克,炙甘草 3 克,大枣 4 枚。水煎服。

加减:腹痛甚者加杭芍 9 克,元胡 9 克。
　　　尿少者加泽泻 9 克,车前草 30 克。
　　　呕吐甚者加蔻仁 9 克,竹茹 9 克,苏叶 9 克。

2. 脾胃虚弱,湿热未尽

主症:纳呆乏力,形体消瘦,低热便溏,夹有伪膜片,舌淡苔白,脉虚数。

治法:健脾和胃,兼清湿热。

方药:参苓白术散(116)加减。

党参 9 克,白术 9 克,炒扁豆 9 克,莲子肉 9 克,茯苓 15 克,炒山药15克，苡米仁15克，陈皮6克，桔梗6克，砂仁9克，黄连4克。水煎服。

加减：若出现脱证，宜用人参9克，熟附子9克，急煎服。

验方：大蒜2至3头，捣如泥后用白开水冲蒜泥温服。日2次，服至症状消失为度。

(包培荣)

6 泌尿系统疾病

泌尿系统急症属中医学中"关格"、"癃闭"、"尿血"、"淋症"，等病症范畴，此类疾病大都由于先天不足，内伤七情，房劳过度，饮食不洁，及邪毒侵袭所致。本章所述内容以"实则泻之"、"虚则补之"为基本治疗原则。

6.1 急性肾功能衰竭

急性肾功能衰竭属中医"癃闭"、"关格"、"水肿"等范畴。

病因病机

因热、毒、瘀、湿、积压外伤等因素，使肾气亏耗，开合无度，气化失常，清气不升，浊阴不降，邪壅三焦而致病。

诊断要点

1. 少尿期：

(1) 在有诱发因素的基础上，特别是休克患者，若体液充分补给、休克纠正以后，每日尿量少于 400 毫升、每小时 17 毫升以下者。

(2) 伴有恶心、呕吐、纳差、头痛、嗜睡，精神恍惚，烦燥不安，甚至昏迷，惊厥，血压升高，或有出血倾向。

(3) 据病因不同尿常规有各种异常表现，尿呈酸性，尿比重可固定在 1.012 左右，血肌酐尿素氮逐日升高，血钾、血镁增高。

2. 多尿期：

日尿量 3 000 毫升以上。此时处理不当可出现脱水，低钾，低钠，软弱无力，恶心呕吐，甚至瘫痪和心力衰竭。

3. 患病早期，因原发疾病引起的失水，休克尚未完全纠正，需与功能性少尿期相鉴别，可做甘露醇静脉注射试验。

4. 发病前，无慢性肾功能衰竭表现。

辨证施治

1. 少尿期

(1) 热毒瘀滞

主症：高热，口渴，少尿或无尿，烦燥，肌肤发斑或吐血，衄血，咯血，尿血，舌质红绛，苔黄舌干，脉滑数或细数。

治法：清热凉血，化瘀利尿。

方药：清瘟败毒饮（117）加减。

生石膏30克（先煎），生地黄15克，犀角粉3克（冲），黄芩9克，栀子9克，知母9克，赤芍9克，玄参9克，连翘9克，丹皮9克，黄连6克，桔梗6克，竹叶6克，甘草3克，生大黄9克（后入），白茅根30克，水煎至200毫升频频而服。

加减：便秘者加芒硝9克（冲）。

出血多者加三七粉6克（冲）。

小便赤少不畅者加猪苓15克，泽泻9克，坤草12克。

(2) 气虚瘀滞

主症：少尿或无尿，全身浮肿，乏力纳呆恶心呕吐，心悸气短，面色苍白，舌质淡，苔白，脉沉细。

治法：温阳利水，益气化瘀。

方药：真武汤（45）加减。

熟附子6克，白术6克，生姜6克，茯苓15克，白芍12克，人参12克，丹参12克，坤草15克，大黄6克，泽泻15克。水煎至150至200毫升口服。

加减：可配合应用中药结肠灌注液1号，大黄12克，黄芪30克，丹参30克，红花15克。水煎至150毫升灌肠，日1次。

瘀血甚者，加桃仁9克，川芎9克，红花12克，水蛭

6克。

肾阳虚者加仙茅9克,灵脾9克。

2. 多尿期　肾阴亏虚

主症:低热心烦,尿量增多,颧红,口干欲饮,头晕耳鸣,腰膝酸软,舌红少苔,脉细数。

治法:益肾固摄。

方药:六味地黄丸(118)合缩泉丸(119)。

熟地24克,山萸肉12克,炒山药12克,丹皮9克,茯苓9克,泽泻9克,麦冬12克,五味子6克,益智仁15克,乌药9克。水煎服。

3. 恢复期　气阴双方

主症:身体虚弱,少气懒言,体倦乏力,小便正常,舌红少苔,或舌淡,脉细无力。

治法:益气养阴,醒脾健肾。

方药:生脉散(26)加减。

太子参30克,麦冬12克,五味子6克,石斛12克,山药30克,砂仁9克,焦三仙各12克。水煎服。

加减:伴有脾肺两虚者,可口服都气丸(120)1丸,日两次。

6.2　尿毒症

慢性肾功能不全所致尿毒症,多继发于泌尿系统疾病,属于中医"虚损"、"癃闭"、"关格"的范畴。

病因病机

由于肾脏久病,正气亏虚而致,湿浊,瘀血,水饮,湿热,郁阻凝聚,阳虚则湿浊上逆,阴虚则肝阳偏亢,甚则邪陷心肝,血热风动。

诊断要点

1. 有原发性肾病史和引起尿毒症的各种诱因。

2. 早期有消瘦，贫血，精神萎靡，嗜睡，食欲不振，恶心呕吐，血压升高，腹痛腹泻，皮肤搔痒及出血现象。晚期症见意识朦胧，烦燥不安，谵妄，抽搐，呼吸深快，有氨味，昏迷加深，可闻心包摩擦音等。

3. 尿毒症早期，C_{Cr} 10 至 25%，血肌酐 2.5 至 5 毫克%，晚期 $C_{Cr} < 10\%$，血肌酐 > 5 毫克%。可作血尿生化检查、肾图等以确诊。

辨证施治

1. 阳虚寒积，浊阴内阻

主症：尿少、尿闭、纳呆乏力，面色晦滞，畏寒怕冷，恶心呕吐，便秘，舌苔薄腻，脉沉弦。

治法：温肾散寒，通腑泻浊。

方药：大黄附子汤（121）加味。

大黄9克，附子12克，细辛3克，枳实9克，黄芪18克，党参24克，白术10克，陈皮9克，清夏9克，甘草9克。水煎服。

加减：呕恶甚者加吴茱萸6克，生姜3克。

阳虚甚者加肉苁蓉9克，仙灵脾9克。

伴阴阳俱虚者，加冬虫夏草9克。

2. 气阴两虚，浊热互结

主症：面色萎黄，神疲乏力，呃恶纳呆，口中尿臭，潮热烦燥，胸中懊憹，头晕耳鸣，舌质红，苔黄腻，脉细数。

治法：益气养阴，清热降浊。

方药：封髓丹（122）加减。

人参9克，天冬18克，熟地9克，黄柏9克，砂仁9克，黄连9克，玄参18克，白茅根30克，大黄6克。水煎服。

加减：瘀血内停者，加丹参30克，桃仁9克，坤草30克。

大便不畅者，加玄明粉9克（冲），川朴9克。

3. 邪陷心肝，血热风动

主症：尿少尿闭，神昏谵语，循衣摸床，牙宣鼻衄，抽搐痉挛，舌质红，苔黄，脉弦细数。

治法：醒神开窍，凉血熄风。

方药：犀角地黄汤（25）合羚羊钩藤汤（22）加减。

犀角 1 克（磨汁冲），羚羊角 1 克（磨汁冲），生地 30 克，赤芍 12 克，白芍 12 克，丹皮 9 克，钩藤 12 克，菊花 9 克，天竺黄 9 克，甘草 3 克。水煎服。

加减：伴有痰浊上蒙者加菖蒲 12 克，郁金 12 克。
　　　昏迷者可选用清开灵注射液，静推或滴注。

6.3　急性泌尿系感染

急性泌尿系感染属于中医学中"淋症"范畴。

病因病机

外感湿热之邪，蕴结下焦，或过食肥甘，脾失健运，积湿生热，湿热下注，或劳欲过度，肾虚，固摄无权，皆可发为本病。

诊断要点

1. 尿频、尿急、尿痛、腰痛、发热寒战。

2. 白细胞计数和嗜中性粒细胞增高，血尿，尿旦白阴性或卄。白细胞＞5 个／高倍镜视野。大量脓细胞，白细胞管型。

3. 清洁中断尿培养及计数，或尿沉渣计数。

辨证施治

1. 下焦湿热

主症：尿痛、尿急、尿频，尿道灼热，发热，腰痛，尿赤，口干喜凉，舌苔黄腻，脉濡数。

治法：清热泻火、利尿通淋。

方药：八正散（15）加减。

车前草 30 克，木通 9 克，瞿麦 12 克，萹蓄 12 克，山栀 6 克，滑石 30 克，大黄 12 克，灯心草 3 克，甘草梢 6 克，连翘 15 克。水煎服。

加减：尿血者加生地15克，白茅根30克，石苇20克。
脓尿者加黄芩9克，苦参9克，土茯苓12克。

2. 阴虚湿热

主症：尿急、尿频、尿少色赤，低热心烦，腰背酸痛，身疲腿软，舌质红，少苔，脉细数。

治法：养阴清热，化瘀通淋。

方药：知柏地黄汤（123）加减。

知母9克，黄柏12克，生地12克，山药9克，泽泻15克，云苓24克，丹皮12克，车前草30克（包煎），坤草30克，白茅根30克，川牛膝9克，石苇30克，淡竹叶9克。水煎服。

加减：若有高热者加双花30克，连翘15克。
伴有咽喉肿痛者，加牛蒡子10克，板兰根15克。

6.4 急性尿潴留

急性尿潴留属于中医学中"癃闭"的范畴。

病因病机

湿热壅结，膀胱气化无力，肾元亏虚，膀胱传递无力，浊精、瘀血、结石内停，阻塞膀胱。肝郁气滞，膀胱气化不利，均可导致本病。

诊断要点

1. 患者有憋尿而解不出的主诉，严重者下腹胀痛，坐卧不安。

2. 耻骨上可触及胀大的膀胱及叩浊。

辨证施治

1. 膀胱湿热

主症：小便点滴而下，热赤不爽，或小便涓滴不通，小腹胀满，口干不欲饮或大便不畅，舌质红，舌苔根黄，脉数。

治法：清热利水。

方药：八正散（15）合石苇散（124）加减。

木通9克，车前子20克（包煎），瞿麦12克，萹蓄12克，滑石30克，大黄9克，山栀6克，灯芯草3克，小石苇30克，赤茯苓20克，甘草梢6克。水煎服。

加减：邪热伤阴，口干少津者，加生地15克，麦冬12克。

少腹胀满，欲尿不得者，加知母9克，黄柏12克，肉桂3克。

热灼肺津，烦渴欲饮，呼吸急促者，去大黄加黄芩12克，桑白皮12克，芦根30克，沙参12克。

2. 浊瘀阻塞

主症：小便淋漓不畅或阻塞不通，小腹胀满疼痛，舌质紫，或有瘀斑，脉涩或细数。

治法：行瘀散结，清利水道。

方药：代抵当丸（125）加味。

大黄9克，当归尾12克，山甲片9克，桃仁9克，丹参30克，川牛膝9克，滑石30克，通草6克。水煎服。

加减：气虚者，加黄芪15克，党参15克。

尿路结石者加金钱草18克，海金沙18克，冬葵子9克。

小腹胀痛加肉桂6克。沉香2克。

3. 肝郁气滞

主症：情志忧郁或多烦善怒，小便不通或通而不爽，胁腹胀痛，舌红，苔薄白或薄黄脉弦数。

治法：疏肝解郁、通利小便。

方药：柴胡疏肝散（106）加减。

柴胡9克，枳壳12克，芍药15克，香附9克，青皮9克，郁金12克，乌药9克，小茴香9克，甘草梢3克，车前子30克（包煎）。水煎服。

加减：可配合沉香粉，琥珀粉各3克，蟋蟀一对共研细末，分两次开水调服。

4. 命门火衰

主症：小便不通或点滴不畅，排出无力，神疲怕冷，面色㿠白，腰膝无力，舌质淡，脉沉细而尺弱。

治法：补肾温阳，益气通窍。

方药：济生肾气丸（126）加减。

附子9克，肉桂9克，熟地12克，萸肉9克，茯苓15克，泽泻12克，川牛膝9克，车前子30克（包），仙灵脾30克，仙茅9克，沉香粉3克（冲）。水煎服。

加减：若年高元气大虚，肾督不振，可加红参9克，鹿角片15克。

针刺取穴足三里、中极、三阴交、阴陵泉等，反复捻转提插，强刺激。体虚者可灸关元、气海，并可膀胱区按摩。也可应用取嚏或探吐之法及外敷法。

（包培荣）

7 内分泌及新陈代谢系统疾病

内分泌及新陈代谢系统急症,属于中医学中"瘿病"、"消渴"、"脱症"、"厥症"等范畴。多由先天不足,后天失养,外感邪毒,七情内伤,及饮食劳倦所致。本系统疾病多以"扶正祛邪"为基本的治疗原则。

7.1 急性化脓性甲状腺炎

(附甲状腺危象)

急性化脓性甲状腺炎,常在甲状腺肿的基础上发生,由血行或局部感染所致,属于中医学中"颈痈"、"锁喉痈"、"瘿病"的范畴。

病因病机

饮食不当,聚湿生痰,肝气不疏,气郁化火,加之外感邪毒内扰,邪毒痰热,相互博结而发为本病。

诊断要点

1. 多为化脓性细菌感染,多见于 20 至 40 岁的女性。
2. 甲状腺剧痛,肿大,发热,有波动感,皮肤发红,伴全身高热。
3. 血白细胞升高。

辨证施治

1. 热毒上壅:

主症:颈项一侧或二侧肿胀,疼痛,表面焮红、发热、伴有身热、吞咽不利,声音嘶哑舌苔黄腻,脉弦滑。

治法:清热解毒,化痰消肿。

方药:牛蒡解肌汤(127)加减。

牛蒡子12克，薄荷9克，连翘15克，山栀9克，夏枯草30克，丹皮9在，玄参20克，石斛12克，公英30克，地丁15克，双花30克，生牡蛎30克，淡竹叶9克。水煎服。

加减：初期可配合应用金黄散（128）外敷；脓成，手术排脓。

2. 邪热伤阴

主症：颈部肿块质软，或有结节，心悸烦热，口干尿赤，舌质红，苔薄黄，脉细数。

治法：养阴清热，化瘀散结。

方药：一贯煎（129）加减。

生地18克，沙参12克，麦冬12克，当归9克，川楝子9克，玄参20克，丹皮12克，赤芍12克，连翘30克，生牡蛎30克，夏枯草12克，黄药子9克。水煎服。

7.2 急性肾上腺皮质机能减退症

急性肾上腺皮质机能减退症，属于中医学中"脱症"的范畴。

病因病机

多为外邪侵袭，因出血、吐泻、外伤及应用肾上腺糖皮质激素不当等原因，致使阴阳大亏，元气衰微，或阴精耗竭、虚阳浮越，或阴寒内盛、气虚阳脱。

诊断要点

1. 头痛，腹泻，厌食，恶心呕吐，神疲乏力，呼吸急促，紫绀，高热或体温不升，意识障碍，皮肤粘膜下广泛出血，休克。

2. 检查可见血浆皮质激素降低，嗜酸性白细胞常在50个／mm^3以上，血钠、氯化物、血糖降低，血清钠／钾比数<30，血钾、红细胞压积、肌酐、酸度增高。

辨证施治

1. 阴脱

主症：高热，头痛，汗多不止，喘促，恶心呕吐，纳呆，口渴喜冷饮，甚者昏迷，舌质红而干，脉数而无力。

治法：益气养阴，生津增液。

方药：地黄饮子（130）加减。

人参9克，生熟地各30克，黄芪30克，天冬30克，麦冬20克，泽泻9克，石斛2克，枇杷叶9克，枳壳9克，甘草9克。水煎服。

加减：昏迷者加安宫牛黄丸1丸（20），可采用鼻饲或肛肠纳药法。

热伤血络者，加用犀角粉1至3克，冲服。

汗多不止者，可加煅龙骨30克，煅牡蛎30克，五味子6克，山萸肉9克。

2. 阳脱：

主症：体温不升，鼻鼾息微，四肢厥冷，口渴喜热饮，或柴绀，昏迷，舌质淡，苔白润脉微欲绝。

治法：益气回阳，扶正固脱。

方药：参附汤（29）加减。

人参9克，熟附子9克，干姜9克，肉桂9克，炙甘草6克。水煎服。

加减：昏迷者加用苏合香丸调服。可配合针刺人中，十宣等穴，中等刺激。

7.3 自发性低血糖症

自发性低血糖症属于中医学中"厥证"、"虚劳"等范畴。

病因病机

多为久病体虚，先天不足，后天失养，房劳过度，致使肝肾不足，精血亏损，清窍不荣脾不健运，聚湿生痰，清阳被蒙而发为本病。

诊断要点

1. 阵发性发作，可在10余分钟后或进食后终止，亦可持续较长时间。

2. 起病急者，有汗出、精神紧张、面色苍白、心悸、脉数、四肢震颤、血压偏高等症状；起病缓者则可见精神症状，如肌肉痉挛、眼花昏倒等。

3. 发作时血糖多在60毫克%以下，但植物神经功能失调者，血糖下降程度不显著。同时可做禁食试验、甲苯磺丁脲试验等，以明确病因作出诊断。

辨证施治

1. 肝肾阴亏

主症：头晕目眩，心悸汗出，面色憔悴，面颧发红，腰膝酸软，男子遗精，女子月经不调或带下，舌质红，少苔，脉细数。

治法：滋补肝肾。

方药：一贯煎（129）加减。

沙参8克，麦冬15克，当归身12克，生地15克，枸杞子18克，川楝子9克，杭芍18克，炒枣仁12克，甘草3克。水煎服。

加减：气血双亏者，方用十全大补汤（151）加减。

心脾两虚者方用归脾汤（50）加减。

2. 气滞痰阻

主症：突然昏倒，喉中有痰，呕吐涎沫，四肢震颤，舌苔白腻，脉沉滑。

治法：行气豁痰。

方药：温胆汤(131)加减。

竹茹9克，枳实12克，半夏9克，橘红9克，云苓30克，菖蒲9克，郁金9克，远志9克，炙甘草6克。水煎服。

加减：视物不清者加决明子30克，可配合针刺人中、十宣、合谷等穴，中等或强刺激，待苏醒后，再辨证用药。

(包培荣)

8 造血系统疾病

造血系统疾病，主要介绍"贫血"、"紫癜"。这些疾病，分别属于中医学中的"脱血"、"血枯"、"虚劳"等范畴。本章疾病多以"急治其标"、"缓治其本"为基本的治疗原则。

8.1 再生障碍性贫血

再生障碍性贫血（简称再障），属于中医学"虚劳"、"血证"等病范畴。

病因病机

本病的产生，发展，转归与心肝脾肾四脏有关，尤以脾肾两脏的亏损为主要症结所在。

诊断要点

1. 发病情况：本病以青壮年多见，男性多于女性。

2. 本病典型发病过程中常有进行性贫血、广泛部位出血、反复感染三大症候群。肝、脾、淋巴结一般不肿大。

3. 实验室检查

(1) 血象：三系细胞均减少，呈小细胞、低色素性贫血，淋巴细胞绝对值降低，此外，网织红细胞绝对数减少。

(2) 骨髓象：骨髓造血组织显著减少，脂肪组织增多，淋巴细胞、浆细胞、网状细胞、组织嗜碱细胞增多。

辨证施治

1. 气血两虚

主症：面色苍黄无华，眩晕眼花，疲乏无力，心悸，口唇爪甲淡白或月经过多，舌质淡苔薄白，脉虚无力或数。

治法：益气补血。

方药：八珍汤（132）加减。

党参15克，白术12克，云苓9克，炙甘草6克，熟地黄12克，当归9克，白芍12克，川芎6克，黄芪18克。水煎服。

2. 脾肾阳虚

主症：面色㿠白或少华，畏寒肢冷，腰膝无力，夜尿频数清长，浮肿，阳萎，遗精，月经不调，气短懒言，食欲不振。舌质淡，舌体胖嫩，苔薄白，脉沉弱或沉缓。

治法：健脾温肾。

方药：四君子汤（133）合右归丸（134）加减。

党参15克，白术12克，云苓9克，炙甘草6克，熟地黄18克，山药12克，鹿角胶10克，肉桂3克，制附子9克，杜仲12克，当归12克。

若出现心脾两虚证，则宜用归脾汤加味，此外尚可加用鹿茸、蛤蚧、紫河车等。

3. 肝肾阴虚

主症：面色萎黄，头晕目眩，两颊潮红，腰膝酸软，胁下隐痛，月经过多，或崩漏不止，或衄血、便血，唇甲淡白，潮热盗汗，五心烦热，舌红无津，少苔，脉细数或弦数。

治法：滋补肝肾。

方药：归芍地黄汤（135）加减。

熟地黄18克，山药12克，山茱萸12克，云苓9克，泽泻9克，丹皮9克，白芍12克，当归15克，五味子9克，水煎服。

4. 阴阳两虚

主症：面色苍白，腰膝酸软，疲乏无力，潮热盗汗，遗清滑精，失眠多梦，或腹冷便溏。脉沉细无力。

治法：阴阳双补。

方药：金匮肾气丸（136）加味。

桂枝6克，附子10克，熟地黄24克，山萸肉12克，淮山药 12 克，丹皮 9 克，泽泻 9 克，云苓 9 克，水煎服。

8.2 急性失血性贫血

急性失血性贫血，属于中医学"脱血"、"亡血"、"血虚"等病证范畴。

病因病机

各种原因引起的出血，导致失血性贫血，皆与心、脾、肝、肾有关。心不主血，脾不统血，肝不藏血，肾不纳气，造成气虚、血虚、阴虚、阳虚等证。

诊断要点

主要查清引起急性失血性贫血的原因。

1. 消化系统：如胃、十二指肠溃疡、胃癌、食道及胃底静脉曲张破裂、钩虫、痔疮等疾病。

2. 血液系统：如血友病、血小板减少性紫癜、急性白血病、再生障碍性贫血及严重出凝血机能障碍性疾病。

3. 妇产科疾病：如宫外妊娠、前置胎盘及其它原因引起的大出血。

另外，各种外伤所致的出血。

辨证施治

1. 心血虚

主症：心悸怔忡，头晕眼花，失眠多梦，面色萎黄，甚者面色苍白，舌质淡，脉细弱。

治法：养血安神。

方药：养心汤（137）加味。

黄芪 15 克，云苓 9 克，当归 12 克，川芎 9 克，炙甘草 6 克，半夏曲 9 克，柏子仁 6 克，五味子 6 克，人参 6 克，肉桂 3 克，大枣 3 枚。水煎服。

2. 脾气虚

主症：食少倦怠，头晕目眩，少气懒言，乏力自汗，大便溏薄，舌淡，脉虚无力。

治法：益气健脾。

方药：参苓白术散（116）加减。

人参 9 克，云苓 10 克，白术 12 克，桔梗 9 克，山药 12 克，白扁豆 10 克，莲子肉 10 克，薏苡仁 15 克，砂仁 3 克，甘草 6 克。水煎服。

3. 肝血虚

主症：头晕目眩，耳鸣胁痛，惊惕不安，手足麻木，甚至抽搐，妇女月经涩少，舌质淡脉弦细。

治法：补血养肝，活血化瘀。

方药：四物汤（138）加味。

当归 12 克，白芍 15 克，川芎 9 克，熟地黄 12 克，水煎服。

若晕眩耳鸣者，加女贞子、磁石、牡蛎各 30 克等以育阴潜阳。

肝胁痛者，可加柴胡、郁金、香附各 12 克等。

4. 肾阴虚

主症：腰酸耳聋，咽痛，颧红，两足无力，视物模糊，舌红绛、少津，脉沉细无力。

治法：补肾养血，滋阴降火。

方药：知柏地黄汤（123）加减。

知母 9 克，黄柏 9 克，熟地黄 21 克，淮山药 12 克，丹皮 9 克，泽泻 9 克，云苓 9 克。水煎服。

8.3 过敏性紫癜

过敏性紫癜又称出血性毛细血管中毒症，属中医"发斑"、"血证"等范畴。

病因病机

多因风湿之邪外袭，与气血相搏，热伤脉络，使血不循经，溢于脉外，渗于肌肤而成。关键在于"血热妄行"，血受热则煎熬成癜。

诊断要点

1. 多见于儿童及青年，前躯症状为：疲乏无力，头昏，食欲不振、低热。一般分如下几型。

(1) 皮肤型：皮肤可见斑丘疹，皮疹周围环绕一苍白区，亦可呈荨麻疹样，可有痒感，散在或融合成片状，多见于双下肢、臀部，呈对称性分布，分批出现。

(2) 腹型：除皮肤紫癜外，尚伴有腹痛、腹泻等消化系统症状。

(3) 关节型：除皮肤紫癜外，伴有关节红、肿、热、痛。多发生在膝、踝及腕关节，呈对称性、游走性。

(4) 肾型：除紫癜外，常伴有脸面浮肿、尿少、蛋白尿、血尿和管型尿。

(5) 混合型：以上几种类型混合称为混合型，但每型均有皮疹出现。

2. 实验室检查

除束臂试验阳性外，其他检查均无明显异常。

辨证施治

1. 风热发斑

主症：发热、微恶风寒，身痛骨楚，斑色红赤，此愈彼发。舌红，苔黄薄，脉浮数。此型多见于早期病人。

治法：疏表清热。

方药：银翘散 (2) 加味。

双花18克，连翘18克，豆豉9克，牛蒡子9克，薄荷6克，荆芥穗9克，桔梗9克，竹叶6克，防风9克，鲜苇根18克，甘草6克。水煎服。

2. 血热妄行

主症：高热面赤，口渴欲饮，斑色紫赤，量多成片，高出皮面；或兼有衄血、尿血，身躁心烦；或兼有关节红肿热痛。舌红，可有瘀斑，苔黄燥，脉滑数。此型多见于中期患者。

治法：清热解毒，凉血活血。

方药：犀角地黄汤（25）加味。

犀角3克，生地黄30克，丹皮9克，芍药12克，知母12克。水煎服。

3. 阴虚火旺

主症：低热不退，倦怠无力，五心烦热，颧红唇干，失眠盗汗，斑疹扁平，色鲜红。舌红少苔，脉细数。

治法：滋阴降火，凉血活血。

方药：滋阴降火汤（139）或知柏地黄丸（123）合小蓟饮子（140）加减。

白芍12克，当归9克，熟地黄6克，麦冬6克，白术6克，酒生地12克，陈皮3克，知母9克，黄柏6克，生姜2克，大枣3枚。水煎服。

4. 脾肾两亏

主症：紫癜反复发作，面色不华，腰背酸痛，头晕乏力，食呆纳差，小便频数，月经量多。舌胖而有齿痕，脉濡。多见于后期患者。

治法：健脾益气，补肾固摄。

方药：补中益气汤（141）合无比山药丸（142）加味。

黄芪15克，党参12克，白术9克，当归12克，陈皮9克，炙甘草6克，柴胡6克，升麻9克，山药12克，五味子9克，杜仲12克，巴戟天30克，泽泻15克。水煎服。

针刺对本病亦有较好效果，常用穴位是足三里、血海、曲池等。

（戴法轩）

9 结缔组织及变态反应性疾病

结缔组织及变态反应性疾病，主要介绍系统性红斑狼疮和变应性亚急性败血症等疾病。这些疾病，中医学没有相似的名称，多属中医学"痹证"、"发斑"、"腰痛"、"发热"等范畴。其治则多为清营解毒，疏风清热，凉血护阴。

9.1 系统性红斑狼疮

系统性红斑狼疮是一种自身免疫性疾病，属中医学"痹证"、"发斑"等范畴。

病因病机

风寒湿之邪伤及脏腑，损伤肌肤，导致气血不通，瘀而化火发斑；瘀而不通，则关节、肌肤疼痛。

诊断要点

1. 多见于青壮年女性。
2. 发热：可为高热或低热。对糖皮质激素治疗敏感。
3. 关节表现：90%以上的患者有游走性多关节酸痛。发作时可有红肿热痛等现象。
4. 皮疹：红斑性皮疹是最常见的皮肤表现，常呈对称性位于面颈或肢体，特别是手指指端和背部、手掌或环绕肘部。
5. 肾脏：50%以上病人可出现狼疮性肾炎。
6. 血液或骨髓涂片，可见"红斑狼疮细胞"，多次检查的阳性率可达80%以上，但并非特异性。

辨证施治

1. 毒热炽盛型

主症：骤然发病，壮热，面颊部蝶状赤红斑疹，关节肌肉酸痛，皮肤紫斑，甚则烦躁口渴，神昏谵语，舌质红绛，苔黄腻，脉洪数或弦数。

治法：清营解毒，凉血护阴。

方药：清瘟败毒饮（117）加减。

水牛角粉 6 克（冲服），丹皮 10 克，生地 30 克，赤芍 10 克，玄参 10 克，双花 15 克，麦冬 15 克，生石膏 30 克，知母 10 克。水煎服。

2. 风湿热痹型

主症：大小关节肿胀酸痛，肌肉痛或伴有低热，舌质红，苔黄糙，脉滑数或细数。

治法：祛风通络，清热和营。

方药：独活寄生汤（78）加减。

独活 15 克，桑寄生 10 克，秦艽 10 克，防风 10 克，生地 15 克，杭芍 15 克，川芎 6 克，杜仲 10 克，牛膝 10 克，当归 10 克，忍冬藤 10 克，丹参 15 克。水煎服。

3. 阴虚内热型

主症：低热，手足心烦热，斑疹暗红，自汗盗汗，关节酸痛，脱发，舌质红，苔剥，脉细数。

治法：养阴清热。

方药：青蒿鳖甲汤（143）加减。

青蒿 10 克，银柴胡 10 克，地骨皮 10 克，胡黄连 10 克，生地 30 克，元参 10 克，麦冬 10 克，石斛 10 克，白薇 10 克，女贞子 15 克。水煎服。

9.2 变应性亚急性败血症

变应性亚急性败血症，属于中医学"发热"、"腰痛"等范畴，此病在临床上不常见。

病因病机

正气不足，风寒湿邪乘虚侵入，流经窜络，阻滞关节，影响气血运行,瘀而化火,形成发热,气血不通,则关节乃至全身疼痛。

诊断要点

1. 长期发热,可间有缓解,热型不定,呈弛张热或稽留热型。
2. 反复出现一过性皮疹及关节疼痛。
3. 白细胞计数增高伴核左移,血沉增快。
4. 多次血培养,无肯定阳性结果。
5. 抗菌素治疗无效。
6. 激素治疗可使症状缓解。

辨证施治

1. 热邪偏盛

主症:骤然发病,壮热,烦躁口渴,大便秘结,胸腹散在皮疹,关节疼痛,汗多,尿黄赤,舌苔黄燥,脉数。

治法:清热解毒,凉血护阴。

方药:犀角地黄汤(25)加味。

犀角粉6克(冲服),生地黄30克,赤芍10克,银花15克,连翘12克,麦冬15克,黄芩12克,天花粉10克,丹皮9克,生石膏30克,知母9克,栀子10克,桑枝9克。水煎服。

2. 湿热蕴蒸

主症:间歇高热,关节红肿疼痛,头胀痛如裹,口渴不欲饮,多汗,舌苔黄腻,脉濡数。

治法:祛湿清热。

方药:独活寄生汤(78)加减。

独活15克,羌活12克,双花15克,桑寄生15克,秦艽9克,防风9克,桂枝3克,细辛3克,当归12克,芍药12克,川芎9克,生地15克,杜仲9克,党参12克,云苓9克,甘草6克。水煎服。

(戴法轩)

10 传染性疾病

传染性疾病，本章主要介绍"流感"、"流脑"、"乙脑"等疾病。这些疾病属于中医学"时行感冒"、"春温"、"暑湿"等范畴。其治则多为"急则治其标"，防患于未然亦是重要措施之一。

10.1 流行性感冒

流行性感冒，又称流感。属于中医学"时行感冒"等病范畴。

病因病机

气候变化多端，寒热失常，人体卫外功能减退，应急功能低下而感受外邪发病。

诊断要点

1. 起病急，局部症状一般较轻，全身中毒症状明显，有高热、全身酸痛、畏寒、头痛、乏力等。流行期间尚可见到以咳嗽、咳痰、胸痛症状为主或以恶心、呕吐、腹泻为主的流行性感冒。

2. 流行季节注意与流行性脑膜炎及麻疹等急性传染病早期相鉴别。

辨证施治

1. 风寒束表

主症：恶寒，发热，无汗，头痛，鼻塞，流清涕，多嚏，咽痒不适，或有微咳，无痰或有少量清稀痰，舌苔薄白，脉浮。

治法：辛温解表。

方药：荆防败毒散（144）加减。

荆芥9克，防风6克，苏叶6克，前胡9克，桔梗6克，淡豆豉9克，水煎服。

肢体酸痛，头痛较甚者，加羌活、独活各6克。

2. 风热犯表

主症：发热，微恶风寒，有汗不多，头痛，咳嗽，痰少黄稠，咽喉红肿疼痛，口干微渴，舌红苔薄白或微黄，脉浮数。

治法：辛凉解表。

方药：银翘散（2）加减。

牛蒡子9克，薄荷3克，连翘9克，金银花15克，杏仁9克，桑叶9克。水煎服。

若扁桃体红肿显著者，加射干9克，山豆根15克。

鼻衄者加山栀9克。

高热不退者，加柴胡9克，黄芩9克。

3. 暑湿在表

主症：夏季感受暑湿外邪，见头痛，四肢困倦或酸痛，身热，有汗或少汗，心烦，口渴，微咳，胸闷脘痞，小便黄或大便溏，舌苔薄黄而腻，脉濡数。

治法：祛暑化湿解表。

方药：新加香薷饮（6）加减。

香薷9克，豆卷9克，连翘9克，银花9克，厚朴6克，藿香9克，佩兰9克，荷叶3克。水煎服。

热盛心烦者，加黄连6克。

若有食滞者，加山楂9克。

10.2 流行性脑脊髓膜炎

流行性脑脊髓膜炎（简称流脑），属于中医学"春瘟"范畴。

病因病机

由于人体正气不足，卫外防御功能减弱，感受温疫时邪而发病。

诊断要点

1. 发病急，高热，畏寒，剧烈头痛，喷射性呕吐，烦躁，惊厥，昏迷，颈项强直，布氏和克氏征阳性，婴儿囟门饱满隆起，角弓反张等。

2. 发病情况：多于冬春发病，病人以学龄前儿童多见。早期即出现皮肤瘀点或瘀斑，直径多在2毫米以上，病后3～5天常有口周与前鼻孔周围的单纯泡疹。

3. 实验室检查

(1) 脑积液：脑积液压力明显升高，外观混浊，呈米汤样或脓性，细胞数增多，以中性多核细胞为主，蛋白增高，糖减少，涂片或培养能找到相应致病菌。

(2) 血培养阳性。血常规中白细胞明显增多，以中性多核细胞为主。

辨证施治

1. 邪在卫分

主症：起病突然，发热不高，恶寒，头痛，呕吐，咽干，口渴，颈项强直（不典型），神情淡漠，皮肤偶有少量出血点，舌苔薄白或淡黄，脉滑数。

治法：清热解毒，佐以疏表。

方药：银翘散（2）加减。

银花18克，连翘15克，大青叶30克，黄芩12克，龙胆草9克，葛根15克，薄荷3克，鲜竹沥10克，半夏9克。水煎服。

2. 气营两燔

主症：高热持续，头痛剧烈，颈项强直，喷射性呕吐，烦躁不安或昏睡，抽搐阵作，皮肤瘀点增多，舌质红，苔黄腻，脉滑数。

治法：清热解毒凉营。

方药：清瘟败毒饮（117）加减。

双花15克，大青叶30克，黄连3克，元参12克，丹皮9克，连翘12克，龙胆草6克，生石膏15克，鲜生地15克，石决明 30 克。水煎服。

热重，抽搐甚，苔黄者，加钩藤 15 克，全蝎 9 克。

高热昏迷，狂躁不安，加安宫牛黄丸，每次 1 粒，日 2 次（成人）。

阴液耗伤者，加麦冬、石斛各 15 克。

3. 邪陷正脱

主症：面色苍白、晦滞、出汗，四肢厥冷，呼吸微弱或不规则，皮肤瘀点迅速发展或融合成片，血压下降，舌质淡，脉微细。

治法：益气回阴固脱。

方药：参附汤（29）加减。

红参 12 克，制附子 18 克，炙甘草 15 克。水煎服。

舌红，脉细数者，加麦冬 12 克，五味子 9 克。

10.3 流行性乙型脑炎

流行性乙型脑炎（简称乙脑），属于中医学"暑瘟"、"暑风"、"暑厥"等病范畴。

病因病机

本病由感受暑热疫疠之邪引起。夏季暑热炎盛，如正气亏虚不能抗邪时，疫邪即乘虚而入发病。

诊断要点

1. 询问预防病史。

2. 在流行季节，遇有突然发病，头痛呕吐，嗜睡或烦躁等现象，并在 2～3 天内逐渐加重者，应首先考虑本病。重证病号可迅速出现昏迷、惊厥或呼吸衰竭等；小儿常见凝视、惊跳。

3. 早期可无特殊阳性体征，2～3 天后常见有脑膜刺激征，幼儿表现为前囟膨隆，部分病人有腹壁反射消失，巴彬斯基征阳性，四肢肌张力增高等体征。

4. 实验室检查

(1) 血液：白细胞增高，以中性多核细胞居多。

(2) 脑脊液：细胞数增高，早期常以多核细胞为主，晚期则以淋巴细胞为主，糖正常或稍高，蛋白质轻度增高，氯化物正常。

(3) 补体结合试验阳性。

辨证施治

1. 邪在卫分

主症：发热，或微恶寒，有汗，头痛，神志清或烦躁，嗜睡，颈项稍强，偶有轻度抽搐，舌苔薄白。

治法：清热透表解毒。

方药：银翘散（2）加减。

双花15克，连翘12克，生石膏30克，薄荷3克，竹叶3克，黄芩12克，大青叶30克。水煎服。

加减：湿邪偏重，见恶心、呕吐、嗜睡、舌苔白腻者，去石膏加藿香9克，佩兰9克，厚朴6克。

2. 气营两燔

主症：高热，头痛，颈强直，昏迷或半昏迷，谵妄，抽搐，两侧瞳孔大小不一，呼吸浅表不规则，舌质红或红绛，苔黄腻，脉数大。

治法：清热解毒凉营。

方药：白虎汤（10）合清营汤（18）加减。

生石膏45克，知母15克，银花15克，连翘9克，板兰根30克，黄连3克，蚤休18克，生地15克，元参15克。水煎服。

加减：热甚、便秘者，加大黄9克。

抽搐不已，加钩藤15克（后入），全蝎9克。

昏迷较甚，酌加石菖蒲6克，制胆星3克。

3. 正虚邪恋

主症：低热不退，午后稍高，面红烦躁，舌质红，脉细数。

见于恢复期。
治法：育阴清热。
方药：加减复脉汤（145）加减。
生地15克，麦冬12克，沙参15克，元参12克，白芍12克，青蒿9克，白薇12克，知母9克，丹皮12克。水煎服。
加减：烦躁不宁，精神异常，舌苔薄黄腻者，加半夏9克，制胆星3克，石菖蒲9克。
阴液耗劫，虚风内动，见手足颤动，舌质红少津，去青蒿加龟板15克，鳖甲15克。

4. 痰瘀阻络
主症：神情呆滞不语，肢体瘫痪，面色苍白，苔腻，可见于恢复期或后遗症期。
治法：益气化痰行瘀。
方药：补阳还五汤（40）合涤痰汤（79）加减。
黄芪15克，当归12克，赤芍9克，红花9克，桃仁9克，石菖蒲9克，法半夏12克，白附子6克，地龙9克，陈皮6克，甘草3克。水煎服。

5. 针灸治疗
乙脑后遗症，如聋、哑、肢瘫等，可采用针灸疗法，效果颇佳。常取穴位是：哑门、天突、尺泽、委中、太冲、曲池、环跳、阳陵泉等穴。

10.4 流行性出血热

流行性出血热是一种自然疫源性急性传染病。中医学称之为"冬瘟时疫"、"疫斑"等，属温病范畴。

病因病机

温热之邪侵袭肺卫，迅速传气入营，而成气营两燔之候，故出现高热不退。热邪内闭或气阴两伤，阳气衰竭，临床出现阴阳离决之候。邪热内盛，肾阴亏耗，出现关格之势，若到病邪渐退，正气不足，肾气不固时，则出现多尿之象。

诊断要点

1. 流行病学资料。
2. 全身中毒症状。
3. 毛细血管中毒征象。
4. 肾脏损害。

辨证施治

1. 发热期

(1) 温邪袭卫

主症：恶寒发热，头痛、腰痛、眼眶痛，颈胸潮红，舌边尖红，苔白薄腻，脉浮数。

治法：疏表清热。

方药：银翘散（2）加减。

双花15克，连翘10克，薄荷3克（后下），桔梗5克，鲜芦根（去节）30克，生甘草6克，白茅根30克，丹皮10克，丹参15克，黄芩10克。水煎服。

(2) 阳明热炽

主症：壮热不恶寒，酒醉貌，口渴烦躁，舌红苔黄，脉弦数。

治法：清热解毒。

方药：白虎汤（10）加减。

生石膏30克，（先煎），知母10克，竹叶10克，山豆根10克，板兰根15克，粳米30克，甘草6克，元参10克。水煎服。

(3) 气营两燔

主症：壮热口渴，目昏烦躁，斑疹吐衄，舌绛苔黄燥，脉弦数或细数。

治法：清气凉营，解毒护阴。

方药：清瘟败毒饮（117）加减。

生石膏60克（先煎），知母10克，生地10克，黄芩15克，栀子10克，犀角5克，丹皮10克，竹叶10克，生大黄6克，元参15克，银花15克，生甘草6克。水煎服。

2. 低血压、休克期

(1) 热厥

主症：手足厥冷，胸腹灼热，面赤心烦气促，舌红苔黄厚而干，脉滑数或沉数。

治法：清热解毒，益气生津。

方药：白虎汤（10）合生脉散（26）加减。

生石膏30克（先煎），知母10克，板兰根15克，人参3克，麦冬15克，五味子15克，升麻10克，白芍10克。水煎服。

(2) 寒厥

主症：畏寒肢冷，蜷卧不渴，气微神疲，面白唇青，脉沉迟而细或欲绝。

治法：回阳救逆。

方药：参附汤（29）加减。

人参10克，熟附子10克（先煎），五味子10克，熟地18克，麦冬10克，丹参15克，炙甘草10克，龙骨30克，牡蛎30克。水煎服。

3. 少尿期

(1) 肾阴亏损，虚火内生

主症：极度衰竭，精神萎靡，嗜睡腰酸，小便涩少，口干咽燥，心烦不眠，舌红苔干，脉细数。

治法：滋肾生津。

方药：知柏地黄汤（123）加减。

知母10克，黄柏10克，生地10克，山药15克，白茅根30克，丹皮10克，麦冬10克，山萸肉10克，泽泻10克。水煎服。

(2) 热陷心包，肝风内动

主症：尿少尿闭，头痛呕吐，神昏谵语，痉厥抽搐，舌绛苔干，脉弦细数。

治法：清心解毒，熄风镇惊。

方药：犀角地黄汤（25）合羚羊钩藤汤（22）加减。

犀角6克，生地15克，丹皮10克，钩藤12克，菊花10克，赤白芍（各）12克，竹茹10克，车前子10克，白茅根30克。水煎服。

(3) 饮邪壅肺

主症：胸满喘急，痰涎壅盛，烦躁不安，舌胖苔白厚腻，脉弦数或滑数。

治法：泻肺平喘。

方药：葶苈大枣泻肺汤（146）加减。

葶苈子15克，大枣10枚，车前子（布包）10克，生大黄10克（后下），白茅根30克，云苓15克。水煎服。

4. 多尿期

(1) 肾气不固

主症：疲倦懒言，口渴多饮，日夜多尿，腰膝酸软，舌淡红苔少而干，脉虚大。

治法：补肾固摄，益气生津。

方药：八仙长寿丸（147）合缩泉丸（119）加减。

麦冬10克，党参15克，熟地24克，山药15克，覆盆子10克，益智仁10克，五味子5克。水煎服。

(2) 肺胃热盛

主症：口干舌燥，烦渴引饮，干咳少痰，多食善饥，尿频尿多，舌红苔黄，脉沉数。

治法：清肺胃热，养阴生津。

方药：沙参麦冬汤（5）加减。

沙参10克，麦冬10克，桑叶10克，花粉12克，玉竹10

克，生石膏 30 克，竹叶 10 克，山药 10 克，益智仁 10 克，生甘草 3 克。水煎服。

5. 恢复期

肾阴亏损，用六味地黄丸（118）加减。

脾阳不振，用参苓白术散（116）加减。

胃阴未复，用益胃汤（173）加减。

10.5 急性重症肝炎

急性重症肝炎，系指急性肝坏死和亚急性肝坏死，约占全部肝炎患者的 0.2%。中医学称之为"黄疸"、"瘟黄"、"胁痛"等。

病因病机

由于脾胃虚弱，或因饮食不慎，酒食不节，以致脾胃功能失常，复感时邪而发病。

诊断要点

1. 发病初情况

起病时与急性黄疸型肝炎相似，但病情发展较快，常有发热、黄疸；同时往往肝脏明显缩小并可于数天内出现肝性昏迷。

2. 急性肝坏死特点

除发病初情况外，突出的临床表现有：如嗜睡或烦躁，谵妄，精神错乱，扑翼性震颤，昏迷，抽搐等。出血现象如鼻衄，呕血，皮肤粘膜瘀点等，以及迅速出现的腹水和肝肾综合征等。本型病人"肝臭"常很明显。

3. 亚急性肝坏死的特点

病程较急性肝坏死病程（大约 2 周）为长。临床表现与急性肝坏死者相似而较轻，主要有黄疸，出血倾向，肝脏缩小，腹水以及中枢神经系统异常等，黄疸常持续不退或进行性加深。

辨证施治

1. 湿热熏蒸

主症：面目及周身俱黄，如桔子色，烦热脘闷，纳呆呕吐，大便秘结或溏，苔黄腻，脉弦滑数或濡数。

治法：清利湿热。

方药：茵陈蒿汤（35）加减。

茵陈 30 克，生山栀 10 克，生大黄 6 克，连翘 10 克，板兰根 15 克，白茅根 30 克。水煎服。

2. 肝气郁滞

主症：胁肋胀痛，脘痞腹胀，恶心嗳气，纳食不香，苔薄，舌质淡红，脉弦。

治法：疏肝理气。

方药：柴胡疏肝散（106）加减。

柴胡 10 克，枳壳 6 克，白芍 10 克，郁金 10 克，甘草 3 克，川芎 6 克，香附 10 克，佛手 10 克。水煎服。

3. 湿邪困脾

主症：胁痛，脘闷腹胀，恶心呕吐，胃纳不佳，口淡不欲饮，身重便溏，苔黄腻。

治法：化湿健脾。

方药：胃苓汤（148）加味。

苍术 10 克，厚朴 10 克，陈皮 6 克，薏苡仁 12 克，白术 10 克，云苓 10 克，猪苓 6 克，车前子（布包）15 克，泽泻 6 克，桂枝 6 克，甘草 3 克。水煎服。

4. 肝阴亏损

主症：胁痛隐隐，低热腰酸，口干苦而燥，手足心热，苔少或无苔，舌质红，舌尖有红刺，脉弦细数。

治法：养阴柔肝。

方药：一贯煎（129）加味。

北沙参 10 克，麦冬 10 克，当归 10 克，生地 12 克，枸杞 10 克，川楝子 6 克，石斛 12 克。水煎服。

5. 热毒炽盛

主症：高热，口渴，烦躁，黄疸迅速加深，腹胀满，大便秘结，小便黄赤，甚则神昏谵语，抽搐，或见便血、尿血等，舌红绛，苔黄腻或黄燥，脉滑数。

治法：清热鲜毒，凉血救阴。

方药：犀角地黄汤（25）加减。

犀角3克（磨冲），黄连6克，山栀10克，生地15克，板兰根30克，茵陈30克，丹皮10克，石斛10克。水煎服。

（戴法轩）

11 中毒及理化损伤性疾病

急性中毒，是临床急症中的常见病。其特点是发病急剧，变化多端，若得不到及时治疗和抢救，往往造成严重后果，甚至死亡，故必须中西医结合采取措施，尽快做出诊断和处理。

中毒在中医学中多分为食物中毒、药物中毒、饮酒中毒和毒虫兽咬伤及秽浊之气中毒等五大类。本章主要介绍"酒精中毒"、"乌头砒类中毒"、"马钱子中毒"等。另外，理化损伤性疾病如"中暑"，亦在此一并叙述。

11.1 酒精中毒

酒精中毒，一般指急性酒精中毒，多见于饮酒过量或误服含有酒精的其他溶液。

病因病机

酒精其性剽悍，入于胃后则胃胀气逆，上逆于胸，内薰于肝胆，令人胸闷志乱，乃至不醒，甚者阴阳离绝，危及生命。

诊断要点

呼吸呈酒味，颜面潮红或苍白，脉搏加速，心悸，汗出，瞳仁缩小或扩大。早期中毒多为醉汉表情，呕吐躁动，严重中毒时则大小便失禁，抽搐，昏不知人。若其毒入肺，涉及华盖，可令人窒息而死。

脉细数或洪数，舌红或红绛或暗，苔薄黄。

辨证施治

1. 涌吐洗胃

(1) 方药：瓜蒂散加味 (100)。

瓜蒂15克，甘草12克，元参30克，地榆5克。水煎服。

(2) 1∶2000 小苏打水溶液洗胃，反复多次。
(3) 自动洗胃机。
2. 排出毒物
(1) 发汗解毒
方药：葱豉汤加味（174）。
淡豆豉30克，葱白15克，甘草15克。水煎，温服。
(2) 泻下解毒
大黄末12克，水冲服；或元明粉6克，水冲服。
(3) 利尿解毒
① 车前草30克，白茅根30克。水煎服。
② 绿豆甘草解毒汤（149）。
绿豆90克，甘草30克，丹参15克，连翘12克，石斛30克，大黄9克。水煎服。
3. 和中解毒法
方药：葛根30克，绿豆30克，黄芩15克。水煎服。
4. 其他应急措施
(1) 注意病人保暖，经常更换病人体位。
(2) 烦躁不安，过度兴奋者可给予利眠宁或安定肌注。
(3) 严重抑制状态而呈昏迷者，可给予呼吸兴奋剂，如半分马钱子煎水服，有胜似利他林之功效。呼吸困难者，可给以氧气吸入。
(4) 低血压者，可给开压药，有脱水者除给以生脉散（26）外，尚可考虑输液等中西医结合措施治疗。

11.2 乌头类药物中毒

乌头中毒即乌头硷中毒。乌头硷系乌头类植物（川乌、草乌、附子）所含生物硷，有大毒，临床多见于误服或因治疗某些疾病误用超量而中毒。

病因病机

服药过剂，反伤正气，致毒邪攻心。

诊断要点

服乌头类药物数十分钟后，口舌麻木，有烧灼感，继之恶心呕哕，口流涎水，轻则头昏眼花，烦躁不安，视力模糊；重则言语不清，吞咽困难，眼睑下垂，肢瘫，呼吸短促，心跳加速，甚则突然抽搐，昏迷紫绀，瞳仁散大，呼吸循环衰竭致死。

辨证施治

1. 生姜30克，双花30克，绿豆90克。水煎服，日二剂。
2. 绿豆90克，黄连9克，黑豆90克，芫荽3克，甘草30克。水煎服，日三剂，直至症状消失后改日1剂，连服1周。

11.3　马钱子中毒

马钱子中毒的主要成分为士的宁和马钱子碱。中毒多由误服或在治疗某些疾病时剂量过大所致。

病因病机

马钱子、苦、寒、有大毒，多外用或入丸剂，一般不入汤剂，若服用过量，则扰乱神明致痉，甚至阴阳离绝，危及生命。

诊断要点

服药数分钟后，先有震颤，胸部有压迫感，知觉过敏，继则咬肌及颈肌抽搐，有时呕吐。痉挛时，神志清楚，脸部哭笑无常，双目凝视，渐至呼吸肌痉挛，全身发绀，瞳孔散大，病人受外界声、光、风等刺激，可立即引起再度强直性痉挛，每次持续几分钟，脉细数，舌紫暗。

辨证施治

1. 立即洗胃，用生姜汁30毫升兑洁水3000毫升左右反复彻底洗胃，最好选用人工洗胃机。洗胃完毕后一次灌入通用解毒粉（活性炭10克，硫酸镁20克）。
2. 高位清洁灌肠，用大黄水或番泻叶水500毫升左右高位灌肠。

3. 芝麻油 30~100 毫升，口服，以保护胃粘膜，加强导泻之功。

4. 药物解毒

黄芩 30 克，甘草 15 克。水煎服，连服数次。

5. 昏厥者，亦可使用安宫牛黄丸 (20)。

6. 其他抢救措施

(1) 针灸二针刺合谷、人中等穴。

(2) 镇静可选用肌注安定。

(3) 建立静脉通道，给以补液，保肝等对症治疗。

11.4　中暑

中暑，中医学按其临床表现和特点分属于"暑风"、"暑厥"、"暑痉"、"阳暑"等证。现代医学称此类病为"热射病"、"热衰竭"、"日射病"、"热痉挛"等。

病因病机

酷夏之日，高温之时，暑邪耗伤气阴而发病。年老体弱者，脾肾已衰，阴津亏乏；产妇暑天分娩损伤元气等，尤易罹患中暑。

辨证施治

1. 暑热耗伤气阴

主症：头晕、心烦，面色苍白，汗出，乏力气短，四肢厥冷，甚则昏迷，舌红苔少，脉细数无力。

治法：益气养阴。

方药：生脉散 (26) 加减。

太子参 60 克（或西洋参 30 克），麦冬 30 克，天冬 30 克，五味子 15 克，花粉 12 克，玉竹 18 克，白芍 15 克，炙甘草 12 克。水煎服。

2. 暑热蒙蔽心包

主症：高热无汗，面目红赤，口干气粗，渴而多饮，烦躁不安，甚则昏迷不醒，舌红苔黄，脉细数或脉伏。

治法：清泄暑热，凉营开窍。

方药：白虎汤（10）加味。

知母 12 克，生地 15 克，生石膏 30 克，粳米 30 克，元参 18 克，黄连 9 克，竹叶 9 克，香薷 12 克，藿香 12 克，西瓜翠衣 60 克。水煎服。

（戴法轩）

12 妇科常见急症

中医治疗妇科急症有自己的特点，这里仅介绍临床最常见的急症，其它疾病可与《妇科学》互参。

12.1 痛经

痛经是妇科常见病之一，属中医"痛经"、"经行腹痛"的范畴。

病因病机

寒湿凝滞，气滞血瘀，气血虚弱，肝肾亏损均可使气血运行不畅而致痛经。

诊断要点

妇女月经前后或经期，出现小腹及腰部疼痛，甚者剧痛难忍。

辨证施治

1. 寒凝血瘀

主症：经前或行经时小腹冷痛或绞痛，喜热拒按。经血量少，色黯红或紫，夹有血块。伴畏寒，便溏，手足不温。舌质青紫，苔白润或腻，脉沉紧。

治法：温经活血。

方药：当归四逆汤（150）加减。

当归12克，桂枝9克，芍药12克，细辛3克，木通6克，炙甘草9克，大枣10克。水煎温服。

加减：寒重者可加吴茱萸、生姜、艾叶、小茴香各10克。

四肢厥冷者加熟附子10克、肉桂6克。

腹痛拒按夹有血块者可加生蒲黄、炒五灵脂各10

克。

全身乏力，面色不华，食少便溏者可加党参、炒白术、茯苓、半夏各 12 克。

2. 气滞血瘀

主症：经前或行经时小腹胀痛，按之痛甚。经血量少或行而不畅，色紫黯有块，血块排出则疼痛减轻。常伴乳房或胸胁胀痛。舌质紫黯或边有瘀点，苔薄，脉沉弦或沉涩。

治法：行气活血，祛瘀止痛。

方药：血府逐瘀汤（32）加减。

当归 9 克，赤芍 9 克，炒桃仁 9 克，红花 9 克，川牛膝 9 克，川楝子 9 克，制香附 12 克，益母草 15 克，川芎 6 克，柴胡 6 克，枳壳 6 克，元胡粉 6 克（分 2 次冲服），甘草 3 克。水煎服。

3. 气血虚弱

主症：经期或经后小腹隐隐作痛，喜温喜按。月经量少，色淡质稀。伴面色苍白，神疲乏力，语言低微，头晕心悸。舌质淡，苔薄白，脉虚细。

治法：补气养血。

方药：十全大补汤（151）加味。

炙黄芪 15 克，党参 15 克，炒白术 9 克，茯苓 9 克，当归 9 克，炒白芍 9 克，肉桂 5 克，炙甘草 5 克，元明粉 6 克（2 次冲服），生姜 2 片，大枣 4 枚。水煎服。

4. 肝肾阴亏

主症：经行量少色淡，小腹疼痛。伴腰膝酸软，头晕耳鸣。舌质淡，苔薄白，脉沉细。

治法：滋补肝肾。

方药：调肝汤（152）加减。

熟地 24 克，炒山药 30 克，当归 12 克，白芍 12 克，山萸肉 9 克，巴戟天 9 克，艾叶 9 克，炙甘草 6 克，阿胶 12 克（烊化)。水煎服。

加减：腰痛重者加杜仲、川断、狗脊各15克。
两胁痛者加香附、郁金、川楝子各12克。
小腹两侧疼痛加小茴香、桔核。

(说明) 对于痛经的治疗，一般应于月经来潮前3～5天开始服药，才可取得理想疗效，同时还需持续2～3个周期，疗效才能巩固。

12.2 急性盆腔炎

急性盆腔炎可见于中医的"热入血室"、"带下病"、"妇女症瘕"、"经水不调"、"经行腹痛"等范畴。

病因病机

平素体弱或经行、产后胞脉空虚，邪毒内侵，客于胞中，而使热毒壅盛致病。

诊断要点

1. 起病急，发热恶寒伴下腹疼痛，分泌物增多。
2. 妇科检查：腹壁紧张，下腹压痛及反跳疼。宫颈及阴道充血，有举痛，宫体稍大，压痛明显，活动性差。两侧附件增厚，有压痛，可扪及包块。若有脓肿形成，后穹窿有饱满及波动感，并有触痛。
3. 白细胞总数及中性增高。

辨证施治

此属热毒壅盛证

主症：恶寒发热，口干欲饮，下腹疼痛拒按。带下量多，色黄如脓，或为血性，秽臭。大便秘结，小便短赤。舌质红，苔黄或黄腻，脉弦数或滑数。

治法：清热解毒，活血化瘀。

方药：银翘红藤汤 (153)。

金银花30克，连翘30克，红藤30克，败酱草30克，薏苡仁12克,丹皮9克,栀子12克,赤芍12克,桃仁12克,元胡9克,川楝子9克,乳香4.5克,没药4.5克。水煎服。

加减:腹痛重拒按可酌加生蒲黄、五灵脂、三棱、莪术、丹参各12克。

大便干结加大黄10克,元明粉6克(冲)。便溏热臭加黄连、黄芩、葛根各12克。

白带多且秽臭加茵陈、黄柏、椿根皮各12克。

若邪毒传里,出现神昏谵语、抽搐等症,可同时急用安宫牛黄丸或紫雪丹1丸。灌服。

12.3 功能性子宫出血

功能性子宫出血属中医"崩漏"的范畴。

病因病机

气郁、湿热、血瘀、血热、脾虚、肾虚,均可使冲任损伤,不能固摄经血,而引起崩漏。

诊断要点

1. 本病多发于青春期或更年期。
2. 主要临床表现以各种不规则的阴道出血为特征。
3. 出血量大或时间较长者可有贫血的各种表现。
4. 妇科检查:无排卵或黄体功能不全。

辨证施治

1. 血瘀崩漏

主症:突然大量下血或淋漓不断,经血紫暗杂有瘀块,少腹疼痛拒按,瘀块排出则疼痛减轻。舌质黯红,边尖有瘀点,苔薄白,脉沉紧或沉涩。

治法:活血祛瘀,理气止痛。

方药:桃红四物汤(154)合失笑散(155)加味。

当归12克,熟地15克,川芎6克,白芍12克,桃仁9克,红花9克,炒蒲黄9克,五灵脂9克,制香附12克。水煎服。

加减：出血量多，改蒲黄为蒲黄炭，再加茜草炭，小蓟炭各12克。

2. 血热崩漏

主症：出血量多，血色深红或紫黑挟瘀块而粘稠，兼口干面赤，渴欲凉饮，烦躁不安。舌质红，苔黄，脉洪数或滑数。

治法：清热养阴，凉血止血。

方药：清热固经汤（156）加减。

生地24克，地骨皮9克，黄芩9克，焦栀子9克，制龟板15克，煅牡蛎24克，生地榆12克，藕节9克，棕榈炭9克，阿胶9克，麦冬15克，沙参15克，甘草3克。水煎服。

3. 气郁崩漏

主症：经量或多或少，淋漓不断，血色黯红或有血块。伴胸胁胀满，乳房、少腹胀疼，纳呆或呕哕吞酸，头痛。舌质黯红，苔薄白或薄黄，脉弦。

治法：疏肝理气，化瘀止血。

方药：平肝开郁止血汤（157）加减。

柴胡12克，当归9克，炒白芍12克，炒白术12克，丹皮9克，生地12克，醋香附15克，茯苓10克，棕榈炭9克，芥穗炭6克，薄荷3克，三七粉3克（冲），甘草6克。水煎服。

加减：伴有呕哕吞酸者加吴茱萸，黄连各10克。

4. 湿热崩漏

(1) 湿重于热者

主症：出血量多或淋漓不止，挟有粘腻水浆，平时白带粘而量多。伴面目浮肿或面色垢黄，头眩胀重，胸脘痞闷，口内粘腻，纳呆食少，少腹胀疼，倦怠乏力，腰重或疼，大便溏薄，小便不利。舌苔白腻，脉濡滑。

治法：祛湿清热，调经止血。

方药：调经升阳除湿汤（158）加减。

苍术12克，炙黄芪12克，柴胡9克，当归6克，升麻6克，羌活3克，独活6克，防风6克，蒿本6克，蔓荆子6克，炙甘草3克。水煎服。

(2) 热重于湿者

主症：出血量多或淋漓不止，血色紫红，腥秽粘稠。伴身热自汗，面色垢腻兼红，口苦而腻，渴而不欲饮，心烦少寐，小腹热痛，按之痛甚，大便秘结或溏泄不畅，小便黄赤不爽。舌红绛，苔黄腻，脉滑数。

治法：清热渗湿，调经止血。

方药：黄连解毒汤（114）加减。

生地15克，黄连6克，黄芩9克，栀子9克，黄柏6克，蒲黄炭6克，艾叶炭9克。水煎服。

5. 脾虚崩漏

主症：大量出血或淋漓不净，血色淡红，质稀薄。伴面色苍白或虚浮，少气懒言，倦怠乏力，胃呆纳少，腹胀便溏，四肢不温。舌质淡红，舌体胖嫩边有齿印，苔薄润或腻，脉细弱无力。

治法：健脾益气，养血止血。

方药：固本止崩汤（159）加减。

党参30克，黄芪24克，白术15克，熟地15克，首乌15克，茜草炭9克，棕榈炭12克，煅龙骨24克，煅牡蛎24克，炮姜炭1.5克，炙甘草6克。水煎服。

加减：若暴崩出血极多，头晕心悸，面唇灰白，四肢厥冷，呼吸低微，甚则昏厥，脉细欲绝或浮大无根，宜急投独参汤，用红参30克，水煎顿服，日2～3次。

6. 肾虚崩漏

(1) 肾阳虚

主症：出血量多或淋漓不断，血色淡红，少腹冷痛喜热敷。伴面色晦暗，精神萎靡，畏寒肢冷，腰膝酸软，小便清长，大便溏泄。舌淡，苔薄白，脉沉细。

治法：温肾助阳，调经止血。

方药：右归丸（134）加减。

熟地 24 克，山药 24 克，山萸肉 12 克，枸杞 12 克，杜仲 12 克，菟丝子 18 克，川断 12 克，阿胶 10 克（烊化），艾叶炭 12 克，附子 6 克，炮姜 3 克，乌贼骨 12 克。水煎服。

(2) 肾阴虚

主症：出血量少或淋漓不断，血色鲜红。伴头晕耳鸣，五心烦热，失眠盗汗，腰酸膝软或脚后跟疼。舌红，苔少或无，脉细数无力。

治法：滋肾养阴，填精固血。

方药：六味地黄汤（118）加减。

生地 30 克，山药 30 克，山萸肉 12 克，丹皮 9 克，茯苓 12 克，泽泻 9 克，女贞子 12 克，旱莲草 12 克，枸杞 12 克，阿胶 10 克（烊化），制龟板 15 克，仙鹤草 30 克。水煎服。

(曹晓岚)

13 儿科常见急症

中医儿科很早就在中医学中形成一门独立的专科，因为小儿生长发育的各个阶段中，在生理病理上、以及疾病的辨证治疗上均有自己的特点，应根据其特点，灵活地加以辨证施治。这里仅介绍临床最常见的急症，其它疾病可与《儿科学》互参。

13.1 新生儿肺炎

新生儿肺炎属于中医学中"乳子喘咳"的范畴。

病因病机

胎内感毒或感受风寒、风热，加之正气不足，无力托毒外出而发为本病。

诊断要点

1. 四季均可发病，冬春为多。
2. 发热，咳嗽，气促，鼻煽为主要特征。双肺闻及细小罗音或捻发音。
3. X线检查，两肺可见点片雾状阴影。
4. 细菌性肺炎白细胞总数升高；病毒性肺炎白细胞和中性粒细胞接近正常或减少。

辨证施治

1. 风寒闭肺

主症：恶寒发热，咳嗽，无汗，鼻煽，口不渴，痰白且稀，舌苔薄白或白腻，舌质不红，指纹青红，多在风关，脉浮紧而数。

治法：宣肺解表。

方药：三拗汤（160）加减。

炙麻黄 3 克，杏仁 6 克，甘草 3 克，前胡 6 克，防风 3 克，桔梗 6 克，苏子 3 克，莱菔子 3 克，白芥子 2 克。水煎后分两次口服。

加减：表里俱寒者加细辛 1 克。

2. 风热闭肺

主症：发热不恶寒，咳嗽，气急，咽红，面赤口渴，舌质红，苔薄黄，脉浮数。

治法：辛凉解表，宣肺化痰。

方药：麻杏石甘汤 (8)：加减。

炙麻黄 3 克，杏仁 6 克，生石膏 15 克（先煎），甘草 3 克，黄芩 6 克，鱼腥草 9 克，连翘 6 克，全瓜蒌 9 克，板兰根 9 克，桔梗 6 克，双花 9 克。水煎后分两次口服。

加减：壮热烦躁者加黄连 1 克，痰热上扰加羚羊粉 0.5 克，勾藤 3 克。

3. 心阳虚衰

主症：面色苍白，口唇紫绀，喘促，四肢厥冷，烦躁，脉虚数或微细欲绝。

治法：温补心阳，开宣肺气。

方药：参附汤 (29) 加减。

人参 2 克，附子 2 克，干姜 2 克，杏仁 9 克，甘草 3 克。

加减：心阳虚衰而见面色唇舌紫绀应加丹参 6 克，当归 3 克，红花 2 克。

大汗淋漓，血压下降者，加煅龙骨、煅牡蛎各 30 克。

4. 正虚邪恋

(1) 阴虚邪恋

主症：低热、口唇干红，多汗，口渴，干咳痰少，舌光红少苔，脉细数。

治法：养阴清肺。

方药：沙参麦冬汤（5）加减。

沙参 9 克，麦冬 6 克，玉竹 6 克，桑叶 6 克，花粉 9 克，地骨皮 6 克，桑白皮 6 克，炙百部 6 克，五味子 3 克。水煎后分两次服。

加减：干咳为主加诃子 3 克。

呛咳严重者加地骨皮 12 克、桑白皮 15 克。

低热明显者可与青蒿鳖甲汤（143）合用。

(2) 肺脾气虚

主症：低热起伏，神疲，面㿠，多汗，咳嗽无力，纳呆，便溏，舌淡苔白滑，脉细。

治法：益气健脾。

方药：六君子汤（161）加减。

党参 9 克，白术 6 克，茯苓 6 克，甘草 3 克，陈皮 3 克，半夏 3 克，炙百部 6 克，五味子 3 克，麦冬 6 克，桔梗 6 克。水煎后分两次口服。

加减：虚汗多加黄芪 12 克。

纳呆加神曲 6 克，山楂 6 克，或麦芽 6 克。

咳嗽重者加紫菀 6 克、冬花 6 克。

13.2　新生儿败血症

新生儿败血症属于中医学中的"邪毒内陷"或"湿毒走黄"的范畴。

病因病机

由于新生儿脏腑娇嫩，形气未充，卫表不固，易为毒邪所侵，邪入营血，化热化火，内陷心包，甚则出现正气虚脱证候。

诊断要点

1. 精神萎靡，面色青灰，体温不升，或壮热烦躁，神昏抽搐。

2. 黄疸进行性加重，肝脾肿大，不吃不哭，或皮肤有散在出血点。

3. 脐部或皮肤粘膜有破损，或口腔粘膜有脓性病灶。

4. 血细菌培养阳性，白细胞计数增高或降低。

辨证施治

1. 邪毒炽盛

主症：发病急骤，壮热烦躁，面目皮肤发黄，肝脾肿大，甚则昏迷抽搐汗出，大便秘结，小便黄，舌红苔黄，脉细数。

治法：清热解毒凉血。

方药：黄连解毒汤（114）加减。

黄连3克，黄芩6克，黄柏3克，双花9克，连翘9克，生地6克，赤芍6克，丹皮6克，甘草1.5克。水煎后，分两次口服。

加减：昏迷者冲服至宝丹。

抽搐者冲服紫雪丹。

2. 气血虚衰

主症：发病缓慢，面色苍黄或青灰，精神萎靡，不吃不哭，气息微，体温不升，四肢厥冷，额出冷汗，身上有出血点，舌淡苔薄白，脉细无力。

治法：益气温阳，扶正祛邪。

方药：参附汤（29）合四逆汤（27）加减。

人参2克，黄芪6克，当归6克，附子1克，干姜1克，甘草1.5克。水煎后，分两次口服。

必要时先扶正，待阳气恢复后，再予以清热解毒之品。

13.3 中毒性痢疾

中毒性痢疾属于中医学中的"疫痢"的范畴。

病因病机

本病的发病原因主要是饮食不洁，湿热疫毒之邪从口而入，蕴伏肠胃与正气相争而发病。

诊断要点

1. 多发于春、秋两季。
2. 常见于 2～7 岁小儿。
3. 发病急，变化快，突然高热达 40℃，面灰肢冷，脉数，开始往往无里急后重，脓血便等。
4. 重者可见神志昏迷，抽搐，皮肤花纹，可见脑水肿征象。
5. 实验室检查：大便培养阳性；大便镜检有大量脓细胞或红、白细胞及吞噬细胞；血白细胞计数增高。

辨证施治

1. 实热内闭型

主症：突然寒战高热，烦躁谵妄，口渴，惊厥，大便腥臭伴脓血，小便短赤，舌质红，苔黄腻，滑数。

治法：清热解毒，开窍熄风。

方药：葛根芩连汤（162）加减。

生大黄 2 克，黄连 5 克，黄芩 6 克，黄柏 6 克，山栀 5 克，葛根 9 克，银花 12 克，丹皮 6 克，赤芍 9 克。

加减：大便见脓血者加白头翁 9 克，地锦 15 克。

高热神昏加紫雪丹冲服。

若热毒深重，不能排毒下达，而无痢疾出现可加之明粉 6 克。

2. 内闭外脱型

主症：除上述突然内闭之证外，见面色苍白，四肢厥冷，冷汗出，脉细欲绝。

治法：回阳固脱救逆。

方药：四逆汤合安宫牛黄丸（20）。

制附子 3 克，干姜 5 克，甘草 2 克。

若阳气外脱为主，则先回阳救逆，安宫牛黄丸暂不用，待阳回厥返，可再根据病情或用凉开醒神，清热开闭之品。还应配合针灸。

(张伟)

14 耳鼻喉科常见急症

中医诊治耳鼻喉科疾病早已形成独立专科，有自己的特点和显著疗效。这里仅介绍几种最常见的急症，其它疾病可与《耳鼻咽喉科学》互参。

14.1 急性鼻窦炎

急性鼻窦炎属中医"鼻渊"的范畴。
病因病机
火热之邪上亢，致肺、胆、脾三经热盛而发为本病。
诊断要点
1. 鼻塞多涕，伴有头痛、发热、嗅觉减退为主要临床表现。
2. 局部检查：鼻甲红肿，鼻道见脓涕引流。
辨证施治
1. 内治
(1) 肺经风热
主症：鼻塞时作，嗅觉减退，涕多黄白而粘，前额及颧部疼痛，兼发热恶寒，咳嗽口干。舌红，苔薄白，脉浮数或浮而滑数。
治法：清热疏风，芳香通窍。
方药：苍耳子散（163）加味。
白芷6克，薄荷9克，辛夷花6克，苍耳子9克，黄芩9克，菊花12克，连翘30克，葛根12克。水煎服。
加减：若头痛剧烈者，巅顶痛加藁本12克，前额眉棱骨痛加蔓荆子15克，双太阳穴痛加柴胡12克。

涕黄浊量多可加冬瓜仁、蒌仁、鱼腥草各30克。

痰多加炒杏仁、桔梗各12克、瓜蒌仁、冬瓜仁各30克。

(2) 胆经郁热

主症：鼻塞，嗅觉差，大量黄脓粘稠浊涕且有臭味。前额或太阳穴或面部颧骨处疼痛剧烈，兼发热、口苦、咽干、目眩、耳鸣耳聋。舌质红，苔黄，脉弦数。

治法：清热泄胆，利湿通窍。

方药：龙胆泻肝汤（97）加减。

龙胆草9克，栀子9克，黄芩9克，柴胡12克，泽泻9克，木通6克，车前子30克(包煎)，当归12克，生地15克，桑叶12克，菊花12克，甘草3克。水煎服。

加减：鼻塞甚者可加白芷、辛夷、薄荷、苍耳子各10克。

(3) 脾经湿热

主症：鼻塞重，嗅觉消失，多量黄浊鼻涕，涓涓长流。伴剧烈头痛，重胀不适。脘腹胀满，肢重纳呆，尿黄。舌红，苔黄腻，脉滑数或濡。

治法：清脾泻热，利湿祛浊。

方药：黄芩滑石汤（164）加减。

黄芩9克，滑石24克，木通6克，连翘30克，茯苓12克，猪苓9克，大腹皮9克，白蔻6克（打碎），石菖蒲12克，藿香9克。水煎服。

2. 外治

(1) 滴鼻：滴鼻灵（鹅不食草、辛夷花）或辛夷滴鼻剂或25%牡丹皮液，每次1～2滴，日3～4次。

(2) 吹鼻：冰连散（黄连、辛夷花、冰片共为细末）少许，吹入鼻腔，日3～4次。

3. 针灸

(1) 肺经风热：取印堂、迎香、合谷、丰隆、通天、列缺，每次2~3穴，强刺激留针10~15分钟。亦可用鱼腥草注射液0.5毫升，肺俞穴注射。

（2）胆经郁热：风池、阳陵泉、绝骨、脑空、行间，强刺激留针10~15分钟。

（3）脾经湿热：取迎香、印堂、上星、合谷、攒竹、通天、风池、足三里，强刺激留针10~15分钟。

14.2 急性扁桃体炎

急性扁桃体炎属中医"风热乳蛾"的范畴。
病因病机
风热邪毒侵袭，引动肺胃火热，上蒸咽喉而致本病。
诊断要点
1. 本病以咽喉疼痛，吞咽不利为主要临床表现。
2. 局部检查：咽部充血肿胀，扁桃体红肿或有黄白色脓点，有的成片状，如白腐假膜，但较易拭去。
辨证施治
1. 内治
（1）风热犯肺
主症：咽喉疼痛、干燥灼热，喉核红肿连及周围。伴发热恶寒、头痛咳嗽。舌边尖红，苔薄白或微黄，脉浮数。
治法：疏风清热，解毒消肿。
方药：疏风清热汤（165）加减。
荆芥9克，防风12克，金银花30克，连翘12克，黄芩12克，赤芍12克，元参15克，浙贝母9克，炒牛子12克，花粉12克，桔梗12克，桑白皮12克，甘草3克。水煎服。

（2）肺胃热盛
主症：咽喉疼痛剧烈连及耳根和颔下，吞咽困难。伴有高热、口渴引饮、口臭、咳痰稠黄、腹胀便秘。舌质红赤，苔黄厚，脉洪大而数。

治法：泄热利咽，解毒消肿。

方药：清咽利膈汤（166）加减。

荆芥6克，防风9克，薄荷9克，山栀子9克，黄芩12克，金银花30克，连翘12克，黄连6克，炒牛子12克，桔梗12克，元参18克，生大黄6～9克（后入），元明粉9克（烊化）。水煎服。

加减：若颌下淋巴结疼痛加射干、瓜蒌、贝母各12克。

若高热加生石膏、天竺黄各12克。

若扁桃体肿疼甚者亦可含服六神丸，每次5～10粒，日3次。

2. 外治

（1）吹药：将少许药末均匀地吹入扁桃体，每日3～5次。咽喉红肿疼痛，轻者用冰硼散，较重或有脓点者可用朱黄散。若疼痛重且全身症状重，或脓性分泌物多，或反复发作者可用锡类散。

（2）含漱：荆芥、菊花、金银花、甘草，煎水含漱，每日3～5次。

（3）含服：铁笛丸或润喉丸，含于口中，渐渐溶化，慢慢咽下，每次1丸，每日数次。

3. 针灸

合谷、内庭、曲池、天突、少泽、鱼际，每次选3～4穴，强刺激，每次留针10～15分钟，日1～2次。亦可取脾俞、肩井内5分、曲池，每穴注射鱼腥草注射液或柴胡注射液各2毫升，日1次。

14.3 急性化脓性中耳炎

急性化脓性中耳炎是耳科最常见急症，属中医"急性脓耳"范畴。

病因病机

肝胆火盛,风热湿邪外侵发为本病。

诊断要点

1. 本病以急性耳膜穿孔、耳内流脓,听力减退为主要临床特点。可伴发热恶寒等全身中毒症状。

2. 局部检查:鼓膜充血,紧张部穿孔,脓液黄稠,闪耀涌出,呈传导性耳聋。

辨证施治

1. 内治

(1) 风热内盛

主症:耳闷胀塞、耳痛。伴发热恶寒、头痛、鼻塞流涕。小儿多有烦躁不安,啼哭不止,甚则呕吐,高热抽搐,神昏项强。舌质红,苔薄白或微黄,脉浮数。

治法:疏风散热,解毒消肿。

方药:银翘散(2)加减。

金银花30克,连翘15克,荆芥9克,薄荷9克,野菊花12克,蒲公英15克,地丁15克,桔梗12克,甘草6克,炒牛子12克,穿山甲10克,皂角刺9克。水煎服。

加减:若小儿伴呕吐、高热、抽搐者可加天竺黄、竹茹、钩藤、蝉衣、白芍各10克。

若出现神昏项强,则宜清营凉血,解毒开窍,请参照第1章"危重征象的急救"中有关部分。

(2) 肝胆湿热

主症:本型以成人为多。耳痛剧烈,听力下降,头痛烦躁,面红目赤,口苦咽干,小便黄赤,大便秘结。舌边红,苔黄腻,脉弦数。

治法:清泻肝胆,解毒排脓。

方药:龙胆泻肝汤(97)加味。

龙胆草9克,焦山栀9克,黄芩12克,柴胡12克,车前子12克(包煎),木通6克,泽泻12克,生地15克,当归12

克，知母 12 克，黄柏 9 克，苍耳子 6 克，辛夷 9 克，甘草 6 克。水煎服。

  加减：疾病初起加穿山甲、皂角刺各 10 克
     脓出之后加用白芷、桔梗、黄芪、川芎等各 12 克。
     大便秘结加大黄 10 克，芒硝 6 克。

2. 外治

(1) 清除脓液：3%双氧水洗涤。

(2) 滴耳：黄连滴耳液、鱼腥草液、银花注射液滴耳，每次 2～3 滴，每日 3 次。

(3) 吹耳：烂耳散（穿心莲 12 克、猪胆汁 10 克、枯矾研末 10 克）或红棉散（枯矾 12 克、干胭脂 6 克、麝香研末 0.05 克)，将耳道内脓液清除后，用纸卷或喷粉器吹入少许药末。

14.4 美尼尔氏病

美尼尔氏病属中医"眩晕"的范畴。因其是"眩晕"中的一种特殊证候，又有人称之为"耳眩晕"。

病因病机

气血不足，髓海空虚使耳窍失养，或阳虚水泛、肝阳上扰，痰浊中阻于清窍，均可致眩晕。

诊断要点

1. 本病以眩晕突然发作，具有旋转性，并有耳鸣耳聋，可伴恶心、呕吐为主要临床表现。

2. 检查：可有自发性眼球震颤。

辨证施治

1. 髓海不足

主症：眩晕发作频繁，发作时耳鸣较甚，听力明显减退，兼腰膝酸软，心烦失眠，多梦遗泄，健忘，手足心热。舌质红，苔少，脉弦细数。

治法：滋补肾阴，填精益髓。

方药：杞菊地黄丸（167）加减。

熟地 24 克，山萸肉 12 克，山药 12 克，丹皮 9 克，茯苓 9 克，泽泻 9 克，枸杞 15 克，菊花 12 克，制首乌 18 克，白芍 12 克，石决明 24 克，牡蛎 24 克，鹿角胶 9 克，龟板胶 10 克。水煎服。

2. 气血亏虚

主症：眩晕发作时面色苍白，神疲思睡，表情淡漠。兼少气懒言，动则喘促，心悸，唇甲不华，食少便溏。舌质淡，苔薄白，脉细弱。

治法：补气养血，健脾安神。

方药：归脾汤（50）加减

黄芪 24 克，党参 18 克，当归 12 克，龙眼肉 12 克，酸枣仁 24 克，白术 15 克，茯苓 12 克，木香 9 克，远志 9 克，炙甘草 9 克，首乌 18 克，熟地 15 克，白芍 12 克，白蒺藜 9 克。水煎服。

3. 阳虚水泛

主症：眩晕时心下悸动、恶寒，肢体不温。兼精神萎靡，腰痛背冷，夜尿频长。舌质淡白，苔白润，脉沉细弱。

治法：温肾壮阳，散寒利水。

方药：真武汤（45）加减。

附子 6 克，茯苓 12 克，白术 15 克，白芍 12 克，生姜 3 片。水煎服。

加减：寒甚者可加川椒 12 克、细辛 3 克、桂枝、巴戟天各 12 克。

4. 肝阳上扰

主症：眩晕每因情绪波动而发，急躁心烦，面赤。兼头痛、口苦咽干、胸胁苦满，多梦少寐。舌红苔黄，脉弦数。

治法：滋阴潜阳，平肝熄风。

方药：天麻钩藤饮（168）加减。

天麻 9 克，钩藤 15 克，生石决明 24 克，栀子 9 克，黄芩 9 克，川牛膝 12 克，杜仲 12 克，桑寄生 18 克，夜交藤 18 克，茯神 12 克。水煎服。

加减：眩晕较剧者，可加龙骨、牡蛎各 30 克。

口苦咽干甚者，可加龙胆草、丹皮各 12 克。

5. 痰湿中阻

主症：眩晕且头额胀重，耳闷、耳胀、耳鸣，听力下降，胸闷不舒，呕吐恶心症状较重，痰涎多。兼心悸，纳呆，倦怠。舌苔白腻，脉濡滑或弦滑。

治法：健脾和中，利湿开窍。

方药：半夏白术天麻汤（169）加减

姜半夏 12 克，陈皮 12 克，白术 15 克，茯苓 12 克，山药 20 克，泽泻 9 克，天麻 9 克，钩藤 12 克，大枣 3 枚，炙甘草 6 克。水煎服。

加减：呕吐频作者加代赭石 30 克、旋复花 10 克、水煎后少量多次频服。

耳鸣耳聋较重加灵磁石 30 克、石菖蒲 12 克。

面色萎黄、气短倦怠者加党参、黄芪、薏苡仁各 30 克。

（曹晓岚）

15 皮肤科常见急症

中医治疗皮肤病早已自成体系有自己的治疗特点和显著疗效。这里仅介绍几种常见急症,其它疾病可与《皮肤病学》互参。

15.1 带状疱疹

带状疱疹属中医"缠腰火丹"、"蛇串疮"、"蜘蛛疮"、"串腰龙"范畴。

病因病机

本病多因肝火郁结,脾湿内蕴,气滞血瘀兼感毒邪而发。

诊断要点

1. 本病多发于春秋季。
2. 可发生于身体任何部位,多见于腰部。
3. 常伴发热、头痛等前驱的全身症状。
4. 皮疹特点:疱疹密集成簇,呈带状排列,绝大多单侧分布,常沿一定神经部位分布,疼痛剧烈。

辨证施治

1. 内治

(1) 肝胆热盛

主症:皮疹潮红、疱疹如粟,密集成片,灼热疼痛。伴口苦咽干,口渴、烦躁易怒,小便黄,大便干。舌红,苔黄,脉弦滑微数。

治法:清热利湿,解毒止痛。

方药:龙胆泻肝汤(97)加减。

龙胆草9克,栀子9克,黄芩9克,生地15克,大青叶15

克,连翘12克,泽泻9克,车前子12克(包),元胡9克,生甘草6克。水煎服。

加减:疱疹发于头面者加菊花12克、钩藤30克。发于胸腰部加柴胡12克。发于上肢者加片姜黄12克。发于下肢者加牛膝12克。

血热明显,出现血疱坏死者加白茅根30克、赤芍12克、丹皮10克。感染重者加金银花30克、蒲公英30克、板兰根30克,大便秘结者加大黄10克。

年老体虚者加黄芪,党参。

(2) 脾虚湿盛

主症:皮损颜色较淡,起黄白水疱,易破溃、糜烂流水、疼痛尤甚。伴腹胀纳呆,便溏。舌质淡体胖,苔白厚或白腻,脉沉缓或滑。

治法:健脾利湿,解毒止痛。

方药:除湿胃苓汤(170)加减。

苍术6克,厚朴6克,陈皮9克,猪苓12克,滑石12克,炒白术12克,炒黄柏12克,板兰根15克,元胡9克,赤芍12克,车前子12克(包),泽泻9克,生甘草6克。水煎服。

(3) 气滞血瘀

主症:疱疹基底暗赤,疱液成为血水,剧疼难忍。舌质暗或有瘀斑,苔白,脉弦细。

治法:活血化瘀,行气止痛。

方药:活血散瘀汤(171)加减。

鸡血藤15克,鬼箭羽15克,红花12克,桃仁9克,元胡9克,川楝子12克,木香12克,陈皮9克,全丝瓜12克,银花藤15克。水煎服。

2. 外治

(1) 有水疱者用雄黄解毒散30克加化毒散3克,混匀后用水调或用鲜马齿苋(或白菜帮)捣烂调合后外用。

(2) 轻度糜烂者用祛湿散,加植物油调后外用。
(3) 后遗神经痛者,用黑色拔膏棍或脱色拔膏棍热贴。

15.2 急性丹毒

丹毒在中医文献中,因其发病部位不同而有不同的名称:发于头面者称"抱头火丹",发于肋下腰胯者称"内发丹毒",发于两腿者称"腿游风",发于胫踝者称为"流火"。小儿丹毒称为"赤游丹。"

病因病机

本病多因血分有热,火毒侵犯肌肤,或因破伤染毒而发。发于头面多为热毒,发于肋下腰胯多挟肝火,发于下肢多兼湿热,发于新生儿多由内热。

诊断要点

1. 发病急骤,常突然起病。
2. 可发于身体各部,但以颜面、下肢为常见。
3. 有恶寒发热、头痛不适、恶心呕吐、关节疼痛等全身症状,局部淋巴结肿大疼痛。
4. 皮疹特点:为略高于皮面的水肿性不规则形红斑,色如涂丹,压之退色,表面发亮,与正常皮肤界线清楚,局部灼疼和触疼,常迅速向周围蔓延。严重者红斑表面可发生水疱、内有浆液或脓液。

辨证施治

1. 内治

急性丹毒辨证为湿热内蕴,热毒炽盛。

热毒内蕴证:

主症:皮损红肿热痛明显,并伴恶寒发热、头痛烦渴,大便秘结,小便短赤,舌红苔黄,脉浮数或洪数。

治法:清热解毒,凉血清营。

方药:五味消毒饮(172)加减。

金银花30克，野菊花15克，蒲公英15克，地丁15克，大青叶24克，蚤休15克，丹皮10克，赤芍10克，板兰根20克。水煎服。

加减：发于颜面者加牛蒡子、薄荷、菊花各10克。发于肋下腰胯者加柴胡、黄芩、栀子。发于下肢者加黄柏、猪苓、萆薢、牛膝。

若伴高烧者加生石膏30克，知母、花粉各12克。

若高热神昏，热入营血者可改用犀角地黄汤（25）或清营汤（19）加减，并配合紫雪丹、安宫牛黄丸，每次1丸，日服2～3次。

2. 外治

（1）如意金黄散30克，化毒散1.5克，混匀后以凉茶水调敷于患处。

（2）新鲜白菜帮或马齿苋或绿豆芽洗净捣烂，调化毒散1.5克、如意金黄散20克，外敷患处。

(曹晓岚)

16　中医急症主要治疗措施

中医治疗急症多采用综合抢救方法，如催吐法、导泻法、针刺按摩法、静脉输液法、雾化吸入法、肛肠纳药法等，可根据不同情况单一选用或联合应用。

16.1　催吐导泻法

催吐法

催吐法是使用能引起呕吐的药物或能引起呕吐的物理刺激，使咽喉胃脘间的有害物质通过呕吐而排出。

1. 适应证

痰涎阻塞咽喉，妨碍呼吸；食物停滞胃脘，胀满疼痛；误食毒物时间不长，尚在胃部，使用吐法较为宜。

2. 常用方药及方法

（1）瓜蒂散（175）

组成：瓜蒂（熬黄）1克，赤小豆1克。

用法：将上两药研细末，混匀，每次1到3克，用豆豉9克，煎汤送服。

功效：涌吐痰涎宿食

应用要点：急性中毒，痰涎宿食，胸中痞闷，烦躁不安，气上冲欲呕，寸脉浮紧有力。

（2）盐汤探吐方

组成：食盐

用法：用2 000毫升 NaCl 饱和溶液口服，服后探吐。探吐方法：将手指洗净、消毒，牙齿间垫上衬物以防咬伤，用棉纤探入咽喉部搅动，使之呕吐。

功效：涌吐痰涎宿食

应用要点：宿食停滞不消，欲吐不得吐，欲泻不得泻，心烦满。

导泻法

导泻法是以泻下药为主组成，以通导大便，排除积滞，荡涤实热或攻逐水饮，寒积等方法。

1. 适应证

热结脘腹，脘腹胀满，大便干结。

2. 常用方药及方法

(1) 番泻叶导泻方：

组成：番泻叶10克。

用法：泡水代茶饮。

功效：通腑泻下。

应用要点：大便秘结，腹满胀痛等。

(2) 大承气汤

组成：大黄12克，厚朴（炙）15克，枳实12克，芒硝9克。

用法：水煎，大黄后下，芒硝溶服。

功效：峻下热结。

应用要点：热结脘腹，大便不通，痞满，腹痛拒按，按之硬，舌苔黄燥起刺或焦黑起刺燥裂，脉沉实。

16.2 针刺按摩法

针刺法

针刺法是用医用针具针刺激患者特定穴位，以达到治疗疾病的目的。

1. 常见适应证

(1) 溺水

治法：开窍苏厥，宣肺清心。

常用穴：会阳，素髎，内关，涌泉。

手法：一般用强刺激。

(2) 中暑

治法：清泄暑热，佐以和胃，佐以开窍固脱。

轻证常用穴：大椎，曲池，合谷，内关。

手法：大椎，曲池，合谷均用泻法；内关用平补平泻法。

重证常用穴：百会，人中，十宣，曲泽，委中，阳陵泉，承山，神阙，关元。

手法：神阙，关元用补法；余用泻法。

(3) 惊厥

发热惊厥

治法：疏调督脉，佐以清热。

常用穴：印堂，太阳，四缝，十宣，大椎，合谷，身柱，曲池。

手法：一般用强刺激。

无热惊厥

治法：疏调督脉，佐以镇痉

常用穴：大椎，筋缩，后溪，阳陵泉。

手法：一般用强刺激。

(4) 休克

治法：疏厥回阳

常用穴：素髎，内关。

手法：持续运针，中强刺激，血压回升稳定后，可间歇运针或留针。

(5) 晕厥

治法：疏厥和中

常用穴：人中，中冲，足三里。

手法：先刺人中穴，短促强刺激，然后刺中冲，足三里。

按摩法

按摩法是医生用自己的手或肢体协助病人进行被动运动的一种医疗方法。具有调和气血，疏通经络，促进新陈代谢，提高抗病能力，改善局部血液循环和营养状态等作用。

1. 按摩手法

(1) 推法：用手或手掌向外推挤患者肌肉，或用力作直线式的按摩。

(2) 拿法：用一手或两手捉拿患处的肌肉，加以挤压，或抓起肌肉后又迅速放手等方法。

(3) 按法：用手指或手掌在穴位或体表某些部位用一定压力，向下或向内压按的方法。

(4) 摩法：用大拇指或手掌在局部或俞穴的部位上反复地摩擦。

(5) 揉法：用大拇指指腹或手掌根部，按压在一定的部位上，以腕关节或掌指关节为主作回旋状的揉动。

(6) 掐法：用手拇指指甲在主治的穴位上，予以一定程度压按的方法。

(7) 搓法：用两手的掌面，紧挟住四肢或腰背部，并带动皮肉，作快速揉搓和上下反复的盘旋动作。

(8) 摇法：用两手固定住某一关节部位的两端，（主要是肩、颈、肘、腕、髋、膝、踝等较大关节），并从两端摇动关节部，作回旋运动，以加强关节的活动能力。

(9) 滚法：手背部的近小指侧部分按压在一定的体表部位上，以腕部作前后上下摆动的方法。

(10) 抖法：用手捏住受伤关节的远端，在向外拨伸时，突然做前后上下摆动的方法。常用于肩部、腰部。活动幅度必须在生理许可范围内进行。

2. 按摩的注意事项：

(1) 对局部皮肤破损或皮肤病患者禁止按摩。

(2) 饱食后不可按摩其腹部；按摩腰腹部前应嘱病人排空小

便。

(3) 对经期、孕期妇女均不宜按摩腰腹部。

16.3 静脉输液法

随着中药剂型的改革,注射输液法已比较广泛地应用于中医急症的抢救和治疗之中。目前,中药注射输液中有中药大输液,中药静脉注射针剂,中药肌肉注射针剂等数种。中药注射输液剂和口服中药相比,具有不受脾胃消化转输功能的影响,且给药方便,剂量准确,作用快,药效高等优点;同西药相比,也具有适应症广,作用时间长,毒、副作用小,不产生抗药性等优点。

1. 丹参注射液

处方:丹参 1500 克,注射用水加至 1000 毫升。

功能与主治:血管扩张药。具有扩张冠状动脉血管与增加冠状血流量的作用。用于心绞痛,心肌梗塞,肾病综合征,腹水,预防脑血栓形成。

用法及用量:肌注,每日 2~8 毫升,每次 2~4 毫升;静注 1 次 4 毫升,用 50%葡萄糖液 20 毫升稀释后,推注 1 日 1~2 次;静滴 10 毫升,用 5%葡萄糖液 100 毫升至 500 毫升稀释后使用,1 日 1 次。

2. 生脉注射液

处方:人参 1000 克,麦冬 3120 克,五味子 1560 克,注射用水加至 1 万毫升。

功能与主治:生脉,醒脑。适用于真阳欲竭,阴不敛阳,虚阳浮越之昏迷及心律紊乱,久病体虚。

用法与用量:肌注,每日 1 次,每次 2 至 4 毫升;静滴加 10%葡萄糖水 250 毫升至 500 毫升,生脉液 10 毫升或 20 毫升。

3. 清开灵注射液

处方:牛黄,水牛角,黄芩,银花,栀子等。

功能与主治:清热解毒,镇静安神。主治热邪内陷,神昏谵语，身热烦躁，抽搐惊厥，对病毒性肝炎，亚急性重型及慢性重型肝炎有较好疗效。同时适用于上呼吸道感染、肺炎、高烧等病证。

用法与用量：静注，每日 20~40 毫升稀释于 10% 葡萄糖液 200 毫升或 100 毫升内；肌注，每日 1 次，每次 2~4 毫升。

注意：如本品产生沉淀或混浊者不能使用。如与葡萄糖液或生理盐水混合后出现混浊者也不能使用。

16.4 雾化吸入法

将适合的药物，用喷雾方法，让病人通过呼吸吸入，以配合咳喘急症的治疗或传染病的预防，叫雾化吸入法。

1. 米醋喷雾

将煮沸的老陈醋 1 斤放入喷雾器，关闭门窗，让病人安坐室内，进行喷雾。具有预防流感的作用。

2. 复方细辛气雾剂

处方：细辛挥发油 50 毫升，冰片 16 克，95% 乙醇 600 毫升，三氯二氟甲烷适量。

功能与主治：宽胸止痛，用于冠心病，心绞痛急性发作。

用法与用量：用时将盛药液的喷雾瓶倒置，喷头圆孔对准口腔揿压阀门上端。喷雾头溶液即成雾状喷入口腔，闭口数分钟。每天喷雾两次或发作时喷雾。

16.5 肛肠纳药法

肛肠纳药法是从肛门进药，纳药肛中，而达到治疗作用的一种治疗方法。肛肠给药可不受病人吞咽功能及上消化道的影响，药物吸收快，作用迅速。如肾功能衰竭的水肿、癃闭、关格等病人用中药：生大黄 12 克，熟附子 12 克，牡蛎 30 克，作保留灌肠，能使血中尿素氮下降。

(张伟)

THE ENGLISH–CHINESE ENCYCLOPEDIA OF PRACTICAL TCM

(Booklist)

英汉实用中医药大全

(书目)

VOLUME	TITLE	书名
1	ESSENTIALS OF TRADITIONAL CHINESE MEDICINE	中医学基础
2	THE CHINESE MATERIA MEDICA	中药学
3	PHARMACOLOGY OF TRADITIONAL CHINESE MEDICAL FORMULAE	方剂学
4	SIMPLE AND PROVEN PRESCRIPTION	单验方
5	COMMONLY USED CHINESE PATENTMEDICINES	常用中成药
6	THERAPY OF ACUPUNCTURE AND MOXIBUSTION	针灸疗法
7	*TUINA* THERAPY	推拿疗法
8	MEDICAL *QIGONG*	医学气功
9	MAINTAINING YOUR HEALTH	自我保健
10	INTERNAL MEDICINE	内科学

11	SURGERY	外科学
12	GYNECOLOGY	妇科学
13	PEDIATRICS	儿科学
14	ORTHOPEDICS	骨伤科学
15	PROCTOLOGY	肛门直肠病学
16	DERMATOLOGY	皮肤病学
17	OPHTHALMOLOGY	眼科学
18	OTORHINOLARYNGOLOGY	耳鼻喉科学
19	EMERGENTOLOGY	急症学
20	NURSING	护理
21	CLINICAL DIALOGUE	临床会话